Seeking Bread and Fortune in Port Said

The publisher and the University of California Press Foundation gratefully acknowledge the generous support of the Constance and William Withey Endowment Fund in History and Music.

Seeking Bread and Fortune in Port Said

LABOR MIGRATION AND THE MAKING OF THE SUEZ CANAL, 1859–1906

Lucia Carminati

UNIVERSITY OF CALIFORNIA PRESS

University of California Press
Oakland, California

© 2023 by Lucia Carminati

Excerpts from the conclusion have already appeared in "An Unhappy Happy Port: Fin-De-Siècle Port Said and Its Connections and Disconnections of Water and Iron," *International Journal of Middle East Studies* 54, no. 4 (November 2022): 731–739, https://doi.org/10.1017/S0020743823000302.

Library of Congress Cataloging-in-Publication Data

Names: Carminati, Lucia, 1984- author.
Title: Seeking bread and fortune in Port Said : labor migration and the making of the Suez Canal, 1859-1906 / Lucia Carminati.
Description: Oakland, California : University of California Press, [2023] | Includes bibliographical references and index.
Identifiers: LCCN 2022059766 (print) | LCCN 2022059767 (ebook) | ISBN 9780520385504 (cloth) | ISBN 9780520385511 (epub)
Subjects: LCSH: Labor—Egypt—Port Said—History—19th century. | Immigrants—Egypt—Port Said—Social conditions—19th century. | Equality—Egypt—Port Said—History—19th century. | Suez Canal (Egypt)—History—19th century.
Classification: LCC HE543 .C285 2023 (print) | LCC HE543 (ebook) | DDC 386/.43—dc23/eng/20230508
LC record available at https://lccn.loc.gov/2022059766
LC ebook record available at https://lccn.loc.gov/2022059767

Manufactured in the United States of America

32 31 30 29 28 27 26 25 24 23
10 9 8 7 6 5 4 3 2 1

To Seba, Elia, and Adele

History is rapidly made in the nineteenth century.
>   —J. STEPHEN JEANS, *Waterways and Water Transport in Different Countries: With a Description of the Panama, Suez, Manchester, Nicaraguan, and Other Canals* (London: E. & F. N. Spon, 1890), 265.

If I am made of dust, then all my countrymen and the world's inhabitants are my kin.
>   —*Il Cosmopolita-Le Cosmopolite—Ο ΚΟΣΜΟΠΟΛΗΤΗΣ—The Cosmopolitan* 1, no. 2 (April 5, 1890).

CONTENTS

List of Illustrations ix
Acknowledgments xi
Note on Transliteration xv
Map of Cited Departure Points and Stepping Stones
of Suez Canal Migrants xvi

Introduction 1

1 · A Universal Meeting Point on the Isthmus of Suez 25

2 · Like a Beehive: Race and Gender on the Suez Worksites 71

3 · A Semilawless Borderland: The Presence of These People
Could Bring Evil 114

4 · Entertainment in Port Said, a Sink of Immoral Filth 153

Conclusion: It Would Be Wonderful If It Were Not Unhappy 195

Postscript 206

Notes 213
Bibliography 309
Index 321

ILLUSTRATIONS

1. Map of the Suez Canal and its surroundings, 1869   *2*
2. Groundbreaking in Port Said, 1859   *5*
3. Route map of the canal of the Isthmus of Suez, 1869   *27*
4. Arab "village" at Port Said, 1862   *28*
5. Fugitive Ethienne Marthoud, 1869   *57*
6. Personnel coming out of the Gouin worksite in Port Said, 1869   *61*
7. Children toiling as earthwork laborers on the Suez Canal worksites, 1862   *65*
8. Unidentified man photographed at a studio on the Quai du Port in Marseille, 1861–1870   *70*
9. Widening of the canal: work on the protection of the banks, 1869–1885   *75*
10. Caricature of Ferdinand de Lesseps by Etienne Carjat, 1862   *102*
11. Hut of *fellah* workers in Ferdane, 1869   *107*
12. Three altars at the canal's inauguration at Port Said, November 16–17, 1869   *118*
13. Examination of passports in Port Said, 1914   *134*
14. Port Said, Quay François-Joseph, 1854–1901   *135*
15. Queen Victoria statue on Port Said's Quay François-Joseph, 1903   *137*
16. "N. Caruana" variously classified as "Maltese," "Greek," and "English," 1893   *142*
17. Peddler of various objects in Port Said, 1880–1890   *155*
18. Commerce Street in Port Said, 1854–1901   *156*
19. Beach at Port Said, 1912   *158*

20. Boulevard Eugénie and the Eastern Exchange Hotel, n.d.   *168*
21. Port Said railway station, n.d.   *199*
22. Port Said's Arab and European quarters, 1911   *203*
23. Statue of Ferdinand de Lesseps at the entrance of the harbor, n.d.   *207*
24. *Construction of the Suez Canal*, watercolor by Abdel Hadi Al-Gazzar, 1965   *208*

ACKNOWLEDGMENTS

This project has led a peripatetic life. It was born of conversations, exchanges, and aspirations I have shared with friends and colleagues in Tucson, Cairo, Jerusalem, Milan, Lubbock, Oslo, and beyond. It would not have been possible for me to write it in isolation, nor would have it been quite as fun. I am indebted to the anonymous readers who have generously provided key suggestions as well as to Niels Hooper and Naja Pulliam Collins of UCP for their sustained backing.

As a graduate student, I dared embark on an ambitious tour de force in archives big and small thanks to the unflinching support of Julia Clancy-Smith, who wondrously morphed from adviser at the University of Arizona to academic champion, *tante*, and beloved friend through the years. I owe much of my passion for academia to her brilliant mind and generous spirit, both of which have made her my role model. Several institutions saw potential in my inchoate ideas and made my travels and doctoral adventures possible: the Social Science Research Council, the Council on Library and Information Resources through its Mellon Fellowships for Dissertation Research in Original Sources, the Zeit Foundation, the American Historical Association, and the Berkshire Conference of Women Historians. Scholars at institutions across the world, whom I fondly consider adoptive mentors, generously read through drafts of this project at various stages and offered feedback: Mario Maffi, who ignited the spark for writing, Elisa Giunchi, who understood my comings and goings, and then On Barak, Beth Baron, Ziad Fahmy, Stacy Fahrenthold, Shana Minkin, Nancy Reynolds, and Mario Ruiz. Others helped at various stages of my research by clarifying obscure words in Arabic, French, Yiddish, and modern Greek: Ahmed Samy and Manal Waly, Eleonora Ballinari and Ivan Toloni, Hillel Cohen, and Anastasia Maravela.

At the University of Arizona's History Department, I was fortunate to share a leg of my journey with exceptional teachers such as Aomar Boum, Linda Darling, Dick Eaton, Kevin Gosner (with eastward detours), Aslı Iğsız, Minayo Nasiali, Farzin Vejdani, and Fabio Lanza; from the latter I learned how academic friendships can be nourished at the dinner table. Benjamin Fortna and Yaseen Noorani at the School of Middle Eastern Studies have always kept their door open for me, for which I am grateful. The apparently unwelcoming Sonoran desert has become a place of the heart thanks to the community of dear friends who have called it home at least for a time: Bill Bemis and Becky Freeman, our children's "American grandparents" who, among other feats, repeatedly crossed the New Mexico-Texas border in the name of our friendship; and the "flying corgis" crew of Sam McNeil, Miriam Saleh, and Murphy Woodhouse, with special guests Dylan Baun, Paul Brown, Ilker Hepkaner, and Nicole Zaleski. Tucson also brought me together with Emma Blake, Amanda Hilton, Sabrina Nardin, Meltem Odabaş, and Francesco Rabissi, who taught me not just the relevance of disciplines other than history but also the value of slowing down and savoring the moment.

In Jerusalem, as a fellow at the Polonsky Academy for Advanced Study in the Humanities and Social Sciences at the Van Leer Jerusalem Institute, I had an opportunity to meet On Barak, extraordinary host Chiara Caradonna, Hillel Cohen in his double identity as historian and unlikeliest and friendliest landlord, Israel Gershoni, and Liat Kozma. I am appreciative of their curiosity and support even while I spent endless hours flying back and forth through the Mediterranean sky to fulfill a two-year home residency requirement that a 2011–2012 Fulbright scholarship tied me to. I am also beholden to Lubbock, where Texas Tech provided an opportunity to get together on campus and in backyards with exceptional friends and academics such as Britta Anderson, Alan Barenberg, Jacob Baum, Paul Bjerk, Zach Brittsan, Sean Cunningham, Stefano D'Amico, Barbara Hahn, Erin-Marie Legacey, Miguel Levario, Daniella McCahey, Johnny Nelson, cherished dinner mate Patricia Pelley, Ben Poole, Emily Skidmore, Aliza Wong, and early modernist, Verona connoisseur, and unrivaled party master Roberto Sisinni. Oslo, finally, gifted me with the wits and last-minute suggestions of Toufoul Abou-Hodeib, Christopher Prescott, and Kim Primiel, whom I thank. I am also indebted to archivists near and far, such as Jasmine Soliman of the Akkasah Center for Photography of NYU Abu Dhabi and Marie-Delphine Martellière of the Centre d'études alexandrines.

Friends in Egypt—Atef el-Sherif, multitalented Carmine Cartolano aka Qarm Qart, Asma' Gharib, Natalie Eiselstein, and Claudia Ruta especially—have generously shared tea, Stella, shisha, and wisdom, compelling me to articulate ideas more clearly. I am also indebted to Michael Reimer of AUC for his gift of a pivotal book and Alia Mossallam for having organized the workshop "Ihky ya Tarikh" in Port Said in January 2016 and proved the present relevance of historical digging. My "Egyptian" community in and out of Egypt has also been a constant source of inspiration and support: Margot Badran, Francesca Biancani, Mohamed Amr Gamal-Eldin, Gennaro Gervasio, Costantino Paonessa, our neighbor to the east Chris Rominger, and Olga Verlato. Finally, I could always count on my comrades in research struggles, international relocations, parenthood, and friendship Teresa Pepe and Estella Carpi. We were brought more tightly together by the tragic assassination in early 2016 of Giulio Regeni, fellow PhD student at Cambridge and builder of bridges, whom I want to commemorate here.

Other kinds of support have enabled me to undertake this project: the perseverance and curiosity my parents Anna Salvi and Franco Carminati passed on to me; the childcare and housework gifted by loving in-laws, aunts, and uncles: Ausilia Fumagalli and Gianni Mussi; Chiara Carminati, Marta Carminati, Paolo Cattaneo, Massimo Fumagalli, and doting Michela Mussi; the proximity of all the friends who were close despite the oceans and the pandemic between us—Marina Begotti; Eleonora Crippa, Anna Gagliardi, Olympia Katsampoula, Dario Longo, Silvia Longo, Maura Maldini, Teo Pozzi, Desiderio Puleo, Alice Sala, Claudia Sala, Alex Tonelli; listeners extraordinaire Eleonora Ballinari, Cristina Biasetto, and Margherita Giorgio, Egypt's fellow explorer ever since our first eventful trip to Cairo in 2007; and last but most important, the decades-long comradeship of my partner Sebastiano Mussi, accomplice in international relocations, companion in the mess and joy of parenting our two beautiful children, reader of drafts, forger of ideas, supporter, ally, and rock. This book is grounded in the world of love and friendship we have created together, for which I am thankful every moment.

NOTE ON TRANSLITERATION

To simplify reading, I have relied on a modified version of the transliteration system of the *International Journal of Middle East Studies*. I have excluded diacritical marks except for the 'ayn (') and the hamza ('). For the placenames and terms with common transliterations in English (e.g., Port Said and imam), I have used that spelling. For proper names that had been transliterated in the sources themselves, I have mostly maintained the original transliteration or included it in the footnotes; for example, El Abbed instead of Al-'Abbad and Moustapha rather than Mustafa.

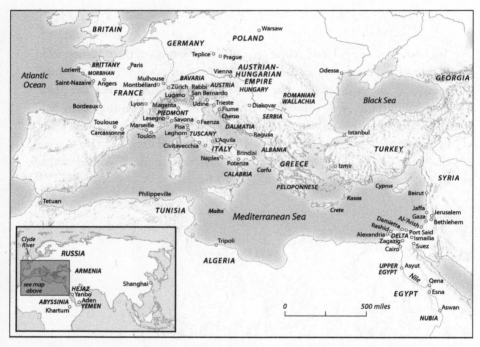

MAP 1. Cited departure points and stepping stones of Suez Canal migrants. Map by Bart Wright.

Introduction

> Port-Said will remain for me the great crossroads of maritime routes where my heart has felt and recorded the pulsation of the arteries of the universal life of our planet. Here, I had the clear vision, the precise feeling of the diversity of human destinies, which snatches the husband from his wife, the son from his mother, the lover from the lover, and throws them violently in space, where they are drawn to more harmonious affinities than those they try to create through familiar ties. Often, our true kinship and our homeland are at the antipode of the place where we come into the world and live as strangers.
>
> —PANAÏT ISTRATI, 1934

ON APRIL 25, 1859, 150 picketers, drivers, sailors, and laborers gathered on the northern Egyptian shore some thirty miles east of the city of Dumyat (Damietta). The group included 125 Egyptian workers. The rest were individuals identified by various sources as Greeks, Austrians, Italians, French, and Maltese.[1] The motley crew was in the hire of the Hardon enterprise, a French contracting firm that executed the first phase of work at the behest of the Compagnie Universelle du Canal Maritime de Suez (Universal Company of the Maritime Canal of Suez; hereafter the Company). The Company, "universal" in name and French in substance, had been conceived in 1854 to undertake the excavation of a waterway across the Isthmus of Suez (see figure 1).[2] As Edward Said once mused, the name of the company eventually created in 1858 "was a charged one and reflected the grandiose plans" that its founder, the French diplomat Ferdinand de Lesseps, harbored for this region soon to enter the world-historical theater.[3] According to the blueprints that representatives of the Company and the Egyptian state arduously drew up, thousands of workers would labor to unite the Mediterranean and the Red Seas, thus fulfilling a millennia-old dream. By bringing down the bridge of land that still united Africa and Asia, the new channel, to be officially inaugurated in November 1869, was meant to abridge the time and space that

FIGURE 1. Map of the Suez Canal and its surroundings, 1869. *Source:* Paris: Lanée, List No. 10599.002, David Rumsey Map Collection, David Rumsey Map Center, Stanford Libraries.

separated Europe and Asia. It would cut in half the time needed to shuttle between Europe and India, or China, or Australia.[4]

Subscriptions to the canal project were first opened to a heedless public in 1858 and only about half were taken up. Hence the Egyptian government agreed to secure the undertaking by buying almost all the rest of shares. For the next decade or so the Egyptian government incurred increasingly onerous financial obligations. Those commitments, coupled with the shock that the domestic economy received from the international depression of 1873, ultimately forced the Egyptian administration to sell its canal shares to the British government at dirt-cheap valuation in 1875, declare bankruptcy in 1876, and thereafter endure Franco-British financial and political control and occupation by Britain in 1882.[5] Neither a formal colony like India, nor a mandate like Palestine, nor even a protectorate as it would become during World War I, Egypt in the last quarter of the nineteenth century occupied an awkward and unique "semicolonial" status.[6] Ironically, Britain had initially resisted the canal project out of fear of French and Russian ambitions in the Mediterranean and in the belief that the territorial integrity of the Ottoman Empire would keep them at bay.[7] Still, shortly after the canal's inauguration British vessels had suited themselves: they were the most active and outnumbered the French, ranking second in activity, five vessels to one.[8]

When workers flung their pickaxes and first hit Egypt's marshy coastal soil on that late April day of 1859, not only did they initiate the so-called Suez Canal, but they also founded the port town of Būr Saʿīd (hereafter Port Said) in the guise of its northernmost labor camp (see figure 2). As the Egyptian official ʿAli Mubarak noted, the city's name originated from the coupling of the French word "port" and the name of the then ruler of Egypt, Muhammad Saʿid Pasha (r. 1854–1863), heir to Muhammad ʿAli.[9] The toponym of the newly created port would forever remind posterity of Saʿid's role in carving a novel waterway for the world. Saʿid had in fact signed the two concessions setting out the terms under which the Suez Canal was to be constructed. The first one, formulated in 1854, prescribed the adaptation of two "sufficient" entries for the canal: one on the Mediterranean and the other one on the Red Sea. It also decreed that the Company could establish one or two "ports" servicing the canal but lacked a clear indication of where exactly they ought to be positioned. The International Commission that gathered in 1856 to discuss plans for the canal clarified Port Said's specific location.[10] This city-to-be was to be emplaced in the bay of Pelusium, also known as the gulf of Tinnis, nestled at the center of a crescent of shorelines comprising the beaches of Egypt, Tripoli, and Tunis in the west and the coasts of Syria in the east.[11] Its proximity to deeper waters and prevailing winds facilitated setting sail. Moreover, this was the point where a longitudinal depression traversing the isthmus encountered the Mediterranean. In the words of a contemporary canal enthusiast, nature had "prepared itself these places for the easy and inexpensive execution of the direct canal between the two seas."[12] Through the roughly 129 kilometers (about 80 miles) separating the bottom of the Pelusium gulf in the north to the uppermost tip of the Suez gulf in the south, the highest altitude amounted to no more than sixteen meters (about fifty feet), and there were two considerable depressions, namely the once-dry Bitter Lakes and Lake Timsah.[13] The excavation of the canal was to proceed southward, from newly established Port Said to centuries-old Suez. Since most tools and provisions would be imported from Europe, this scheme promised to control costs and prevent delays. As the digging made strides, the artificial waterway advanced, and supplies could be more easily transported to the bridgehead marching into the isthmus desert.[14]

The canal excavation and the erection of Port Said created both a fresh interface between Egypt and the Mediterranean and untrodden ground for confrontation among the Egyptian administration, the Ottoman government, and Western powers. They also engendered a novel urban environment, new employment opportunities, an unprecedented migratory trajectory for job seekers in and out of Egypt, and a peculiarly unequal migrant society. In Port Said, everybody was a newcomer. As On Barak pointed out, it was "initially a tabula rasa with no one then 'native' to it."[15] Yet not all new arrivals began in the same way. Throughout the nineteenth century, Egyptians were conscripted, drafted for public works away from their homes, prevented from leaving them, and struggling to get away if they so wished.[16] Some did manage to move about Egypt in search of work by, for example, leaving their native Upper Egyptian hometowns, heading for Suez, and ending up in the Nile Delta.[17] Meanwhile, foreigners poured into Egypt, where they would settle not just in Alexandria and Cairo but also in Port Said, Ismailia, and Suez, among other locations.[18] To European individuals on the move, Eastern Mediterranean ports were more accessible than many European or US destinations because the use of passports and nationality documents was still not standard, and even when such documents were required, the law was not systematically applied.[19] Immunity was granted to those foreigners hailing from countries that had negotiated so-called capitulations, agreements

FIGURE 2. Groundbreaking in Port Said, 1859. *Source:* Marius Etienne Fontane and Edouard Riou, *Le canal maritime de Suez illustré; histoire du canal et des travaux* (Paris: Aux bureaux de l'Illustration A. Marc et Cie., 1869), 24–25.

between the Ottoman sultan and the European powers dating back to the sixteenth century that dictated that foreign powers had a prerogative over their own subjects residing in Ottoman lands.[20] In the nineteenth century capitulatory privileges remained more extensive in Egypt than anywhere else in the Ottoman empire.[21] These privileges included the very ability to cross the Egyptian borders and move about the country. All Europeans were admitted without distinction and allowed entering, settling, working under mostly no restrictions, and enjoying the additional benefits of exclusive consular jurisdiction and exemption from taxes.[22] For the states and societies on the Mediterranean's southern shores, as noted by historian Julia Clancy-Smith, these displacements were neither inconsequential nor necessarily benign.[23]

How did the Egyptian administration cope with the arrival of Egyptian nationals and foreigners in the Isthmus of Suez? How did bureaucrats from the Company's ranks deal with the swelling and peripatetic isthmus population? How did the modes and options for mobility change all along the canal when the British unfurled their occupation army into Egypt in the summer of 1882? This book addresses Cairo- and Paris- or London-centered perspectives, but it also capsizes them to grasp at new angles and ask: How did isthmus-bound individual and collective trajectories differ from one another? Who were those who moved to, dug, erected, and inhabited Port Said and the other fledgling towns along the nascent waterway? How did newcomers respond to the authorities' dictates on and off the canal worksites? How were different groups of migrants incorporated into the isthmus's labor regimes? What licit or illicit behaviors did they partake in? What options did men and women from different migrant groups have for their wherewithal or leisure? How did their relationships with Company cadres, Egyptian bureaucrats, and British colonial officials evolve? Finally, is it even possible to identify these workers on the move as a homogenous group of people and label them all as undifferentiated "migrants"?

If, as Ulrike Freitag and others argue, the life of migrants can be read "as a text that is rich in detail about the whole of society," then the mobile population of Port Said and the Isthmus of Suez can potentially illuminate the multifaceted Egyptian context through which they moved.[24] At the same time, this book breaks apart that apparently uniform category of "migrants" and accounts for differences in the types, strategies, and implications of their displacement, comprising travel, nomadism, purposeful relocation, and flight, among others. If approached expansively, the notion of "migrant" accommodates everyone—merchants and mamluks, saints and shaykhs, lumpen proletariat and high rollers—and satisfies none.[25] Similarly to what Zachary Lockman argued about "workers," such generic labels divert attention away from local subjectivities and the meanings of individuals' actions.[26]

Mobility is "a resource that is differentially accessed," geographer Tim Cresswell has claimed. The act of moving between locations ought to be unpacked and approached as a productive nexus of meaning and power. Mobility, according to Cresswell, becomes "socially produced motion," at once comprising the empirical reality of movement, ideas about it, and subjective experiences thereof.[27] These are racialized and gendered. Moreover, as illustrated by geographer Doreen Massey, differentiated mobility is not just about who moves and who stays still. Different individuals and social groups are placed in distinct ways in relation to these uneven flaws and interconnections. Some people are more in charge of this somehow differentiated mobility than others. Some initiate flows and movements; others are effectively constrained within undesirable, undocumented, crowded, and dangerous options of mobility.[28] As acknowledged by the so-called mobility turn in the social sciences, the concept of mobility or "mobilities" ought to embrace large-scale movements of people, objects, capital, and information across the world but also more local, daily, and mundane transactions, as well as instances of fixity, stasis, and immobility.[29] Both mobilities and moorings do not just happen in places, presumed to be fixed, given, and separate from those passing through.[30] They are constitutive of specific arrangements of power and space.

Seeking Bread and Fortune argues that the differentiated mobility of a diversified workforce and the formation of an unequal migrant society produced Port Said and enabled the realization of the Suez Canal project.

Between the 1850s and the 1900s three different but at times overlapping managerial elites and their subordinates—the French Company, the Egyptian government, and after 1882, the British-controlled Egyptian state—opposed one another in claiming control over the canal region. They all substantially failed to single-handedly impose the social order they envisioned over the unruly and elusive isthmus population. But the measures and practices they enacted gave way to a profoundly unequal migrant society, one in which supposedly ethnic or racial differences and gendered notions of respectability dictated uneven access to relocation, employment, lawful behavior, and leisure. This study examines the disparate sets of norms and practices of rule that, through five decades, different institutions attempted, chiefly in Port Said. At the same time it highlights the actions undertaken by migrant individuals and groups to counter the obstacles in their way. Overall, it takes stock of external interferences—the state's or others'—imposed on such actions and of the different and unequal ways in which people experience either domestic or cross-border movements.[31]

In other words, this work overcomes the dichotomies between "structure" and "agency" and institutional "strategies" and individual "tactics" that Michel De Certeau identified in his classical study on people's appropriation of quotidian life.[32] This Port Said–grounded history rejects binary frameworks by showing that neither institutional representatives nor migrant groups appeared homogenous or acted coherently. Port Said's residents were neither completely subject to controlling authorities nor fully autonomous in navigating their relations. At the same time, they were not unfailingly engaged in systematic or revolutionary subversion. Some of them appropriated the modes of action that were being imposed on them to advance their own interests, often at the expense of others in comparable circumstances. In the vein of other recent work on the history of labor in the late Ottoman context, this study privileges a complex social order rather than a scenario animated by idealized workers and mechanistic capitalism.[33] Moreover, it does not isolate women from other mobile workers. Often depicted as "unfree" and thus denied broadly construed autonomy, female migrants add to the complexity of labor and migration history because they set out for distinct reasons and followed migration patterns that differed from men's.[34] In Port Said and elsewhere on the isthmus, individual migrants pursued different and at times contradictory goals, while the proponents of social control that theoretically governed their lives were most of the time unwilling to follow or incapable of implementing unified agendas.

Seeking Bread and Fortune also argues that the creation and sustenance of an apparently peripheral spot such as Port Said altered circuits of mobility within Egypt and the Mediterranean. Not only did this brand-new hub play a novel role in connecting the Mediterranean and the Red Seas and provisioning passing ships, but it also forged a newfangled arena of connectivity with locations farther down on the canal banks. This arena was fed by flows coming from the rest of Egypt and abroad, but it also functioned independently from them. Recasting the Isthmus of Suez in this light shows how this region was at once self-contained and connected to sites elsewhere.[35] Moreover, Port Said operated as both inlet and egress for things and people that were both documented and unaccounted for. It was traversed in all directions by substantially unhindered movements weaving together the Mediterranean, the canal, and beyond. Stowaways from Mediterranean and Red Sea ports, for example, resurfaced from steamship steerage crannies in Port Said, their often unintended destination. Customs-free, contraband, or stolen stuff circulated in and out of the canal area in ways that went unsupervised. Although surveillance measures, such as Customs, came to be established in Port Said and made it into a checkpoint intended to serve both Egypt and the canal, "sans-papiers" and goods still found ways to enter and exit with ease. They turned Port Said into a unique living contradiction as both a gateway and a getaway spot, a chokepoint and a conduit, where "fixity" and "flow" converged and lay bare constantly shifting power relations.[36]

Port Said created its own orbit and timeline. If observed from this spot on the Isthmus of Suez, the period from the late 1850s to the early 1900s appears to have embraced both momentous change and substantial continuity. Around 1859 this quiet strip of land, theretofore inhabited by scattered fishermen and nomadic Bedouins, witnessed the unprecedented arrival of people and goods from the rest of Egypt and farther away. As a seemingly forlorn outpost on the eastern Mediterranean coast of Egypt, Port Said struck many as isolated and detached from the Egyptian interior. To the eyes of the French consul dispatched there, this town was in 1874 "part of Egypt only in name" because of its position and isolation. It produced nothing, had no outskirts, and lacked "a sufficiently easy and quick way to communicate with the rest of Egypt, with the [Egyptian] government willing to do nothing about it."[37] Its apparent seclusion was meant to cease in 1906, when a standard-gauge railway promised to connect it to the Egyptian railway network and, by extension, to the rest of the country more efficiently than in times past. The tide turned in other ways as well. In 1907 the country registered an economic crisis that burst just before Evelyn Baring or Lord Cromer, Britain's consul general of Egypt since 1883, resigned from his post in March that year. A years-long financial speculation had then come to an end.[38] That was also when the number of foreign residents of Egypt reached its peak: excluding the Anglo-British-controlled Sudan, they totaled 221,000 or 2 percent of the population. In 1907 foreigners made up roughly one-third of Port Said's population, and the number dropped afterward.[39] By 1908 what consular personnel termed the "financial crisis" had reportedly produced throngs of unemployed workers. Egyptians themselves were out of work, and there was no demand for foreign labor whatsoever throughout the country.[40]

Taking Port Said's foundation in 1859 and the advent of the standard-gauge railway into town in 1906 as benchmarks, my work explores the ironies embedded in political histories and demographic data. It shows that an incessant movement in and out of the isthmus constantly interlaced, albeit with ebbs and flows, this port town with the isthmus region south of it, the rest of Egypt, and sites outside of Egypt throughout the first half century of its existence. Inspired by what Adam McKeown has called "global mobilities from below," this book challenges the prevailing emphasis in migration history of transatlantic European mobility and 1914 as the end of the age of mass migration.[41]

ZOOMING IN ON PORT SAID

I focus on the history of a specific place rather than that of predetermined migrant groups in order to explore the relational dimension in the formation of social configurations.[42] By concentrating on one locale, I welcome the diverse actors entering this single frame and probe their tangled experiences.[43] I thus join the fold of those historians of late Ottoman port cities who attenuate the previous emphasis on economic structures and dependency theories and boost instead individuals, groups, and their interactions.[44] Meanwhile, by paying attention to the Suez Canal and Port Said's sundry migrant communities in its first fifty years of life, my work diverges from the scholarship on mobility in modern Middle Eastern history that either traces the history of specific groups within Egypt, or treats the migration of Americas-bound individuals from turn-of-the-century Mount Lebanon, or examines the movements of certain refugees across the late Ottoman Empire.[45] It builds on the empirically rich work authored by Zayn al-'Abidin

Shams al-Din Najm in 1987 and takes it past its chosen benchmark of 1882. It also converses with the trailblazing study authored by Valeska Huber in 2013 on the forms of mobility that the Suez Canal accelerated or restrained, by providing her global history with a physical street address. My scholarship remains deeply indebted to her insight that the openness of movement through the canal was highly dependent on the status of the individual traveler, even at times of allegedly passport-free travel, and that it impinged on slowdowns elsewhere.[46] Like hers, this history also stares at the canal, but it plants its feet in the swampy isthmus ground as solidly as possible. It aims to offer a localized historical account of broader interconnections.

I embrace the notion that a city is not a "simple material product" but rather "the production and the reproduction of human beings by human beings."[47] As I have argued elsewhere, scholarly research on the Suez Canal region has so far mostly excised ordinary people.[48] On the one hand, Egyptian historians have formulated unforgiving views of the Company's record of political deceit and economic exploitation.[49] On the other hand, scholars in especially French circles have produced mostly sympathetic accounts of the technological and entrepreneurial challenges faced by the Company and its eventual success.[50] In the scholarship by the latter, the feats of foreign technicians and officials especially in matters of architecture and urbanism throughout the isthmus have taken pride of place. But they have supplanted actual dwellers. Some studies in this vein have acknowledged that the canal cities brought multiple actors together into the city-building process.[51] Nonetheless, they have still approached these locales as yet other places where Europeans erected many notable buildings and achieved stylistic homogeneity and beauty, thus revealing the long-term influence of previous works aimed at celebrating the canal's venture as a technical and political triumph of "French genius."[52] Far from being empty or inert backdrops, however, the nineteenth-century urban creations along the canal and Port Said affected interactions among individuals, shaped their lived experience, and determined possibilities for authorities' intervention. This was a place where demarcation lines between one group and another were in flux. At the same time, social stratification came to manifest itself right away. From a technical standpoint, the canal undertaking ushered in the creation and the transformation of Port Said. But by peeking from the angles offered by a social and cultural history of this infrastructure, it becomes apparent that Port Said and the heterogenous population all along the canal shaped the course of the enterprise.

I bring this population to life by extrapolating fine-grained details from a myriad of arguably unremarkable cases. In aggregate, they construct the world of some of the workers who made the canal project possible. This world included gender relations and the labor of women, who had always been present and yet went unacknowledged by most contemporary and later observers. My goal in ferreting out "trifles" has been "to elucidate historical causation on the level of small groups where most of real life takes place and open history to peoples who would be left out by other methods."[53] I thereby borrowed from microhistory, requiring the historian to imagine the world as seen by her subjects and examine the terrain on which they traveled, worked, disobeyed, and reveled.[54] But the historical record of migration is by definition scattered and fragmentary. Only a handful of subjects left significant traces among the wide spectrum of sources consulted for this research. It was hardly feasible to reconstruct individual lives in a traditionally conceived microhistorical fashion.[55] Therefore, I juxtaposed multiple individuals in a narrative that combines the insights of microhistorians, customarily more attracted to idiosyncratic figures than ordinary folks and relying on "exceptional" sources, with the intuitions of writers of everyday life, who privilege ordinary individuals and the material circumstances of their daily existence. They all hail the idea that history's protagonists were simultaneously its objects and subjects.[56]

Through a microspatial perspective, I have first prioritized a close analysis of sources and of human action across multiple connected contexts and harnessed the specificity of a local context to reassess general questions.[57] I have given preeminence to flesh-and-blood individuals and unraveled dehumanized, teleological, and triumphalist narratives. In the historiographical landscape of the modern Middle East, microhistory is especially poised to displace the centrality of political history that ensued from the publication of Edward Said's groundbreaking *Orientalism* in 1978.[58] Second, when it comes to investigating how power relations work in context, details that place subjects in time and place are key. What is more, small scales of analysis make women's often opaque lives more visible.[59] Third, minutiae from everyday life shed light on how members of different communities within nineteenth-century Egypt partook in similar daily routines, thus balancing assessments of how segregated and seldom intermarried they were.[60] Finally, in the realm of migration history, a shift from state-level policies to the arena of mobile individuals brings us beyond the typical spatial migration patterns, which differentiate only between short-term and long-term migrants. It illuminates variables at the macrolevel of whole societies and the global economy, at the mesolevel of regional economies and cultures, and at the microlevel of neighborhood and family networks, as well as individual characteristics such as age, gender, or access to information.[61] But different analytical scales are not necessarily incompatible or isolated from each other. Conceptualizing historical processes as being located at distinct scales or "levels" may actually be counterproductive; it attributes fixed characteristics and rigid possibilities of knowledge to each layer of observation.[62] The history of migration is particularly poised to show that microhistorical and world-historical approaches flow into one. Straddling the local or the national, the transnational, and the global, both domestic and border-crossing migrants trek across one or more of these levels. They continue connecting them while going about their daily lives and carving out a place for themselves in their new surroundings.[63]

TIME AND SPACE IN MODERN EGYPTIAN HISTORY

Following up close the migrant laborers who, starting at the close of the 1850s, began journeying toward the Isthmus of Suez from the rest of Egypt as well as other countries brings about two main shifts in perspectives on time and space in modern Egyptian history. As for chronology, the years 1869 and 1882 are here placed in the middle of my narrative, unlike many historical works on the Suez Canal that often treat these two moments as beginning or ending points.[64] This exposes the threads of both continuity and interruption across these two apparently insurmountable divides. In so doing, this study offers new insights into the often unexpected ways the canal's official inauguration and the British occupation changed or left substantially untouched the existence of the isthmus population. Arguably, this waterway's premiere inauguration took place as early as August 13, 1865, when a cargo of coal proceeded southward while a Marseille-bound shipment of Indian and Arabian products crossed the canal in the opposite direction.[65] In 1870 the seamen who were interrogated by the French consul in Port Said acknowledged that, even if the passage from one sea to the other was possible, it nonetheless remained very difficult.[66] The last diggings proper were completed no sooner than April 15, 1871. Rocks below water level at three different points along the canal endangered the large trading boats that were supposed to shuttle between Europe and India.[67] While political history and economic history lend themselves to sequential narrative analysis, in which events can be linked into a consecutive sequence, social and cultural changes appear in the historical record less as events than as processes. The history of migration in particular resists standard narratives of historical turning points in either "sending" or "receiving" countries. Its chaotic timetable is contingent on individual decisions to depart as well as the resolutions of legislators.[68] Therefore, my aim is not to suggest an alternative chronology of the canal. Rather, I want to highlight how, for instance, a migrant worker may have conceived of the year 1869 as more or less significant depending on whether he/she got to keep his/her job as the completion of the infrastructure loomed near. Conversely, the import of 1882 may as well be downsized if we consider, for example, Port Said's smuggling business and the loopholes in harbor surveillance that persisted even after Anglo-Egyptian authority set in.

A view centered on Port Said's everyday life can also alter conventional perceptions of space in Egyptian history. The historiography has mainly approached modern Egypt as a highly centralized country, in which change descended from the top down and radiated from the city to the rural masses. In either Arabic or English, histories tend to be centered on Cairo or, at any rate, the Egyptian north.[69] Scholars of modern Egypt have joined efforts to begin "decentering" Egyptian history, while emphasizing that this is no simple geographic move: it implies dislodging the assumed centrality of Cairo and its archives, embracing non-elite voices and experiences, and revisiting the notions of "center" and "periphery" themselves.[70] Hence, writing the history of Port Said is not just about filling in the northeastern chasm within a lacking historiographical panorama. It also aims at reconstructing Port Said as an unequaled and unequal city built on and around people's movements and shaped by them. As such, Port Said proved to be neither part of a homogenous whole nor simply a local reflection of general phenomena. This is not to say that it did not change along with its broader context. For instance, Egypt's centralizing bureaucracy reverberated in Port Said, where new local governance and technologies of population control were introduced. Further, Port Said was part and parcel of the metamorphosis of the Eastern Mediterranean into a point of passage, a line of transit, or a canal-oriented funnel.[71] In a way, Port Said became "a link in the maritime chain" joining Britain with its Asian and Pacific possessions.[72] The nineteenth-century Mediterranean came to be dominated by British and French navies, armies, merchant fleets, and telegraph lines, but it did not become just a "colonial sea," nor did it wash up "seemingly insignificant social realities." Rather, as

Clancy-Smith has pointed out, this liquid space rippled under the effect of the "countervailing winds of the unexpected, unintended, and undesired."[73] In a word, this book insists that transformations in Port Said were not simply "the passive outcomes of changes in the social whole," as Henri Lefebvre would put it, but rather steered the history of the Suez Canal, modern Egypt, and Mediterranean mobility.[74]

EGYPT IN WORLD HISTORY

This Egyptian port city did share with its immediate Mediterranean and Ottoman coastal neighbors some elements of novelty. Both Port Said and Izmir or Smyrna, for example, were sites of great institutional innovation as well as uncertainty in the last part of the nineteenth century.[75] Like Algiers and Tunis, chiefly studied by Clancy-Smith, Port Said became a critical hub in Mediterranean-wide migratory currents that included other port cities and islands, such as Minorca, La Valletta, Marseille, Toulon, and Bastia, among others.[76] According to historian Charles Issawi, Port Said was a "heterogenetic" seaport of the likes of Casablanca, Oran, Algiers, Tunis, Tripoli, Benghazi, Alexandria, and Jaffa-Tel Aviv.[77] At this time, ports all across the Mediterranean basin experienced transformation and renewal. They provided a refuge to communities excluded from nation building, individuals of indeterminate identity, or marginal groups guilty of supposedly immoral or criminal behavior. At the same time, they became sites where the ordering power of the modern state and the reach of the international economy converged.[78] Relatively more is known about the latter aspect. Port Said's emergence as a strategic port for major European shipping powers has in fact received some attention.[79] On the heels of Wallerstein's world-system approach, the historiography of Ottoman port cities has often confirmed that "port-cities in the periphery emerged as the privileged locales of contact in the world capitalist economy."[80] While this premise has successfully shifted the focus of Middle Eastern urban history away from inland cities, it has also diffused an overly structural analysis that exaggerated the colors of the world economy while only sketching the local. As suggested by Ziad Fahmy, much less gravity has been given to "the cultural significance and the micro dynamics of this borderland environment."[81]

Port Said begged to differ from other Mediterranean or Middle Eastern port towns. Born as the backdrop of massive infrastructure requiring cheap, temporary, and skilled as well as unskilled labor who preferably would toil hard and whine little, everything had to be built from scratch. It did not originate as a conglomerate of old and new neighborhoods like Cairo and Alexandria, other Ottoman port cities, or even the metropolises of French North Africa. Unlike in other migratory contexts, people reaching the Isthmus of Suez could not rely on preexisting networks or resources. Neither could they be subjected to established administrative norms or practices. Considering the initially significant role played by the Company in Port Said's urban management, this center could be described as a "company town" designed to "facilitate production in an isolated area and manage labor and resources effectively with the expansion of industrialization."[82] Rather than being like its neighbors Alexandria and Damietta in the west, Port Said would appear closer to Djibouti or Port Sudan, colonial port cities created from scratch by French and British officials in 1888 and 1905, respectively, to attain control of hoped-for increases in coaling and shipping.[83] Our gaze could venture farther away. Port Said's history also bears a resemblance to Karachi, Dakar, Singapore, or Hong Kong, whose ports were being built anew.[84]

Contemporary observers offered apparently wild comparisons among far-flung corners of the globe. Indeed, various continents in the course of the nineteenth century experienced a similarly invasive penetration of cabling, railway, and global shipping.[85] As a newspaper article suggested in 1850, the attention of the "ancient world" was focused on a navigable channel through the Isthmus of Suez at the same time that the "Western world" was engaged in connecting the Atlantic and the Pacific via railroads and channels.[86] At the same time, some accounts were laced with fantasies about the world's frontiers and relied on a circulating set of platitudes. French sculptor Auguste Bartholdi, for example, on his 1869 business trip to Egypt (before visiting the United States for the first time in 1871), combined testimony and imagination by describing Port Said as a town "with all its houses, huts, or shacks lined up along large streets on the sand" that aptly resembled what he fancied the towns built by "American pioneers" must have looked like.[87] A visitor passing through around 1929 commented that one of its entertainment venues, the Eldorado Cinema, was, "in construction, almost exactly like those breath-taking dancing saloons in films about the Klondike gold rush."[88] Egypt as a whole frequently reminded contemporaries of the American frontier.[89] Indeed, Port Said's fortunes could be compared to those of Dawson City, founded in 1896 in Yukon territory during the so-called Klondike Gold Rush. Like Dawson City, after all, Port Said bore a distinctly international flavor, was geographically distant from supply centers, and demanded that families and institutions be adapted to suit local conditions.[90] In its status as a "desert town" that did not have to account for existing urban structures and practices, Port Said may also have evoked Nevada's Las Vegas during the construction of the Hoover Dam in the 1930s.[91] To others, who rebaptized Port Said the "Cayenne of the Desert," Port Said's lack of drinking water and exilic character reminded them instead of French Guiana, a south American outpost and France's depository for political and common criminals beginning in the late eighteenth century.[92]

Finally, this Egyptian port town's history jibed with that of the Atlantic seaport of Colón (formerly Aspinwall), founded in 1850 as the Caribbean terminus of the Panama Railroad.[93] The Isthmus of Panama, de Lesseps proclaimed in 1858, was "in the same condition as the isthmus of Suez," nothing but an obstacle to be overcome via either railroad or channel.[94] De Lesseps would himself become involved in an effort to build a trans-isthmus canal at Panama from 1880 to 1889 that failed spectacularly, dishonored its champion, and ushered in US control over the newly created Republic of Panama starting in 1903. Colón, which would become the Panama Canal's northern gateway and a crucial basis for that undertaking from 1904 to 1914, was, like Port Said, built on a swamp, lacked clean water, imposed preposterously high rents, and provided workers with entertainment options.[95] According to a late nineteenth-century journalistic account, their populations were alike arguably because both towns attracted "the scum of all nations" and disputed "the palm for degradation."[96] Whether factual or fictional, these scattered endeavors at comparative inquiry suggest that Port Said was changing along with its counterparts elsewhere on the planet, thereby countering stereotypes of the Middle East as inherently exceptional or isolated.

Through an urban history that interweaves multiple scales of analysis, I try to account for the confluence of a host of local, regional, and global variables in the formation of a new urban community.[97] A full-scale study of the entire canal region into the twentieth century would have distorted my lens. I did investigate Port Said's connections to other canal cities, but I kept a tight focus on this particular spot, which had much to say precisely because of its smallness and apparent irrelevance.

A WORLD OF SOURCES

In tracing the multiple trajectories that traversed Port Said and later came together in this book, I was compelled to step across national boundaries and traverse long distances like the migrants who animate these pages. I trekked disparate and sometimes unexpected paths across Britain, Egypt, France, and Italy. The archives of mobility are scattered across various towns and states.[98] Such archives also come about in a variety of languages and shapes. The multisited nature of my fieldwork—disorienting at times, occasionally miserable, always feverish—has itself steered the course of this book, for three main reasons. First, all research agendas are shaped on the ground by the researcher's positionality, external constraints beyond the historian's will or capacity, and random events.[99] In my case, as I have explored elsewhere, conducting research in Egypt in the fraught 2015–2016 political juncture affected the questions and conversations I developed within and outside the archives.[100] Second, with all due consideration to how twenty-first-century mobility differs from its antecedents, being displaced while examining displacement forced me to approach the experience of migration as subsuming multiple scales. This happened in part because historians of migration, as suggested by Donna Gabaccia, need to function simultaneously as historians of the world, of several nations, and of the particular loyalties that sustain and sometimes motivate migration.[101] But it also occurred because my own nomadism made me question the conventionally siloed historical categories of the microscopic, the national, the transnational, the regional, and the global. Third, my very ability to move across borders brought others' immobility into sharper relief. When dealing with fellow travelers of past and present times, I could not help but grasp how some mobile individuals were recorded, others slipped through, and yet others received a disproportionate share of often unwanted attention. I realized that my status as a frequent traveler "easily entering and exiting polities and social relations around the world, armed with visa-friendly passports and credit cards," was hardly universal. And with mobility being socially differentiated, so is the often privileged research-related mobility of US-based academics.[102] In other words, as Anna Lowerhaupt Tsing suggests, "How we run depends on what shoes we have to run in. . . . Coercion and frustration join freedom as motion [that] is socially informed."[103]

With this awareness, I broadened and diversified my source base as much as possible, welcoming state records and private correspondence, the official and the intimate. My goal became to look at Port Said as if through a kaleidoscope so that I could parcel the making of the Suez Canal into microscopic details and, conversely, assess how microscopic details are key to our understanding of world-historical processes. Alas, despite my frenzied search for the "particles" of ordinary lives, driven by "an energy all the greater for their being small and difficult to discern," I unintentionally replicated in this work some of the very unevenness encountered on my archival trails.[104] Egyptian migrant workers are less conspicuous and vocal than I wished.

Seeking Bread and Fortune relies on a wide variety of archival and published sources. Foremost among them are materials from the Egyptian National Archive (Dar al-Watha'iq al-Qawmiyya) in Cairo, the Suez Canal Company series currently housed at the Archives du Monde du Travail in the French town of Roubaix, the French diplomatic archives in Nantes and Paris, the National Archive in Kew-London, and the archives of the Italian Ministry of Foreign Affairs in Rome (some of which, those called "Vecchio Versamento," are still uninventoried). Also crucial, albeit less abundant, have been finds at the Propaganda Fide archives in the Vatican, meant for the Roman Papal Curia to gather information from its missionaries abroad, and at the motherhouse of the Catholic female order of the Bon Pasteur, whose nuns disembarked in Port Said as early as 1863. These documents show that while the experiences of ordinary people on the move across polities tend to be elusive, the records of their surveillance are rich.[105] It is indeed possible to disinter people's stories even in those instances in which creators and guardians of archives never meant for such excavation to take place.[106]

Several caveats are in order. First, these sources illuminate some aspects of migration while obscuring others. State records such as, for instance, population registers, tend to only concentrate on the numbers of visitors or resident aliens and the timeframe of their sojourning. Another example is provided by those requests for clarification that the Company received from families or consular authorities about the whereabouts of specific individuals who had presumably traveled to the isthmus to find work. These do shed light on the provenance of migrants but do not elucidate their trajectory, motivation, or destination. Only exceptionally do they go into detail, such as when they describe one Elias "the Greek" as a tall man of white complexion and black moustaches, wearing black woolen "big Greek trousers" and a so-called Greek beany. Most of the time, interpellated supervisors from assorted campsites simply shrugged off requests and replied that the individual was "unknown."[107] Second, the records of religious institutions were chiefly meant to document their representatives' successes and struggles overseas. As a result, as Beth Baron has pointed out, the populations among whom they worked seemed to have no names, no voices, and little to say about their experiences.[108] As argued by Ussama Makdisi, histories of mission work have often juxtaposed the indispensability of missionaries with the "irrelevance of actual native history and agency."[109] The Bon Pasteur nuns may have legitimately believed that many "lost sheep" needed "saving" in Port Said, but they appear to have been more stirred by the ungodly character of these souls' plight than by the alleged suffering of their bearers.[110] Third, records such as police reports or consular court records, attesting to the separate judicial system that existed for foreigners from capitulatory countries, do resound with the voices of mobile workers but mostly reproduce faint and distorted echoes. They ought to be approached as products of a dialogue with the authorities, one that was first mediated and later bluepenciled by state recordkeepers. Even though, as noted by Ehud Toledano, it was interrogators who enabled (or forced) the interrogated to step out of silence, the latter's language remains muffled. In addition, the only stories we hear are those of people who had brushes with the law.[111] Overall, it is an abundant and yet fraught and selective set of sources. The archives of surveillance do offer information about migrant society, while also revealing that those charged with counting, retrieving, "rescuing," and questioning people on the move were seldom up to the task.

The memoirs, travel guides, and newspapers authored by foreigners residing in or passing through Egypt are expedient as well as deceptive. Despite their claims of truthfulness, they stand as records of their authors' highly subjective impressions of the isthmus and its population.[112] They tend to leave workers, and especially Egyptians, out of the picture. The latter, in those rare instances in which they are featured, are presented through racist and Orientalist lenses (as showcased in most photographs and postcards of the time).[113] As for memoirs in particular, they often repeat sheer hearsay or platitudes found in previous publications.[114] Moreover, they are mostly gendered interpretations. Works penned by men, in fact, came to dominate the market for memoirs on Suez. After all, foreign female visitors were ridiculed as late as 1969, when a British journalist scoffed at them as the "inevitable gentlewomen" of the "tribe which has haunted the Middle East" since the early nineteenth century.[115] As for guidebooks, they provided snapshots of tourist-catering institutions at different points in time, thereby sketching a convenient chronology of the changing urban geography. At the same time, however, they tended to standardize travel experiences and homogenize the narratives about certain destinations. Moreover, they reduced Egypt as a whole to its antique sights. As noted by Huber, guidebooks still recommended a tour of the isthmus but remained unimpressed with the canal cities and tried to circumvent them as far as possible.[116] Finally, as for foreign newspapers as sources, their political orientation and intended readership clearly affected the ways in which their pages described Port Said and the canal works. For example, *L'Isthme de Suez: Journal de l'union de deux mers*, printed in Paris from June 25, 1856, through 1869 (and later continued as *Le canal de Suez* in 1870–1871 and *Le canal de deux mers: Journal du commerce universel* in 1872–1875), was the Company's mouthpiece and a tool in de Lesseps's hands to promote the "Egyptian canal issues" as he saw fit.[117] Some of the articles appearing in the press, as Kenneth Perkins posited in regard to the coverage of early twentieth-century Port Sudan, included flights of fancy that would have provoked guffaws from construction workers and administrators alike had they ever had the opportunity to leaf through them.[118]

Whenever possible, I have privileged migrants' own perspectives. Accidents of discovery came to the rescue regarding especially international migrants. The great majority of nineteenth-century workers on the move were unlettered. As shown by Clancy-Smith in her study of nineteenth-century Algiers, not only were state officials and Western memorialists imposing filters on them, but so was migrants' illiteracy.[119] However, some mobile individuals or family members looking for them did occasionally commit their lives to paper, either in their own hand or through a scribe. Whenever the postal carrier could not reach the intended recipients, either in Egypt or abroad, these missives remained ensnared in the consular bureaucracy. Both contemporary administrators and later archivists thus ended up thwarting these attempts at communication. Undelivered letters authored by or addressed to migrants remained ignored and unindexed, casually slipped between unrelated papers or filed away in misleadingly titled folders. The historian thus has become the only recipient of these papers, never producing their longed-for outcomes. Their addressees did not read them, the money they beseeched was never delivered, the messages of love and despair they carried stirred no feelings. Their pledges and their threats became stranded on their way to or from Egypt. In mine as in Lucie Ryzova's research, "the 'logic of preservation' is intimately linked to the 'logic of loss.'"[120] The presence of these happenstance records in the archive is predicated on their absence from their intended destination, on their being marooned and separated from their contemplated receivers. This estrangement is itself a source of historical information on the workings of consulates abroad and the frequent alienation of their subjects. Further, these missives are enriched by what only apparently curbs them. In fact, their authors, whenever capable of writing, often did so in a faltering handwriting that captured words and parts of speech as they sounded to their ears and with no knowledge of, time, or patience for grammar or punctuation. The Italian locution for "another paper" thus became the phonetically correct grammatical jumble "unal tro foglio" (rather than "un altro foglio"). The French turn of phrase for "what she desires" turned into "ce quil desire" instead of "ce qu'elle désire." In between the lines, notwithstanding their grammatical and syntactical heterodoxy, claims are loud and urgency is pressing. As argued by historian of eighteenth-century France Arlette Farge, these are precious auditory memories, striking reminders of the role of intonation in speech, and unique traces of the author's voices.[121]

OVERVIEW

Seeking Bread and Fortune progressively closes in on Port Said by first landing on the Isthmus of Suez, then moving on to its multiplying worksites, later approaching the town's surroundings, and finally mooring in its streets and bars. It maps one cycle in the life of the port town, with each chapter depicting it under a different light: from a swelling labor camp to a node in an isthmus-wide regimented labor regime, on to lawless borderland, and finally on to a hotbed for leisure and vice. After all, places do not necessarily have single, essential, or static identities.[122] As the chapters try to capture Port Said's shimmering forms, they follow a heterogenous ensemble of migrants (see map 1) as they moved, settled down, strove to find and keep a job, fell in love, caroused in drinking establishments, fought in bar brawls, left the isthmus, or ended their days there. As such, this book has different moods. It takes the reader on a ride through melancholic and joyous times.

The first chapter centers on the isthmus-bound migration of laborers hailing from the rest of Egypt as well as European countries and Ottoman lands. It begins moments after the Egyptian government signed a concession in 1854, bestowing upon Ferdinand de Lesseps the exclusive powers to constitute the Universal Company of the Maritime Canal of Suez. The chapter examines the Company's reliance on forced Egyptian labor between 1861 and 1864 and explores the overlap between different labor regimes. It connects the end of the forced-labor regime to the recruitment of workers in locations near and far. It also shows that controlling the mobile workforce on the isthmus was a critical drive in French, Ottoman, and British diplomatic maneuvers at midcentury. The chapter argues that the excavation of the Suez Canal and the construction of Port Said redirected labor flows toward the Egyptian shores while also tapping into existing circuits of mobility. The second chapter focuses on the employment opportunities available at the destination to the novice migrant population. It explores the canal worksites and their surroundings to outline what jobs both male and female would-be laborers in the second half of the 1860s could access for their wherewithal. The chapter discusses the ethnic and racial categories and hierarchies as well as the gender-based expectations and norms that both the Company and the Egyptian government enforced to manage the isthmus population. At the same time, the chapter analyzes the ways in which migrants circumvented or harnessed such hierarchies and norms to their own benefit. They all, but women in particular, resorted to mobility as a convenient strategy. As the excavation-related bonanza continued through the second half of the 1860s, migrant workers moved, switched occupation, and changed residence frequently. This chapter not only emphasizes the existence of female labor migrants but also stresses their role as active participants in the migration process.

By the time the 1860s came to a close, the atmosphere had begun to change; the third chapter registers a downward mood shift. It investigates conditions in Port Said and in Egypt after the canal's inauguration in 1869, when thousands lost the jobs they had held in and around the isthmus's worksites. A perception of increasing delinquency and growing precariousness characterized isthmus life in the 1870s. Port Said in particular appeared as utterly out of control. The traffic going through its port and waterfront seemed unmanageable. A growing stream of people and goods was outlawed. Authorities began defining the terms of lawful versus unlawful forms of mobility through town and enforcing attendant regulations. This chapter lingers on the changes ushered in by 1869 and examines the transformations in isthmus governance that occurred after the British took over the Egyptian administration in 1882. This chapter also takes into consideration what remained substantially unaltered. In fact, it contends that throughout the 1870s and 1880s the Egyptian police, the prison, and foreign consulates constantly failed in their attempts to impose their vision of social order upon the port's populace. Law and order in Port Said and the canal area, even if increasingly sought, appeared as elusive as ever. Meanwhile, immigration into Egypt continued even at a time when jobs had become rare. Taken together, these processes help conceive both 1869 and 1882 more as transitions than as ruptures. The fourth and final chapter, spanning the 1880s and the 1890s, focuses on the leisure options available on the isthmus and zooms in on Port Said and its drinking establishments. The chapter continues the discussion of the employment opportunities that the port city offered to female migrants, including those available in the city's brothels and private residences. The chapter argues that Port Said's entertainment scene catered to a range of budgets and inclinations. It highlights how activities such as newspaper reading or attending sacred and profane commemorations tended to separate individuals along national or religious lines. Drinking establishments, instead, appealed to everyone. Occasionally these venues turned into stages for the definition and the performance of group boundaries. Gradations of drunkenness and expressions of masculinity merged unto Port Said's drinking venues, loathed and yet popular well into the last decade of the nineteenth century and the onset of the twentieth. Finally, the conclusion discusses the chosen periodization (1859–1906), coinciding with the time in which Port Said was supposedly disconnected from the rest of Egypt and yet plugged into local, regional, and global routes. It explores the ironic twists of Port Said's at once promising and disappointing history.

Overall, this study probes Port Said's relationship to the Suez Canal and highlights the import of workers' mobility on the realization of this much-celebrated infrastructural project. Delving into the everyday life of one diversely populated port town, it examines the daily and often mundane negotiations among the population in Port Said and the institutions in power. Through four different prisms, this book explores the varied forms and meanings of mobility in late nineteenth-century Egypt. While the first chapter embraces international and internal migration and comprises back-and-forth movements, transiency, and running aground, the second chapter zooms in on the circuits that migrants created and followed within the isthmus by chasing better living conditions. The third chapter analyzes the different kinds of flows that entered and exited the Isthmus of Suez almost undiverted, especially through the canal's northern terminus at Port Said. The fourth chapter simultaneously embraces the worldwide and the microscopic. It examines Port Said's regional and global interconnections through the trajectories of desired commodities and peripatetic people, while focusing on the leisurely comings and goings of its population through its thriving drinking establishments. On the whole, this book provides a ground-level perspective of many of the most significant processes that took place in late nineteenth-century Egypt: heightened domestic mobility and immigration, intensified urbanization, the transformation of urban governance at both central and provincial levels, and growing foreign encroachment. Following Panaït Istrati, "proletarian intellectual of the traveling bohème," *Seeking Bread and Fortune* also likens Port Said to the pulsating core of "the universal life of our planet."[123] It is into this throbbing spot on our globe that this book dives.

ONE

A Universal Meeting Point on the Isthmus of Suez

The men of the deserts, the men of cultivated regions, attracted by the seductive charms of this good work, will come to us like fecund rains; and the wonders of their industry will fondle us.

—RIFAʿA AL-TAHTAWI, 1856

This small nook, remote and neglected, surrounded on all sides by salty water, lake or sea, and sand, was nothing, and it has become the greatest worksite of the world, while looking forward to becoming its most beautiful port. There have never been and will never again be a similar assemblage of living and mechanical forces.

—HENRI P. C. BAILLIÈRE, 1867

SOMETIME IN THE EARLY 1860S, Emmanuel Dot, not yet twenty years old, left for Egypt. A lemonade maker of unspecified nationality, he headed toward the Isthmus of Suez, where he roamed, unsuccessfully practiced the profession of hunter, and neglected to send news of his whereabouts to his increasingly distressed family. In 1862 an imam by the name of Mustafa relocated to the Al Guisr worksite, where he tended to the spiritual needs of the Egyptian drafted laborers toiling there while also voicing their remonstrations to Company cadres.[1] A Mr. Audibert, after having worked in the French navy, later found employment on the isthmus as a Company employee. He eventually reinvented himself as a storekeeper in Port Said during the day, while he allegedly liked to shoot guns in the dead of night. He must have done well enough to rent out his property-turned-brewery around 1878. But the Egyptian tax authorities foreclosed on his property for unpaid taxes a decade later.[2] Around 1866 a woman who went by the last name of Conti also settled in Port Said, where she resided out of wedlock with a Mr. Petanseau, a trader in dry goods at a store called Botte d'Or. She had originally relocated to the isthmus with her brother, who later moved farther south and settled at the worksite of Al Firdan.³ In 1867 an engineer called Giacomo Lepori from the Swiss town of Lugano likewise landed on the Isthmus of Suez. He would become one of the most "active and intelligent collaborators" of the excavation undertaken by the Company. He later turned into a contractor of public works on behalf of the Egyptian government. Having made a remarkable fortune for himself, he withdrew to his native land but regularly spent winters in Cairo with his family.⁴

Altogether, the trajectories of each of these individuals elucidate how the Isthmus of Suez had become a "universal meeting point."⁵ They demonstrate that the creation of the Suez Canal, initiated in newly founded Port Said in April 1859, engendered new isthmus-bound trajectories for which this port functioned as both a terminus and a transit point. They also show that the canal region gave rise to a relatively autonomous circuit for the circulation of bodies and objects. The Company played an important role in forging this new scenario of migratory movements.⁶ Yet it also tapped into existing networks and orbits of mobility. The Eastern Mediterranean in the late nineteenth and early twentieth centuries was part of larger labor markets. The Company alone did not affect migration toward this spot on the Eastern Mediterranean. Economic downturns and political changes in countries of origin, as well as regulations adopted by the Egyptian and Ottoman governments, shaped migrants' decisions to move to this area of Egypt.

This chapter traces the multiple paths trekked by individuals who voluntarily and forcibly left home, journeyed, and finally reached the isthmus. It connects Port Said to both single nodes of mobility and trade, such as Damietta, Alexandria, and Marseille, and to Egyptian and European regions that people left behind for manifold reasons. Finally, it links Port Said to the rest of the isthmus as it follows migrants all along the length of the waterway in the making. Ultimately, it stresses the reliance of the canal undertaking on a heterogenous mass of migrant laborers. It privileges the mundane dimension of migration, thus undermining the dichotomies between local and international mobility or between permanent and temporary relocation.⁷ Finally, it eschews attempts to pigeonhole types of migration or outline a hierarchy of migrants, because it confirms that labor migration includes both voluntary and coercive elements, which are difficult if not impossible to disentangle.⁸

By April 1860 ten functioning encampments dotted the trace of the canal-to-be. First, up north was Port Said. Proceeding southward, one would

FIGURE 3. Route map of the canal of the Isthmus of Suez, 1869. *Source:* Marius Etienne Fontane and Edouard Riou, *Le canal maritime de Suez illustré; histoire du canal et des travaux* (Paris: Aux bureaux de l'Illustration A. Marc et Cie., 1869), 108.

encounter Qantara, Al Firdan, Bir Abu Ballah, Timsah (renamed Ismailia in 1863), Al Fowar, Al Mourra, Tossoum, Serapeum, and Jabal Janifa.[9] In the following years more worksites emerged at Ra's al 'Ish or "Raz-el-Ech" (between Port Said and Qantara), Al Guisr (south of Al Firdan), and Chalouf (south of Serapeum and north of Suez) (see figure 3). The "French" associated with the Company set up foundries and workshops, collected machinery, arranged stores of all kinds, and laid out "plans of towns" to organize "everything most completely."[10] They also attempted to divide the migrant population and assign groups to different parts of the budding towns. Port Said, among others, saw the creation of a separate European or Ifrangi "quarter" and an Arab so-called village, respectively located on the canal's eastern and western banks. Originally these two sections were separated by the progressively deepening ditch.[11] In 1862 the Arab quarter was repositioned west of the European quarter; the canal no longer kept them apart. Nonetheless, a physical separation between these two areas remained in place, as did the belittling sobriquet of "village" for the formally Arab part of town and the persistently worse condition of its services (see figure 4).[12] Before the 1860s came to a close, the string of campsites and budding towns along the canal offered a number of employment opportunities. Migrant trajectories traversed and often went beyond fledgling Port Said.

FIGURE 4. Arab "village" at Port Said, 1862. Illustration by A. Gusmand. *Source:* P. Merruau, "Une excursion au canal de Suez (1862)," *Le Tour du monde: Nouveau journal des voyages VIII,* 2nd semester (1863): 25.

WORKERS MOVING BETWEEN STIPULATIONS AND ISTHMUS REALITIES

The very first workers to be employed in the Suez Canal works came from the area around Lake Manzala, on the northern Mediterranean shore of Egypt, adjoining the site where Port Said eventually came to be.[13] Even the Suez Canal Company's founder, Ferdinand de Lesseps, admitted that their task was tough, as they were standing, naked, in water up to their knees, excavating the bed of the lake with axes. With their bare hands they lifted blocks of earth, which were then passed on from worker to worker until they reached the edge of the projected waterway. The mud dredged up from the bottom of the lake gave off "an offensive smell."[14] Albeit unacknowledged, forms of local knowledge came in handy. These workers were lake fishermen who came to the Company engineers' assistance by showing them how to handle the mud as if it were heavy dough. They first pressed it against their chests to remove excess moisture and then shaped it into slabs that they carried on their backs to a spot where they left them to dry rock-solid in the sun.[15]

The concession that the Suez Canal Company and the Egyptian government agreed to in January 1856 stipulated that four-fifths of the labor force employed by the Company was to be made up of Egyptian workers. The clause clarifying the composition of the workforce did not specify the practical ways in which the four-fifths quota would be achieved.[16] Rather, it seemed to accomplish several diplomatic goals. In part, the stipulation promised to satisfy the Company's desire to facilitate its search for manpower, hardly enhanced by the desert prospects. Incidentally, it saved the Company the costs associated with hiring a mass of free workers. According to de Lesseps, the Egyptian government recognized that it had a "duty" to provide labor because the Company was theoretically forbidden to recruit abroad and unable to enlist workers locally.[17] At the same time, however, the requirement that four out of five workers be Egyptian did not constitute an utter capitulation by the Egyptian government to the Company's whims. In fact, it assuaged Egyptian concerns about the "influx" of a considerably copious "European" populace into the country.[18] Moreover, by committing to provide most of the necessary manpower, the Egyptian government strove to retain a degree of leverage over the Company. Egyptian officials did in fact later use the withdrawal of Egyptian labor as a lever to recuperate the lands they had initially conceded to the Company. By 1863 they would want both Egyptian workers and lands out of the Company's hands.[19]

The Egyptian government enshrined the project of the Suez Canal as one among other comparable governmental public works. On July 20, 1856, in the "Decree and regulation for the fellah workers" (*fellah* means "peasant" in Arabic), the Egyptian government eventually committed itself to providing the workers for the canal, paying them each 2.5 to 3 piastres per day and feeding them (for an unspecified span of time). Workers under age twelve would receive only 1 piastre per day but full rations of food. Those defined as "skilled workers" (*ouvriers d'art*), such as masons, carpenters, stonemasons, and blacksmiths, were to receive pay in compliance with what the Egyptian government normally paid for such works. The decree also established that the Company would provide all workers with abundant fresh water at its own expense. It would also accommodate them in tents, hangars, or other "suitable" lodgings. Moreover, the Company was supposed to erect a hospital for sick workers. The decree also established that the Company would pay those workers who were unable to work due to sickness a salary of 1.5 piastres during recovery. Finally, it called for the Company to cover travel expenses for workers and their families to get to the worksites.[20]

At least on paper, this regulation pledged to better the lot of those Egyptians forcibly recruited to work for the government—that is, if we believe those testimonies claiming that customarily each government-mandated corvée (forced labor) shift took sixty days, during which men left their fields uncultivated and let their families go hungry. If drafted by the Egyptian government, workers were supposed to bring their own provisions and tools; each man was to take his own basket and each third man a hoe. When exposed to bad weather, they faced the risk of freezing to death. Reportedly they received neither pay nor food. Those who could afford it could pay for a substitute worker and his foodstuffs.[21] There is evidence that the Egyptian state did pay the men it conscripted, even if reluctantly.[22] Conversely, the Company may have not kept its word. According to Egyptian statesman Nubar Pasha, the canal workers either returned home with no pay and under the burden of having had to pay for food or were only paid one-tenth of what had been initially promised, which was one franc per day.[23]

Both the concession and the "decree and regulation for the fellah workers" stipulated in 1856 established the principle that forced labor could be employed on the Suez Canal worksites. In the middle decades of the century, the corvée was still a common practice in Egypt.[24] The government relied on it to carry out most of the major public works it undertook, including railway construction, improvements to the irrigation system, and the cleaning of the main canals to ensure they could continue to carry water during the summer when the Nile was at its lowest. In many cases, men were coerced to work far from home.[25] Among other projects, the massive undertaking of the Mahmudiyya Canal connecting Alexandria and Cairo (1817–1820) stood out as the Suez Canal's infamous herald. In order to achieve that undertaking, the then ruler Muhammad 'Ali had forcibly drafted as many as 315,000 workers. Exhaustion, starvation, disease, dehydration, accident, cold, and heat claimed a death toll as high as 100,000.[26]

In 1858 the supporters of the righteousness of forced labor on the Suez Canal harnessed this historical antecedent. Why, they asked, since the former ruler Muhammad 'Ali had not hesitated to coerce into and sacrifice hundreds of thousands for the digging of the Mahmudiyya Canal, make a fuss about the grandiose "Canal of the Two Seas"? From the Company's standpoint, the use of forced labor posed no quandary. Appealing to the fact that the Egyptian government had long relied on it for useful public works in agriculture, navigation, and commerce, they argued that it was now France's turn to enjoy the resources that Egypt had to offer.[27] On their end,

British authorities in Egypt did not object to forced labor in principle, but they did criticize what they saw as the abuse of forced labor on the isthmus. They were particularly concerned by the idea that as many as twenty to thirty thousand Egyptians could be forcibly conscripted to serve French interests.[28]

To Egyptian administrators in the country's capital, Egyptian workers may have embodied the means to attain control over the sparsely inhabited Suez region. Multitalented Rifa'a al-Tahtawi (1801–1873), an Azhar scholar, the imam of the first mission sent by Muhammad 'Ali to study in Paris (1826–1831), the director of the new School of Languages and translation bureau, the planner of Egypt's new educational system, and a prolific writer, penned a poem to celebrate the undertaking of the Suez Canal. In it he praised the moveable workforce of "men from the deserts and cultivated regions" alike who, seduced by the charms of that "good undertaking" (*bienfait*), blessed Egypt "like fecund rains" and brought to it the "caressing wonders of their industriousness."[29] As Khaled Fahmy noted in regard to the Mahmudiyya Canal, "the sheer size of the labor force involved as well as the cost and duration of the project testify to the ability of the Pasha's administration in Cairo to tap and control the human and material resources of his province."[30] If managing the Mahmudiyya Canal epitomized Egypt's move toward a more centrally controlled bureaucracy, so could the Suez Canal a mere four decades later.[31]

Ambitions to control the Suez territory and to harness the workforce disseminated on it came to inextricably inform the outlook of both the Egyptian and Ottoman administrations on the canal project. They shared similar fears about a veritable invasion of Egypt by a "swarm" of foreign workers, "scattering through the country and bringing about disorder and even subversive ideas which the government could not harness" by virtue of the protection that the regime of consular jurisdictions afforded foreigners.[32] In particular, the Egyptian ruler Sa'id worried about the concentration of thousands of foreign workers on Egyptian soil and feared they could constitute "a threat to the independence of the country."[33] However, the Ottoman and Egyptian administrations had very different positions in respect to Egypt's contribution in land and manpower to the Suez Canal project.

First, the Ottoman government found that its own territorial sovereignty had been violated. In spite of the viceroy's explicit assurances that activities on the isthmus ought to involve only preliminary examinations and not the piercing itself, operations had nonetheless continued. Istanbul found that the undertaken works were no simple preparatory surveys but "of more compromising nature."[34] Moreover, the Ottoman ruler was vexed by Egypt's reliance on forced labor.[35] The corvée had been formally abolished throughout the Ottoman Empire since the beginning of the reign of Sultan Abdülmaçit (1839–1861) as part of his Tanzimat reforms. The Porte was in no position to sanction it in its autonomous yet subordinate Egyptian province. The pursuit of the "progress and the civilization in the Orient," the grand vizier wrote to the viceroy of Egypt, was incompatible with a system of labor that had been severely condemned by all "civilized" nations, in whose fold he implicitly included their own.[36]

On March 16, 1859, the Egyptian government forbade the presence of workers in the area of the projected canal's northern opening. But Company representatives and laborers must have brushed aside that injunction because, just one month later, ground was broken at Port Said, inaugurating it as the first worksite of the canal in the making. On July 22, 1859, the government further instructed Ja'far Pasha, the governor in Damietta, to go to the spot where "de Lesseps's workers" (with the latter rendered as *'ummal wa-shughghala*) could be found, round up all the "Arabs of Egyptian nationality" (*awlad al-'arab al-misri al-jinsiyya*), bring them back to Damietta, and report their total number to Cairo. From that moment, no more Egyptian workers were allowed to engage in the works.[37] Both the governor of Damietta and that of the neighboring province of Daqhaliyya received a directive to stop the influx of workers and monitor those who had already made it to the isthmus. Finally, Ja'far Pasha was also secretly instructed to import no more food or water to that area. By the summer of 1859 it had become clear to foreign consulates in Egypt that the Ottoman sultan had enjoined the Egyptian viceroy to stop the works right away and wait for a *firman* (imperial order), and that the Egyptian government had consequently entered a state of great embarrassment.[38]

But neither works nor workers on the move stopped there and then. After the Egyptian government had hastily forbidden its subjects to participate in the excavation works in the spring of 1859, de Lesseps resorted to gathering all the Europeans that he could find in the country. The viceroy reacted by requiring European consuls to address their nationals and urge them to leave the canal area.[39] Some foreign authorities did try to comply, but most of them appeared either unwilling or unable to intervene. Even those consuls who were keen on curtailing the power of de Lesseps, the French champion of the canal, could not formally ban their nationals from taking part in the works. The Austrian consul general in Alexandria, for example, argued that the sites offered "an honest way to make a living in a free country." He added that one hundred Austrian mariners were transporting materials on Nile boats. Had these mariners been barred from this task, he claimed, they would have been left with no jobs whatsoever.[40] Even those consular authorities who did command the evacuation of the workshops were ineffective. The consul general of France, who had fully adhered to the Porte's requests, issued orders to the French vice-consul in Damietta to interrupt the works right away, while lamenting that de Lesseps's actions were arbitrary and prone to attract costly annoyances for him and the viceroy.[41] However, the Company's director of works in Port Said, Félix Laroche, decided to act in open defiance and made all French workers stay put. "No Frenchman," one contemporary observer proudly quipped, "quit the battlefield." In line with the Company's desiderata, Paris repaid the coherence of the French consul general with dismissal.[42]

THERE IS NO NEED TO COERCE THESE MEN

Despite the pleas of the viceroy and foreign consuls alike, non-Egyptian workers kept venturing to the isthmus. Already in 1858, as later in 1859 and 1860, contractors from the Italian peninsula, France, and Slavonia, in the Hungarian portion of the Habsburg Empire, were offering to recruit hundreds of men willing to leave for the isthmus.[43] The supposition de Lesseps himself often uttered during his campaign in favor of the Suez Canal was that "the Greeks of the islands, the Bedouins, and the Syrians would hasten to his appeal, encouraged by relatively small salaries ... with regard to the cost of working days in Europe, but amply remunerative for them."[44] His wishful assumption proved to be partly right. In 1860 the subordinates of the trained engineer heading the worksite at Jabal Janifa, for example, comprised skilled workmen from various countries, including "a sprinkling of Algerian soldiers and colonists."[45]

Egyptians similarly challenged their government's ban and continued to toil on the canal project. They constituted the majority of Port Said's population at the time, which amounted to about 150 individuals, including the high-ranking Company employees.[46] By December 1859 the presence of 30 so-called Arab workers, often but not necessarily always identifiable as Egyptians, was recorded in Port Said. They had been freely recruited in Damietta, a coastal town to the west, with no apparent opposition from the governor. Other similarly unrestrained engagements had also taken place elsewhere. The Company dispatched individuals who could speak Arabic and "knew local customs" to recruit peasants to those areas of the Nile Delta that neighbored the isthmus.[47] In 1861, on behalf of the Company, the entrepreneur Alphonse Hardon, the principal contractor for the canal works, also hired two Greek contractors in Cairo, Jean Palidi and N. Costa, and promised they would "receive a sort of bounty upon each labourer brought to the works." For the first one thousand men, contractors would get 10 paras; for another thousand men, 12 paras, and so on exponentially.[48] In early 1861 de Lesseps gave orders to have a notice publicizing job opportunities on the canal worksites translated into Arabic, printed in Cairo, and distributed in the capital as well as in the main villages of Middle and Lower Egypt by either Company agents or other trusted individuals. Notices in Arabic were also to be affixed to mosque gates, in railway stations, and at the entrances of police stations. Recruiting efforts thereby appealed to workers directly, to both the literate and the ones within earshot, but also promised advantages to those village heads who would send out workers, thus promising to empower them.[49] Such exertions yielded only mildly promising results. De Lesseps hoped that by March 1861 five or seven thousand more "Egyptian" workers would join the three thousand who were currently on the worksites. However, at the end of the first trimester of 1861, only 200–250 "free indigenous workers" dwelled in each of the six worksites scattered on the central isthmus plateau known as "Seuil," the sill.[50]

That paucity of workers partly explains why the Company's recruiting efforts reached the Syrian coast and interior, touching on both Beirut and Bethlehem. De Lesseps was beset by an additional set of concerns, perhaps the same that determined his decision to target specifically Egypt's Copts for recruitment.[51] He tried to muster between fifteen hundred and three thousand Syrian Christians to guarantee the project's progress while Muslim workers fasted during Ramadan (in March 1861). He instructed Company agents to entice to the Suez worksites all the Syrian Christians they could find: men, but also children, elderly men, and women (in this context, children were individuals under sixteen and the elderly included anyone above forty years of age). While men between sixteen and forty years of age were promised at least 1 franc a day for their employment, the others would be paid proportionately to their labor. From the Company's standpoint, other incentives included residing in separate villages and, for those who owned camels, renting them out to the Company for 2.50 francs (more than their owners would have earned had they toiled themselves, attesting to the key role of animal labor in the enterprise). The Company even offered to cover the costs met by Syrian Christians to reach the isthmus whether via land or sea and to hand out small recruitment bonuses.[52] Overall, Syrian workers were present on the Suez worksites as early as 1861. That year the presence of five to six hundred "Arabs from the Syrian frontiers" was registered in Qantara alone. However, only a relatively small number of Syrians seem to have been effectively recruited for the Suez worksites. Rather than from the Syrian interior, they mostly came from the surroundings of coastal Al-'Arish and Ghazza (Gaza).[53] After residing for some time along the canal, many may have moved on to Cairo and Alexandria in pursuit of better economic conditions.[54]

Company representatives declared they had done nothing to lure those who were flocking spontaneously to the Suez excavation sites from the rest of Egypt, Syria, or, among other places, Greece, Italy, and Dalmatia, asking for work. They claimed that the canal had a naturally "strong force of attraction that radiated in a wide circle and exerted a pull over workers from the whole of the Mediterranean basin."[55] They asserted that "Arab workers" perfectly understood "the advantages of the [work] system" and appreciated being handled with gentleness, justice, and respect for their "customs" and "beliefs," as well as receiving regular and high wages. The Company founder boasted of having allowed all Egyptian workers, "*even* the women," to appoint one *shaykh* as a "judge of the peace" who would be in charge of executing those measures adopted by the Company or its subsidiary companies that applied to said workers.[56] "There is no need to force these men," de Lesseps proclaimed to the Company General Assembly in May 1861, "because they come running to our worksites."[57] At this time, both the Egyptian government and the Company manifestly portrayed workers as mere volunteers. The former made open profession of having simply given de Lesseps permission to recruit voluntary personnel. De Lesseps, on his part, emphasized that the orders the Egyptian government released to provincial governors simply facilitated the recruitment of *fellahin*.[58] And yet, starting in May 1861 the governors of the provinces that neighbored the isthmus may have actually begun drafting hundreds and then thousands of workers to be dispatched to the sites of the canal proper, such as Al Guisr and Qantara, and the sweetwater canal in the making.[59] By June 1861 British sources reported that eyewitnesses had seen men brought up in gangs from their villages and handed over by government officers to the Company's agents. These agents, stationed in several districts, rigorously inspected every batch; none but "picked" men were "passed." Although some contrived to escape from the train, they were recaptured in a matter of a few days and sent back to their taskmasters.[60] By September 1861, even if Company representatives still considered the recruitment service scarcely satisfying, numbers had increased and the workers' contingent was 5,771 strong.[61]

Despite its boastful claims, the Company actually saw its initial efforts to attract a sufficiently numerous and diversified workforce for the canal frustrated and may have attempted various approaches to secure the labor it needed. For example, twenty-five Maltese who had been engaged in August 1859 fled as soon as they had disembarked in Alexandria after having traveled at the Company's expense and received one month's pay in advance.[62] In November 1860 the number of workers on the isthmus reached a mere seventeen hundred: roughly six hundred were north, in Port Said; nine hundred were either in Damietta or working on the communication canal between the Nile and Lake Manzala; one hundred were in Qantara farther south; and one hundred were distributed in the central isthmus between Maxama, Bir Abu Ballah, Timsah, and Seuil.[63] In early 1861 Mougel Bey, an engineer at the service of the Egyptian government, announced he would prepare a deal with some contractors (*entrepreneurs*) to provide ten to twelve thousand workers.[64]

In sum, the arrivals of would-be workers on the worksites remained scattered and the various attempts to attract a privately recruited work force were unsuccessful. Neither the Company nor entrepreneurs were able to secure sufficient workers from the rest of Egypt, European countries, or neighboring Syria. The situation was dramatically transformed, however, by a major change in Egyptian state policy.

FORCING OLD AND NEW ON THE CANAL WORKSITES

In January 1862 the Egyptian administration officially instituted a systematic scheme of corvée on the canal's worksites. A representative of the Egyptian ruler called Isma'il Bey went to Seuil to ensure numbers were sufficient and remained stable.[65] At that point, corvée workers were already present in relatively small numbers in two points within the Seuil section: at Al Firdan and at a worksite identified as no. 6 (adjacent to the northern edge of Lake

Timsah). But their presence, from the standpoint of the engineer in chief, was no guarantee of stability or productivity.⁶⁶

By blending established practices and novel elements, the system of forced labor on the Suez Canal ensured that a sufficient workforce would supply the needs of the canal project. More substantial contingents, swelling the ranks of the Canal force to twenty thousand men overall, reached the isthmus in February 1862. They hailed from Upper Egypt. In March 1862 the canal worksites brought workers from Upper Egypt and Lower Egypt. Sources specify that workers from Upper Egypt would eventually be drafted in the Minya, Asyut, Qena, Esna, and Nubia areas. Workers from Lower Egypt departed the central delta areas of Monufiyya and Qalyubiyya, Buhayra in the western delta, and Daqhaliyya (Mansura in particular) and Sharqiyya in the eastern delta.⁶⁷

Reportedly, at any given time up to twenty thousand men were on their way to the works, twenty thousand were digging, and twenty thousand were commuting back home.⁶⁸ These were rough estimates, considering that drafted workers occasionally brought their families with them to the sites.⁶⁹ Their engagements were supposed to last one month, excluding travel time to the destination. The trip from Upper Egypt took ten to twenty days.⁷⁰ For those hailing from adjacent areas, the average number of days needed to reach the isthmus could drop to six.⁷¹ Even if workers followed often circular trajectories, the consolidation of the forced-labor system on the canal worksites still forged new paths of isthmus-bound mobility that connected the rest of the country to this northeastern strip of land.

On the excavation sites, Egyptian forced laborers were each given a small pickaxe and a basket made of palm leaves, in which they placed the dirt they excavated.⁷² They toiled under the threat of the *courbache* (whip; *Kurbaj* or *kurbak* in Arabic and *kırbaç* in Turkish), a whip made of hippopotamus hide. While de Lesseps publicly declared that all forms of corporal punishment were prohibited on the worksites, Western observers of the time insisted that the whip could ensure the discipline, obedience, and overall submissiveness of Egypt's "native" bodies, including those of children (see figure 7 later in the chapter).⁷³ Drafted laborers were placed under the authority of both *shaykhs* (headmen), figures to whom the government routinely entrusted the selection of men for the corvée and military service, and European foremen, like the Turin-born engineer Edoardo Gioia, who was in charge of as many as twenty thousand drafted men at the Al Guisr worksite alone.⁷⁴ The *shaykhs* occupied an ambivalent yet powerful position, since they were both tasked with enforcing the Company's measures and presenting workers' claims.[75] For the managers of especially large foreign enterprises, contracting labor through *shaykhs* was advantageous, as it enabled them to rely on supervisors who were familiar with local conditions, could be held responsible for the completion of the work, solved potential problems, adjusted the fluctuating demand for labor, and ensured a steady inflow of manpower while preventing the formation of a stable workforce with demands for higher wages or better working conditions. One *shaykh* called Sidi Ahmed Maghoule, for example, traveled to Al Firdan in May 1861 with a contingent of 327 men and then again in July with 300 more.[76]

The ways in which Egyptian contingents reached the canal area demonstrate the complementarity of Egypt's budding canal and railway network. Drafted workers approached the excavation sites via the brand-new railroad connecting Cairo and Suez, which proved instrumental to the advancement of the canal project. Completed in 1858, the Cairo-Suez tract extended the one laid out between Alexandria and Cairo between 1851 and 1856 under the direction of British engineer Robert Stephenson.[77] Comparatively fast, the railway enabled travelers to depart at 7:30 a.m. from Cairo and arrive in Suez by 6:30 p.m. that same evening.[78] Depending on the task they were assigned to, workers reached either Suez or Zagazig, at the canal's southern tip or near its middle tract, respectively. Company agents then guided laborers to the worksites and camels carried their luggage. At the Company's insistence, the Egyptian government even decided to build a small railroad that branched east from Zagazig, where the Company used to store some wares.[79] Moreover, in the 1860s, on a three-hour-long leg of the journey between Timsah-Ismailia and Suez, the train stopped in Serapeum and Chalouf for the exclusive benefit of the Company.[80] Even after a sweetwater canal between Zagazig and Timsah-Ismailia had been completed in January 1862, travelers still needed to rely on both tracks and water: they first reached Zagazig via train and then embarked on a barge headed to Timsah-Ismailia for the daily evening service established by the Company. Overall, it took one or one and a half days, a relatively long time, to cover this distance, depending on the direction taken.[81] Yet in the long run, the Suez Canal may have posed a French threat to the British interests vested in the Alexandria-Suez railroad.[82] While both river steamboats and the Egyptian State Railway depended on the canals' and Nile's waters, they did eventually enter into direct competition.[83] Man-made canals and railways, after all, had been at odds before and elsewhere due to the latter's superior speed and greater punctuality.[84] However, the converging paths followed at this time by the canal workers show that Egypt's railway and canal projects functioned, at least for a while, in synergy.

SURVIVAL IN THE WORKSITES

At the onset of the enterprise, only a few workers would deliberately try their luck in an area that could become so hot that the sun alone could cook an egg and roast fleas in ten minutes' time if infested shirts were laid out on the sand.[85] In 1862, as the summer approached, more and more workers were increasingly suffering from sunstroke.[86] Recorded average August temperatures at the time in Suez reached 39.9 degrees Celsius.[87] Either with threats or pleas, workers were at times forced to toil day and night with no rest. While in the first section of the canal, at the beginning of the works, each worker was assigned the feasible task of digging forty meters of simple sand and with embankments that were only four meters high, the pace of work had to slow down as the digging progressed. As the canal's breadth and depth increased, it became impossible to complete the given task within the scheduled shift. Workers were constantly late. They were forced to stay on until the arrival of the following contingent, thus raising the complaints of the *mudiriyat*, the "provinces," they came from (for example, in May 1863 protests from the administrators of the Monufiya, Qaliubiya, and Sharqiya provinces were recorded).[88]

Moreover, some of the *shaykhs* subjected workers to extortion. According to one observer who fell in line with the Company's perspective, workers, in groups of eight or ten to make the transaction easier, received their pay directly in their hands and not through their *shaykhs* (as had been claimed).[89] However, the archival record shows that, in at least one instance, a man called Mustafa, the above-mentioned imam who had accompanied drafted workers to the Al Guisr worksite, having attended to numerous testimonies, requested the chief of the encampment to withhold the accounts of the *shaykhs* so that the money they had previously extorted could be given back to the claimants. He also wrote to higher bureaucrats connected to the Company so that they could inform the *Mudir*, the Egyptian governor, who could take punitive measures against the "Grand Chek," presumably a *shaykh* in charge of all other *shaykhs* on the Isthmus.[90]

Life on the Isthmus was made even more unpalatable by the scarcity of potable water at the digging sites. Water's retrieval governed the existence of its thirsty laborers as well as the success of whole enterprise. Each of the workers on the worksites was said to need up to ten liters per day.[91] The excavation of a canal for drinking water had been in the plans since the Egyptian government granted a first concession to the Company in November 1854. In January 1856 a second concession rehashed the contents of the first one and added that a freshwater canal would be dug from the Nile River to the midpoint of the Suez Canal through the *wadi*, the valley, called "al-tumilat." From there, two branches would be built northward and southward.[92] After an initial back and forth to determine who would pay the bill, it was decided that the Company would take up the costs.[93] The Company thereby came to virtually own the sixty thousand hectares extending along the two branches of the drinking water canal extending toward Port Said in the north and to Suez in the south.[94] But the swampy and salty nature of the lands through which the drinking water canal ought to be cut posed a serious challenge.[95] Meanwhile, water had to be transported to the isthmus from elsewhere. To Port Said specifically, drinking water was initially transported by camel or boat from Damietta in iron boxes; it was brackish and had to be rationed.[96] A ditch carried water from the wadi to Bir Abu Ballah, located west of Lake Timsah. From there, a terracotta pipe carried water until the worksite of Al Guisr. A lock established by the Egyptian government in Zagazig in 1861 allowed Nile boats to navigate the *wadi* canal until Qassasin, in the Sharqiya province, halfway between Zagazig and Timsah. It was not until April 1861–January 1862 that a canal was hurriedly excavated, from Qassasin to Nefiche, on the western edge of Lake Timsah, mainly responding to the urgent need for a communication line. Between November 1862 and December 1863 a branch was accomplished between Timsah and Suez.[97] This canal network was meant to provide the Company with seventy thousand cubic meters of water daily to quench people's thirst, water lands, guarantee the functioning of machines for the canal's maintenance and exploitation, irrigate the seedlings and the dune plantations, and stock the boats passing through the canal.[98] A water-made network came to connect the center of the isthmus to Cairo as well as to adjoin it, via the capital, to both Lower Egypt and Upper Egypt.[99]

In the summer of 1859 two steam-operated devices to distill and desalinate salty water were installed in theretofore underserved Port Said; each desalinated five thousand liters daily.[100] By December 1860 French Company officials declared that a third distilling machine was to be installed near the area in Port Said known as "Arab village," so that water could be distributed more evenly throughout the town. Moreover, by the end of that year a water reservoir was ready.[101] As argued by On Barak about Middle Eastern arid environments in the last third of the nineteenth century, desalination conveyed "political power and sovereignty." Managing to extract salt from seawater meant capturing and maintaining a foothold on otherwise impenetrable territories.[102] Desalinated seawater, though, had a yellow color and an irony taste. Water brought over from the Nile, even if lukewarm, remained a better-tasting alternative. By 1863 Port Said residents, worn out by thirst, still clung to the hope that other options to get water more easily, vaunted since at least 1862, could be found.[103] But it was not until April 1863 that two cast iron pipes were laid down by the Lasseron company to pump drinking water to Port Said, thanks to machines installed in Ismailia.[104] This water, however, tasted like tar. Years later, Nile water was still the beverage of choice.[105] By the end of Port Said's first parched decade, public fountains could be found throughout the city's quarters and were constantly surrounded by men, women, and children of twenty different nationalities.[106] Still, water provision would remain a bone of contention in Port Said, leading to clashes among residents about who could access the water and how its price ought to be regulated.[107]

The poor quality of the water contributed with several other factors to the overall sickly condition of the workers. According to contemporary medical personnel, health on the isthmus was affected by, among others factors, temperature variations, ground humidity, the spiking cost and low quality of staples, abuse of alcoholic beverages, sleeping in the open, and lack of hygiene.[108] As for the quality of staples, Company sources argue that drafted workers were provided with biscuits, onions, oil, lentils, and rice (the value of which exceeded, per man, 1 piastre daily, which equaled around one-third of an adult's day's pay and a full child's day's pay).[109] But others say laborers were only given coarse dark-colored biscuits, hard as wood, and often of bad quality. The cost of this food was deducted from the pay of the workers, who had to bring whatever else they needed from home.[110] Initially, a Cairo contractor provided Port Said with fresh beef and mutton.[111] Greek boats carried cattle from Cyprus and Syria.[112] In general, however, meat, vegetables, and even fish were hard to come by. The only vegetable available in town was the salad that arrived from Alexandria, which since it took about ten days to arrive was no longer green but black.[113] Another similarly unappealing option on the isthmus menu was preserves, half-jokingly said to date back to the Crimean War.[114] The cattle murrain or epizooty (plague) that struck Egypt in 1864 made meat scarce. Finally, fish, apparently appreciated by the Greeks in particular, was at first an oddly rare delicacy; presumably, fishermen did not sell it to the public. Overall, staples were exorbitantly expensive. Especially during winter, food may have been scarce given that landings became rare.[115] Apparently only in the second half of the 1860s did fish become more available and the main staple of the urban poor's diet.[116]

As for sleeping arrangements, according to British observers, no shelter could accommodate such a large crowd. Corvée workers made hollows in the sand; piled a few stones on the windward side; lit a fire when the desert supplied a few bushes; and crouched together, drawing over themselves a camel-hair rug.[117] Other sources, however, suggest that the Company provided both agents and workers, whether indigenous or foreign, with tents, some mattresses, and coverings.[118] In Port Said, in April 1861 pillared constructions coexisted with tents and shacks (see figure 4), some of which were aligned while others had been erected pell-mell.[119] Medical assistance and hospitals were available to those workers who fell ill, but smallpox, fever, dysentery, gastric fever, and ophthalmia kept thinning their ranks.[120] In Port Said alone, between the summer of 1864 and the summer of 1865, 850 people suffered from wounds, contusions, and burns, while 102 were afflicted by some form of cancer, gonorrhea, and syphilis. On a seemingly seasonal cycle, weak digestion struck the most, followed by ophthalmia and diarrhea. These were all especially acute during the summer months. Enteritis and dysentery caused several deaths, while two individuals perished of chronic diarrhea and two more of rickets.[121]

Egyptian drafted laborers reportedly hated the isthmus works more than any of the tributary works owed to the ruler, in which, it was claimed, at least food was abundant.[122] Their salaries may have been calculated depending on how much dirt they excavated in a day: they received 40 or 50 centimes per cubic meter.[123] By March 1864 they received the sum of 2.5 or 3 Egyptian piastres per day, equal to about 0.70 franc (3.8 Egyptian piastres equaled 1 franc in 1870).[124] The claim that Egyptian earthwork laborers earned a daily sum ranging from 1.10 to 1.30 francs was perhaps blown out of proportion.[125] At the end of the month of their forced permanence on the isthmus, Egyptian workers' trip back home was not covered, and they had accumulated an average of only 5 piastres, two days' pay or less.[126] Moreover, the timing of their convocation might have damaged their wherewithal. While the Company claimed that workers were conscripted only in those times of the year when the fields did not demand their attention, other observers argued that the *fellahin* were drafted regardless of their need to attend to the harvesting or the sowing of crops.[127] All in all, workers might have had few incentives to leave their villages voluntarily and little need for a salary so small it would have scant effect on their well-being.[128] An Egyptian scholar starkly judged the forced recruitment of workers for the Suez Canal in 1902: "The work of the peasants in digging the Isthmus clearly differed from the normal corvée; nothing remained of its elements save the coercion of workers to come to work."[129]

Discontent with the canal enterprise had begun festering within the Egyptian administration as well. The ruler Isma'il, who had succeeded Sa'id when the latter died in January 1863, did not look favorably on the fact that large tracts of territory had been assigned by his predecessor to the Company.[130] Moreover, he was hostile to the forced labor scheme and favored its demise.[131] In fact, he may have opposed the fact that the Suez Canal excavation sites drained the agricultural manpower needed to cultivate cotton during the boom years of the American Civil War and its attendant cotton shortages. By the mid-1860s, 40 percent of all fertile land in Lower Egypt had been converted to cotton farms, and Egyptian rural cultivators had quintupled their cotton production. The labor previously used by the Suez Canal Company was now to be diverted to the cultivation of that lucrative staple.[132] For the whole of 1863, however, Isma'il hesitated. In mid-1863, for example, Nubar Pasha, the Armenian director of Egypt's foreign affairs, suggested that the viceroy, rather than dismissing the forced laborers altogether, should engage in providing a mere six thousand men a month. According to Nubar, this would have allowed the Egyptian government to maintain its leverage with the Company.[133]

In April 1863 the forced workers' ranks reached a peak when 22,480 individuals arrived at the canal sites.[134] That is around the time Istanbul took a firmer stance. On April 6, 1863, the Ottoman government addressed a note to the sultan's representatives in Paris and London expressing regret that the "unauthorized works" on the isthmus were still advancing. The note pointed out that the Porte had never approved the plan for the canal in the first place. Notwithstanding the little respect that the Company had shown theretofore for the rights of the sultan, the Ottoman suzerain was still willing to show goodwill toward the enterprise if the Company scrapped all unacceptable clauses in the concession and fulfilled some "indispensable" guarantees. First the Ottoman government demanded the drafting of international stipulations that would declare the complete neutrality of the canal, as in the Dardanelles and the Bosporus cases. Then it insisted that forced labor ought to cease on the canal worksites. Istanbul took issue with the fact that twenty thousand men were forced, every month, to abandon their jobs, return home at their own expense, and mostly travel great lengths to do so. On top of the thousands who were snatched away from agriculture, industry, and commerce each month, thousands more were on their way or preparing to leave. Finally, Istanbul opposed the Egyptian practice of concession of lands to the Company. The Porte feared that the Company would take possession of these lands and that Port Said, Timsah-Ismailia, and Suez, as well as the whole border with the Syrian province, would pass into its hands. The Company, the Porte noted, would thus acquire the power to create "semi-independent colonies" on this "important point of the territory of the Ottoman empire." No government with any sense of independence and duty "would subscribe to such transaction." The note emphasized that the Egyptian viceroy was nothing but a "high official" who governed under the sovereignty of his sultan and under the condition of obtaining the latter's sanction. Could Paris and London seriously expect the Ottoman government, the note wondered, to let a joint-stock or incorporated (*anonyme*) company such as the Canal Company establish itself on its imperial territory and claim dubious rights issued by its Egyptian underling?[135] Dispatched to Istanbul in August 1863, Isma'il's representative Nubar committed Egypt to buying back the land concessions at whatever price was required and to negotiate the abolition of the corvée on the canal. De Lesseps prepared to react aggressively. He confided that the prospective changes would not cripple the enterprise but was nonetheless prone to sell the lands as dearly as possible at the highest possible cost.[136] The Company's management maintained that the land grants were indispensable for transforming the desert into a fertile region. They indeed constituted one of the sources of revenue that had been promised to its shareholders.[137]

By the spring of 1864 the average monthly contingent had dwindled to twelve or thirteen thousand workers. It kept decreasing until May 1864, when all contingents of workers were withdrawn by the Egyptian government. The corvée system officially lasted on the canal works until 1864.[138] Earlier that year the Egyptian viceroy had agreed to de Lesseps's proposal that they submit their differences to the arbitration of the French emperor, whose wife Eugénie happened to be de Lesseps's cousin, and who had to support the Company lest he discredited himself in the eyes of the public.[139] In July 1864 Napoleon III issued a decision that unsurprisingly disfavored

Egypt. For the annulment of forced labor, the Egyptian government would disburse to the Company an "indemnity" of 38 million francs. For the retrocession of lands, the Egyptian government would pay 30 million francs.[140] Conventions agreed upon in January and February 1866 eventually settled the dispute, with the participation of an Ottoman commissioner and of representatives chosen, respectively, by the Egyptian viceroy and de Lesseps. Italian diplomats in Egypt praised the viceroy for liberating a huge portion of land and claimed that the Company was in a far from prosperous situation.[141] However, the French emperor's arbitration, by sanctioning the fact that the Company no longer owned the lands along the canal but could now only use them, actually still deprived Istanbul and Cairo of the possibility to determine their width. Therefore, the new situation still infringed on the sultan's rights.[142] On the whole, these transactions ratified a "confiscation masquerading as equity" that may have been even more vexatious and costly than earlier abuses.[143]

AN EGYPTIAN KLONDIKE

After the official end of the corvée in 1864, attempts intensified both locally and internationally to recruit the scores of laborers that the canal project still needed to reach completion. The Company dispatched emissaries to the Egyptian delta, Syria, and the Greek archipelagoes, but only a few reinforcements were mustered. The need for workers erected practical and moral obstacles to the momentum of the Company's works.[144] In this new predicament, the sustenance of the Suez Canal project proved once again to both depend on preexisting circuits of recruitment and generate novel migratory routes toward the isthmus worksites.

After 1864 the thousands of Egyptian *fellahin* who had been, in spite of themselves, previously connecting locations in Upper and Lower Egypt to the Isthmus of Suez were no longer forcibly brought to the canal worksites. Yet some apparently chose to relocate there. Individuals from both Lower and Upper Egypt found work in the coal-heaving business and in the transportation of goods.[145] Upper Egyptians were disproportionately motivated to settle in the canal region, for they had both transportation options and economic incentives to do so. In fact, in the second half of the nineteenth century southern Egypt saw the expansion of sugar cane cultivation and the extension of a railway system that allowed easier connections, but that area remained much less economically developed than the more prosperous and populous north, where the Nile Delta had been converted to perennial irrigation and increased cotton cultivation.[146]

The canal works, compounded with the cotton boom of the early to mid-1860s, drew to Egypt a growing multitude of foreigners as well. Immigrants had been heading to the country in previous decades.[147] But now "the works on the Canal attracted a mass of desperate people to the Orient, the refuse of several countries, who fetched up in Egypt when the American war turned this country into an Eldorado."[148] In 1864 an article in the *Times* declared that Europe "finally understood the immense future of Egypt and is eager to develop its budding resources. Every steamer is pouring a new population and a golden stream on our shores; energy and capital are taking possession of the land." Alexandria and Cairo were "receiving so great an influx of inhabitants that, although whole quarters are rising on every side, house room is still insufficient, and rents are always increasing, but the inland towns and villages are overrun."[149] According to *Le Bosphore égyptien* (established in Port Said in 1880 and transferred to Cairo in 1881), the country provided Europeans with a bustling commercial traffic, the chance to establish various industries, and prospects of jobs in the state administration.[150] In 1895, on the pages of *L'Imparziale*, a Cairo-based paper published in Italian since 1892, a journalist described the Egypt of decades past as a place where easy money could be made, everyone thought "that they were saved from need for ever," and people rarely cared about putting savings aside.[151] Reportedly an average of twenty thousand foreigners per year disembarked at the port of Alexandria between 1847 and 1856. After the canal works took off, more people kept coming: 33,000 foreigners, including visitors, arrived in 1862; 43,000 in 1863; 56,500 in 1864; and 80,000 in 1865. The steady crescendo paused and fell off to 50,000 when cotton prices collapsed in 1866.[152]

The Egyptian census recorded the doubling of those Europeans residing outside of Cairo and Alexandria in the span of twenty-five years: from 15,000 recorded in 1846, there were 29,271 in 1869, mostly concentrated on the Isthmus of Suez.[153] To prospective migrants, this eastern Egyptian-Ottoman region may have conjured up a blend of fascination with the unknown, freedom, prestige, and desire for adventure. Its growing cities may have promised a life away from the drudgery of the countryside that offered no modern housing, entertainment, or even a modicum of free time.[154] Finally, the mirage of remuneration on the canal worksites, with vaunted salaries of 5, 6, or even 7 francs per day, reportedly tipped the scales for the undecided. Calabrians, for example, were apparently inclined to cross the Mediterranean because they had never handled such sums before.[155]

In 1874 a Catholic priest dispatched to the isthmus attempted to describe its precariously established new parishes to his superiors in Rome by comparing his surroundings to California.[156] According to contemporary and later observers, seekers of gold "from every corner of Europe" were descending en masse upon Egypt "as on a new California" and their movements now amounted to "a real Klondike."[157] They thereby compared Egypt-bound migration to the so-called gold rushes of 1848, when gold was first "discovered" in California, and 1896, when migrants departed for the Klondike River in what was to become western Canada's Yukon territory.[158] The combination of the Suez works and the cotton boom in the mid-1860s had thus turned Egypt into the successor of the Californian frontier and transformed the Suez Canal into the harbinger of the Klondike.

Changing Egyptian and Mediterranean Migratory Paths

Migration toward the isthmus changed both Egyptian and Mediterranean trajectories and options for mobility and in the process affected some key urban nodes along its path. Suez, the canal's namesake located at the southern tip of the isthmus, enjoyed a brief renaissance thanks to the arrival of foreigners in town, the establishment of commercial agencies, and brisk building activity. Pierre Lebret, for instance, who was employed in the inspection of materials for one of the Company's contracting firms in 1866 in Port Said, left that town to go reside in Suez.[159] West of Port Said, the port city of Damietta began suffering the consequences of Port Said's growth as early as the late 1850s. This came on the heels of another blow Damietta had received in prior decades, when Alexandria and its role in Egypt's revitalized trade with Europe had displaced Damietta from its previous status as a key commercial mediator between Egypt and the Levant.[160] In the 1860s Damietta still counted sixty thousand inhabitants and was roughly six times larger than Port Said. Nonetheless, it was allegedly "disappearing" given that so many of its denizens were relocating to Port Said. Some turned to trading food and other goods and made good money, especially in their dealings with foreigners. Many were artisans. Others found jobs in the service of foreigners' houses or shops.[161]

Farther west on the Egyptian coast, Alexandria also became entangled in swelling Port Said by means of the isthmus-bound migration. "European outcasts," barbers, and tailors out of work hailing from Alexandria were among the "free" workers who had first showed up for recruitment but were deemed unfit and swiftly dismissed.¹⁶² Others stayed on. One man called Antonio, for example, was a coach driver in Alexandria before relocating in 1865 or 1866 to Ismailia, where he later had an extramarital affair with a laundrywoman. Around 1867, twenty-six-year-old Joseph Mayond spent two months in Alexandria, where he hoped to recover from sustained fevers that had afflicted him for some time. While he had previously resided in Port Said, whence he wrote to his parents in August 1864, he relocated to the Chalouf worksite three years later. The following year the two Calamita sisters also moved from Alexandria to the isthmus when they were recruited to sing and dance at a music hall to be opened in Ismailia.¹⁶³

Finally, Marseille, on the southern French coast, became a key node in the circulation of isthmus-bound goods and bodies. Even though it was in competition with other Mediterranean ports such as Genoa, Trieste, and Izmir, Marseille was a growing and bustling port city through which immigrant workers and refugees came and went.¹⁶⁴ It also became the main provisioning center of the isthmus; tiles, bricks, lubricating vegetable oils for engines, marine rope, carpentry elements, and rustproofing were produced in Marseille to then descend into the Mediterranean and point east towards Port Said.¹⁶⁵ It was in Marseille that the Company charged an official with disbursing payments to those employees who disembarked from Egypt and again headed back after their vacation was over.¹⁶⁶

International and Domestic Recruiting Efforts

In the summer of 1864, right after the corvée was officially terminated on the isthmus, the Company weighed its options for active foreign recruitment. In May and June 1865 intermediaries in Hong Kong offered to procure up to twenty thousand "coolies" or "Chinese workers." Without specifying whether these laborers would be indentured or not, middlemen vaunted their excellent attitude in the works of earth-leveling and removal (*terrassement et deblaiment*) and promoted them as docile, intelligent, and tireless. It was reasoned that since they were moving in great numbers to the Western Indies, Peru, California, and Australia, among other places, they could just as well make it to Egypt.¹⁶⁷ On the canal, Chinese workers would be capable of the "patient endurance" needed to turn the desert into a farmland. Supposedly attractive traits also included their lack of interest in religious fanaticism, their being "more used to liberty than Muslims," and their convenient habit of feeding on rice, a staple of which nearby Damietta could provide quantities in "superior quality." In sum, middlemen praised Chinese workers' intelligence and industriousness and claimed that they embodied "the Greek of the Pacific Ocean," a metaphor that blended the familiar and the exotic in Western minds.[168] This marketing pitch suggests that Company cadres, to whom the offer was supposed to appeal, must have been susceptible to both Orientalist tropes about Muslim fanaticism and penchant for authoritarian rule and to the stereotypes about the meek and productive Chinese "coolie" laborers that also circulated in nineteenth-century North America.[169] Further, even if it is unclear whether the Company ever got to recruit Chinese workers, Company echelons still may have considered relying on yet another form of indentured or forced labor even after the Egyptian sovereign had forcibly terminated the corvée of Egyptians along the canal.

In its hunt for manpower, Company management was vested in banking on mediators to simplify its services and reduce personnel and expenses. It occasionally relied on the aid of French consular authorities overseas, as in the case of the French vice-consul in the Moroccan port of Tetuan who, in August 1864, endeavored to hire "Moorish" laborers for work at the Suez Canal.[170] Indeed, there is evidence that the Company succeeded in enlisting Moroccan workers.[171] In general, Company management was not willing to deal directly with railroad or navigation companies for the transportation of workers and food staples from the ports of Trieste, Brindisi, or Marseille to Port Said. It was reluctant to take on the responsibility of recruiting workers, bringing them to the isthmus, and eventually repatriating them, all the while having to worry about finding the mechanics and other specialized laborers it needed.[172]

Disparate ideas circulated among Company cadres about the different ways in which international recruiting ought to be organized. Alphonse Couvreux, a well-known French entrepreneur who had been entrusted with fifteen kilometers of embankment (*terraissement*), suggested that the Company ought to cover workers' travel. According to Couvreux, the Company also should grant them full freedom to work at whatever worksite they chose throughout the isthmus, which they could reach for free on the barges floating up and down the portion of the canal in the making (if they decided to move to a different worksite, however, they would have to pay for passage). Couvreux insisted that workers ought to be set free from contracts that bound them to toil at the identical chore, in the same spot, for a uniform salary, or for a determined time.[173] The self-styled impresario Joseph Casanova in Civitavecchia, a Tyrrhenian port in central Italy, hypothesized even more guarantees. He similarly expected the Company to pay for migrant workers' passage to Egypt. But Casanova also suggested that the Company should disburse their salary starting from departure day, provide them with suitable lodging, guarantee them food, and ensure an ambulance service. Those sick workers who did not need hospitalization would get half a day's pay. In Casanova's plan, if activities languished or stopped altogether due to the Company's responsibility, workers would be paid as if they had been working or, alternatively, they would be conducted back to the point of their arrival. If, because of climate, one or more workers could not bear living in Egypt, the Company would cover repatriation costs. Unlike Couvreux, Casanova proposed that the Company fix the hours of work and the daily pay for each type of worker (earthwork laborers, wheelwrights, carpenters). Salaries would be maintained as long as work proceeded under the Company's control (*à regie*) rather than by piecework (*à la tâche*). Finally, the Company ought to pay for migrant workers' passports.[174]

In 1866 the Company struck a compromise. But this appeared to apply only to workers recruited in France. It would pay them starting on the second-to-last day before their departure for Egypt. It would cover their transportation costs. It would provide them with lodging and an indemnity equivalent to fifteen days of work upon arrival. However, it would withhold 10 percent of their salary for repatriation costs up to 300 francs, which would later be reimbursed at a yearly interest of 5 percent after their return to France. French workers left home on a three-year contract.[175]

Much more haphazard appeared to be the arrangements worked out for other migrants, such as those departing from Calabria, a region at the southernmost tip of the Italian peninsula. Brokers or contractors (*imprenditori*) had indeed recruited them by the dozen in both 1865 and 1866. In October 1866 more than eighty Calabrian migrant workers landed in Alexandria with the "English steamboat" coming from Naples, after undergoing ten days of quarantine. When it came to paying for travel to the isthmus and their later repatriation, however, contractors left migrant workers stranded, thus shifting the onus onto their consular representatives in Egypt.[176] Theoretically, contractors were supposed to help workers navigate the complicated procedures necessary to obtain passports, change currencies, distribute incoming mail, and build trust. They also galvanized migrants' hopes about their departure.[177] However, when the southern Italian port cities of Naples and Bari were flooded daily by hundreds of perspective migrants from the surrounding provinces, brokers apparently limited themselves to negotiating their transportation with navigation companies and made sure to withhold generous sums for each of the recruited men. Sometimes migrants would be carried to a different destination than the one they had originally subscribed to.[178]

Experience in port construction and railroad work, compounded by a proclivity to displacement, was a desirable asset in this job market. Contractors pitched the ideal Company worker as someone who had toiled in the tough worksites for the Adriatic ports and the Italian railroads and who could "expatriate easily." They then offered the toolkit they had described as desirable. Casanova proposed, for example, to bring to Egypt two or three hundred hard-working men originally from L'Aquila, a town in a mountainous area of central Italy, who had toiled on the leveling (*terraissement*) of the Rome-Civitavecchia railroad. It was preferable to extradite them while they were still displaced, Casanova added, because it would be difficult to make them leave once they returned home. He himself had been employed as the storeroom guard in the Rome-Civitavecchia railroad works. Others, such as Johann Steiner from Diakovar in Slavonia, offered to recruit 150 or 200 workers who had been employed in the construction of railroads for the Austrian monarchy and who would have been available for two or three years.[179]

Established recruitment networks appear to have been in place by mid-1866. Within Egypt, the Company relied on intermediaries. The *tâcheron* (subcontractor or taskmaster) Emile Girardeau, for example, trusted his associate "Hadju Ahmet," an Egyptian subject, for the regular provision of men to be employed at the excavation sites. When, in 1869, the latter failed to fulfill this task, Girardeau pleaded with the governor of the isthmus to summon "Hadju Ahmet" to the Diwan, the governmental office, in Ismailia.[180] Contractors (*entrepreneurs*) were entrusted with the task of finding workers outside of Egypt as well. Within each country, scouting often occurred at specific sites and repeatedly. In the French town of Saint-Nazaire, situated in Brittany at the Loire's estuary end, for example, men were recruited both in 1865 and in 1866. Saint-Nazaire, in the nineteenth century, was well known for its Chantiers de l'Atlantique shipyards.[181] It could perhaps provide the specialized workers the canal worksites and workshops needed. A certain Jules Menard left Saint-Nazaire at the beginning of 1865 and was employed for a while at the worksite of Al Guisr by the Couvreux enterprise. Another man called Chenet, in June 1866, was also recruited in Saint-Nazaire and was bound to leave from Le Havre on a mid-June Sunday. Saint-Nazaire also functioned as a recruitment center for nearby locations; multiple presumably Breton workers were mustered there, for instance, from nearby Lorient, farther northwest on Brittany's Atlantic coast. Joseph Le Pesquer of Lorient left Saint-Nazaire for the canal works in 1865. Eugene Warri, born and living in Lorient, left his hometown in 1866 and was employed at Serapeum at least until September 1868, the date of the last letter he sent to his family back in France.[182]

Some of the workers who were traveling back and forth seem to have acted willy-nilly as recruiters. Those expatriates who fashioned themselves into middlemen could capitalize on their prior knowledge and potential influence within their local communities.[183] In 1866, for example, an entrepreneur enlisted a few Calabrian laborers to go back to their country to recruit yet more workers. Their travel expenses were covered, and each man they brought over to the Isthmus would guarantee them an allowance of around 5 francs. A few weeks later, two hundred Calabrian workers landed in Egypt.[184] In 1867 one subcontractor successfully recruited, in Izmir and other unspecified sites, a certain quantity of Georgian and Armenian earthwork laborers (*terrassiers*). Disembarked at Port Said, they were later conducted to the worksite of Seuil to the south.[185] But workers continued to approach the Isthmus of Suez independently as well, either as isolated individuals or in groups, such as the mechanics who together left the town of Mulhouse in eastern France to be hired in the canal works in 1868.[186]

The Millions to Be Made on the Canal Works

Word traveled that there was good money to be earned on the isthmus. Information on available jobs or possible accommodations along the Suez Canal circulated quickly through publications, word of mouth, or even hearsay. Newspapers around the world had been trumpeting the millions that could be made on the canal works.[187] Casanova, the abovementioned self-styled contractor in Civitavecchia, eagerly devoured the "journal of the piercing of the Isthmus of Suez," presumably *L'Isthme de Suez: Journal de l'union des deux mers*. Casanova may have chatted about his reading with the workers he met daily on the railway construction sites. He boasted that skilled journeymen (*maîtres compagnons*), wheelwrights, ironsmiths, carpenters, stonecutters, and platelayers pressured him to know when they could finally set off for the isthmus.[188]

Whether it was through literate readership or word of mouth, perspective migrants knew about the canal works and had been considering the Isthmus of Suez as early as 1858. That year crowds of workers in Montbéliard, in eastern France, daily assailed a navigation company agent because they believed he could enroll them to work on the piercing of the isthmus.[189] A great number of Italian nationals, for instance, reached Alexandria in 1866 and preposterously claimed that they had been engaged for the Suez works. Theoretically, as argued by a high-ranking Company official in 1866, the Company was "always ready to maintain the contracts it had subscribed with foreign workers" but refused to unconditionally enroll those men who arrived at the worksites haphazardly (*à l'aventure*) and with no preliminary arrangements.[190] This stance suggests that migrants had indeed been arriving at the isthmus either on prearranged contractual bases or autonomously on the basis of passed-on knowledge. It does not clarify whether or not they thought signing a contract could have placed them in a better position. At any rate, they may not have enjoyed the luxury of having that choice to make. Lacking guarantees, some foreign migrants may even have chosen Egypt due to rumors they had heard. It has been hypothesized that the Maltese, for example, up until the First World War, preferred to be less well-off in the southern and eastern Mediterranean than to venture to the Americas and Australia, believed to harbor strange and deadly diseases.[191]

Perspective isthmus-bound migrants sought to gather information from those who had already reached their intended destination. Responding to one of these requests, for example, a man named Antonio Candida replied from Egypt in April 1865 to a (male) "dearest friend" of his that surely employment was to be found, although he could make no guarantees about the pay. He even pressed his correspondent to leave right away.[192] As argued by migration scholar Leslie P. Moch, people have historically been the most effective conduits of information about relocation possibilities. The unevenness of human relations and networks can help explain why migrants did not act as perfectly rational decision makers, possessed incomplete knowledge, and concentrated on particular destinations and specific occupations.[193] Letters from fellow villagers may have contributed to raising exaggerated expectations and exciting vain hopes. Maria Salvagno, writing from the then Habsburg Island of Cres/Cherso off the Croatian coast, made a few attempts at asking her husband, already in Egypt, to find a job for one of their relatives. While she did so at first rather timidly, she later reported to her husband that the man she had written about got very angry for not having received any tip. He was willing, she reported, to leave right away. "We are bust," she added. "God has seemingly forgotten these places" (*pare che Dio si abia scordato di queste parti*).[194] Here, too, few migrants made choices only as isolated individuals. Decisions were made in the context of information and assistance obtained from relatives and village members who had already ventured further afield.[195]

If reliable information about several destinations was available, migrants may have opted for those places where relatives or friends already lived and could provide shelter for the first nights or access to jobs and property commensurate with the newcomers' skills or means.[196] Women who reached Egypt from central Italy to work as nannies, for example, often joined their mothers, sisters, or female townsfolk, who provided them with the necessary money, information, and professional opportunities.[197] Family members appeared to be on the isthmus at the same time, although not necessarily in the same exact location. They may have undertaken their Egypt-bound voyage together but parted ways upon arrival in order to diversify their options. For example, it was the brother of a Greek navvy (*terrassier*) who had been killed in a brawl who collected the deceased's unpaid salary. The brother of a French restaurant keeper in El Guisr, doing the same job in Ismailia, closed down the canteen his sibling had abandoned due to a state of alleged "mental alienation."[198] As in similarly remote destinations, forms of so-called chain or network migration may have encouraged migrants to settle near one another, stay longer, and in turn, invite others back home to join them.[199]

But obtaining the papers required to cross international borders was no easy feat. While many Ottoman citizens could obtain domestic travel documents that permitted them to travel to Egypt (after the Ottoman Empire regulated internal passports in 1887), non-Ottoman immigrants experienced complicated, lengthy, and confusing procedures. Those departing from the Italian peninsula were a case in point.[200] First they had to request a "no-objection statement" (*nulla osta*) from the mayor of the town in which they resided; only afterward could they request a passport from the "authorities of public security." Before 1888, when restrictive policies on emigration still prevailed in Italy, the poorest citizens may not have had a chance.[201] At every step of the way, aspiring migrants were subject to fines and bribes in one office after another.[202] Finally, Italian passports could be denied on multiple grounds: familial obligations (if, by departing, individuals were presumably abandoning their families), absence of fatherly agreement, absence of tutor's agreement for minors and those proclaimed "unfit," and in the case of "girls," presumed intention to practice prostitution.[203] Apprehension about the latter would motivate Greece to pass legislation in 1912 forbidding women under twenty-one from traveling abroad without a special permit.[204] Women intending to relocate abroad thus often had to bear with narratives of victimhood and shame, imposing protective and yet restrictive policies on their ability to relocate.[205]

Those unwilling or unable to navigate the bureaucracy may have crossed borders clandestinely. In 1869 alone, for instance, there were at least twenty-two recorded cases of "illegal" Italian migrants who had departed for Africa.[206] Italian citizens could, for example, apply for those passports to be used "internally" that municipal administrations released easily. Then they could cross over to France, where having a passport was not required to obtain a residency permit. There they could secure a forged passport and board ship in either Marseille or Le Havre. In Marseille, a "semi-clandestine" agency run by Italian subjects provided paperless migrants with "more or less falsified" documents.[207] Forging was a rather florid industry in Egypt as well. By the mid-1870s, not only was falsified gold and silver currency, even in relatively smaller denominations of 5 francs, circulating in Port Said, but procuring a fake passport appeared to be quite a feasible feat.[208]

Departing might have been a familial strategy aimed at social mobility. Recent migration scholarship has emphasized the family as a fundamental arena of decision-making.[209] However, breadwinners may have migrated in order to escape from unhappy situations rather than to abide by decisions taken with full household consensus. Some of the men recruited by the Company's mediators along the Atlantic coast of France in 1865 and 1866 seemingly abandoned their families and left them destitute. The wife of the migrant worker called Chenet, for instance, declared she was utterly surprised by the letter with which her isthmus-bound husband informed her he had left to go find work, all the while consigning to her the responsibility of rent and an acute form of migraine. Consequently, she was compelled to seek out a public scribe's services to request that a reasonably sufficient sum be withheld from his pay and sent over to her, while pondering whether she should chase him at Le Havre before his imminent departure. Joseph Le Pesquer's wife, "a poor woman in misery" with two young children to support and no means of subsistence, decided instead to turn to the local police.[210]

Flight from military conscription and criminal prosecution also appear as frequent triggers for emigration. Egypt apparently teemed with Romanian deserters and defaulting soldiers.[211] Citizens of the newly unified kingdom of Italy, where the compulsory draft lasted at least seven years, also may have swelled the ranks of deserters on the run within the country.[212] The Ottoman Empire in general, North Africa, and Egypt had been harboring European revolutionaries since the 1820s, offering asylum to Italian Carbonari and to participants in the 1848 revolutions fleeing the repression spreading in western, southern, and Balkan Europe.[213] On the whole, there might have been several reasons for fleeing from home. In 1868, for example, a man by the name of Raffaele Marendi used a fake passport to escape from the Italian town of Faenza after killing a man. He had been in Suez for just a few days, with no means of subsistence, when he was assassinated, perhaps in revenge. The French, eighteen-year-old Ethienne Marthoud and the Swiss, twenty-four-year-old Emile Schoerr had similarly found refuge in Egypt, perhaps in Chalouf. Marthoud (see figure 5) had stolen from his previous employer, a stockbroker in Lyon, while Schoerr stood accused of falsification and embezzlement in his prior post as teller at the Federal Bank in Zurich.[214] Since the 1820s, as highlighted by Julia Clancy-Smith, political upheaval, boom-and-bust capitalism, demographic imbalance, and environmental degradation converged as valid motives for setting out. Reasons ranging from criminal pursuits, vendettas, kinship obligations, marriage patterns, trading networks, flight from undesirable spouses, to debts could all potentially prompt people's displacement throughout the Mediterranean in the nineteenth century.[215] They muddle the artificial distinction between voluntary migration and forced relocation.

Migratory movements were also facilitated by the cheap steamship lines that served the southern Mediterranean.[216] In turn, new forms of mobility dictated changes in transportation. Women willing to relocate from central Italy to Egypt, for example, knew they could embark on the Florio-Rubattino vessels anchored in Ancona, a conveniently located port city.[217] European firms had been running regular steamship lines since the early 1850s. In particular, the French state had passed a convention in 1851 with a company called "Messageries nationales" (later known as "impériales" and then as "maritimes") to establish a Mediterranean postal service, soon to be adapted to human transportation.[218] In a rush to outcompete one another, different navigation companies entered a race to grab the opportunities offered by the canal in the making and lure the canal-bound migrants. Companies adapted their Mediterranean routes accordingly. Those in the business of recruiting

FIGURE 5. Fugitive Ethienne Marthoud, 1869. *Source:* CADN, ARI, C9, Photograph attached to letter from Alexandria, 4 December 1869, French Managing Consul to Geyler, French Consular Agent in Ismailia.

and transporting workers attempted to entice Messageries Impériales in Civitavecchia, Maison Bazin in Marseille, and Austrian Lloyd in Trieste and Brindisi to function as potential carriers.[219] In 1865 it was indeed possible to reach Port Said directly from Marseille thanks to the services of the Maison Bazin. Moreover, once prospective migrants reached Alexandria, they could count on the regular transportation service that this company had established between that port city and Port Said. Starting in 1866, the Fraissinet Company also began stopping in Port Said three times a month.[220] In 1874 the Russian Company of steamship navigation also publicized its biweekly service between Alexandria and Port Said at the cost of 42 francs for first class, 30 for second class, and 10 for deck passage. The price for the latter kind of travel may have fluctuated considerably in response to the degree of competition prevailing among shipping companies.[221]

Migrants could embark with approximately sixty kilograms of luggage each.²²² The relatively short trip on a steamer from Alexandria to Port Said took up to eighteen hours. Similarly, passengers from Jaffa to Port Said had to endure twelve to fourteen hours on board.²²³ The trip to the isthmus could take a toll on voyagers' health, especially on those who were sickly to begin with.²²⁴ Travelers of different means were accommodated in dissimilar fashion: the wealthy and high ranking had nothing in common with poorer migrants, be they Syrians who were off to America or destitute Italians shuttling from and to Mediterranean ports. Those in third class slept either on straw beds arranged on deck and protected by a tent or on two levels of shelves built into the sides of the boat, on mattresses tossed next to one another, and in a melee of bodies. Men, women, and children all laid down together.²²⁵ Lactating mothers who were seasick with the boat's rolling motion could hardly nurse their babies, and it is unlikely those among them who were in third class could rely on artificial milk either.²²⁶

Upon disembarkation at an Egyptian port, different border crossers were expected to produce disparate kinds of identification papers. Until the early 1870s, while identification documents eased passage, they were not necessary for it. The white, male, well-dressed travelers traversing Egypt and the Ottoman Empire with an 1872 edition of the *Help-Book for Travellers to the East* in their pockets ought to be provided with them, but passports were not often called for—at least from them.²²⁷ In theory, those workers who, upon their arrival at the canal works, had a passport with them had to leave it either at the chancellor's office of the closest consulate or with their employer. Bespeaking ambitions for closer surveillance of the workforce, a regulation drafted in 1862 prescribed that Company workers would receive camping items or lodgings only after leaving their passports or residence permits (issued by consulates to those residents of Egypt and the Ottoman Empire who could establish foreign protection or subjecthood) with the head of the worksite (*chef de campement*). Moreover, workers would receive their departure tickets only after the approval of said boss and the restitution of all they had previously received.²²⁸

Some isthmus-bound migrants followed trajectories that were circular rather than linear. As had been the case for domestic mobility in their own countries of origin, some of them relocated in temporary or cyclical fashion.²²⁹ For example, the overland Syrian migrants who stepped onto the isthmus via Al-'Arish, a coastal town east of Port Said, may have shuttled back and forth with ease. This was an apparently seamless borderland, where a properly demarcated boundary only came to be in 1906 when Egyptian, Ottoman, and British delegates drew a line from the Mediterranean to the Gulf of Aqaba and placed Al-'Arish within Egyptian territory.[230] In 1869 alone, 5,574 people were recorded as immigrating into Egypt via Al-'Arish, with 4,104 recorded as trekking in the opposite direction.[231] For some, such as the Maltese, the possibility to easily and frequently return must have made the North African and Egyptian shore look appealing, as did its familiar languages and customs.[232] Other needs may have dictated migrant workers' trajectories. Jules Morand, for instance, after having worked for many years at the arsenal of the state in Toulon as a boilermaker, embarked in 1863 for Egypt "for the piercing of the Suez Isthmus." He stayed there until 1865, when he returned to France to regain his health. Once recovered, he left again for Suez.[233]

Not only did migrants travel back and forth, but they also maintained strong connections to the places where their journeys had begun via, for instance, flows of money or correspondence. The husbands of Maria/Marieta Salvagno, Maria Baraga, and Ana/Anna Cobau had all migrated to the Suez Isthmus with other countrymen from the Istria region, then part of the Habsburg Empire. These women, writing from the island of Cres/Cherso and the neighboring seaport of Rijeka/Fiume, received the meager sums of money sent by their husbands from the hands of returning countrymen.[234] In turn, women entrusted their letters to the laborers who were about to depart again, thus hoping to find alternatives to the often unreliable postal system. The husbands of Maria/Marieta, Maria, and Ana/Anna did not make it back home, or at least not during the time frame covered by the one-sided correspondence housed in the archives. These women waited to spot their spouses among the many returnees from Port Said. Beleaguered by looming rent payments and their children's hunger, they kept on writing, begging for money, lamenting their husbands' lack of attention, scolding them for how little they sent over, and even threatening to depart for Egypt by themselves to ferret them out.[235] Through letters and remittances, migrants strove to maintain relationships with those back home and keep the door open for their return. Arguably, going back constituted an ever-present option, both for those migrants who eventually decided to head back or for those who continued to walk those steps only in their thoughts and dreams.[236]

WOMEN AND FAMILIES SETTING OFF
FOR THE ISTHMUS

Initially the population in Port Said seemed to be mostly composed of men who had left their families in their countries and "had come to Port Said exclusively driven by the goal of getting richer."[237] In Ismailia, a square had been renamed "square of unmarried men" (*place des célibataires*) after the Company had assigned all its surrounding houses to single males. In Port Said, an entire neighborhood where young and playful unmarried men resided had been ironically nicknamed Charenton after the French lunatic asylum in Charenton-Saint-Maurice.[238] After 1865, with the intensification of the international recruiting of labor, the arrival of whole families in Port Said became especially notable.[239]

But women, even if faintly discernible between the lines or along the margins of the archival record, holding a ladle and a small cup (see left side of figure 6), also undertook migration to the isthmus right at the onset of the canal project, either with males or by themselves, sometimes reaching the location where their kin had already, if precariously, settled and at other times fending for themselves. The official record is reticent; it mostly resists categorizing women according to what work they performed on the Isthmus of Suez in the second half of the nineteenth century. Nonetheless, it clarifies that this region did not just welcome the male and unaccompanied prototypical migrant disfavored in terms of marriage and reproduction.[240] It was a much more capacious home.

De Lesseps explicitly recalled that, by mid- to late 1860, Christian workers and their families already dwelled in Port Said, and Egyptian men and women were also in the process of settling in the isthmus.[241] The first foreign immigrant women who disembarked in Port Said were Greek ladies from the Aegean Island of Kasos who found employment in domestic labor.[242] As early as April 1860, some of the one hundred European workers settled in Port Said had "their wives with them."[243] At the onset of the 1860s, for example, the wife of Alphonse Goëlo, a Company foreman who had been previously employed in Brittany's department of Morbihan, accompanied her husband to the isthmus. When he died there of dysentery in 1862, she was left with four young children and her own precarious health to attend to. She had no family savings to draw on, since all had been spent on their travels to and from Egypt. She may have had no choice but to remain on the isthmus and eke out an existence there for herself and her little ones. In 1863

FIGURE 6. Personnel coming out of the Gouin worksite in Port Said, 1869. A female worker is visible in the bottom left corner. *Source:* Bibliothèque nationale de France, Paris, Justin Kozlowski, *Canal maritime de Suez: Photographies d'après nature* (n.p.: n.p., 1869), 5.

Manuel Lepin's pregnant wife joined her husband, who had been in the hire of the Company since 1860. After his request to move as a couple to a free room was denied, she had to bear the hardship of living in a tent with Lapin and several other men while gravid.[244] In 1864 the wives and families of 150 English and Scottish workers, who had done dredging work before on the Clyde River with the Ayton company, landed all together in Port Said.[245]

Company management encouraged the arrival of families and their on-site settlement in hopes that workers would take up residence in a relatively permanent fashion and stay on for the entire duration of the works.[246] De Lesseps declared that by 1861 the Company had created villages "especially for Arabs all along the path of the works to bring their families along." He bragged that each village included a mosque and that a canal would provision all villages with Nile water.[247] The Company resorted to similar strategies when it came to enticing Greek workers to come with their families and stay long enough on the isthmus, sponsoring the construction of a "chapel" in Port Said and founding there a school with a Greek male teacher. Indeed, numerous Greek workers did relocate with their families and founded the first nuclei of the Hellenic communities of Port Said, Ismailia, and Suez.[248]

Almost all workers on the canal worksites reportedly had a "small, more or less legal, household arrangement" (*ménage*).[249] Some migrants looked for spouses in their home countries, either in person or through the offices of the mediators (*sensali*) who were active in Egypt at the time.[250] Men may have had two families at once, such as a Mr. Fontane, a lemonade maker in Port Said. He "was married in Toulouse where he left his wife and daughter in poverty to go, they say, in America." But the grapevine suggested he was really in Egypt, where he lived with his partner and their two children. Similarly, Giuseppe Giammugnai, who got engaged to a young lady in Port Said, was already married and had children, they said, in the Italian town of Savona.[251]

Marital unions and adulterous liaisons also formed on-site among members of different ethnicities or classes. Within dense urban neighborhoods, the physical proximity of members of different communities provided a context for relationships cutting across supposedly impermeable group boundaries.[252] "Similarly to others hailing from the places of our birth," Italian Company employee Ludovico Alessandrini wrote in 1870 to his consul in Cairo, he now wanted to fulfill his promise of marrying the Greek subject Regina Serali, with whom he had lived in the Suez "encampments" for many years already.[253] In 1877 a widow in Port Said asked her French consular representative to authorize the marriage of her daughter, born in Marseille, with a man of Austrian origin.[254] De Lesseps's strenuous defense of his ill-informed conviction that no Egyptian peasant woman could ever bear the children of a European man actually suggests that such unions were not just possible but also frequent.[255] Foreign officials and notables residing in Egypt publicly declare that "mixed" marriages were very rare but ended up admitting that a few Italian men had indeed married indigenous women.[256] Even if infrequent, examples of "mixed" relationships and affairs challenge the notion that discrete and mutually exclusive religious and social strata have inhabited the Egyptian past.²⁵⁷

On the Isthmus of Suez, as elsewhere in Egypt, conversion functioned as either a prerequisite or a consequence of interfaith marriages. It enabled nuptials or sanctioned already existing households. A Muslim woman, for example, was baptized "Maria Giuseppa Emilia" so that she could wed the Dubrovnik-born Catholic Giovanni Ivankovic, with whom she shared affection, resided, and had children (after having abandoned her blind Muslim husband).²⁵⁸ Francesco Talamasi, residing in Egypt since at least 1861, was an Italian subject who lived in Port Said and was known as "Mohammed Abdallah." Upon his death, he left a Muslim widow and daughter as well as other "collateral" heirs—presumably conceived in previous marriages—who resided in Alexandria. Eugenia Brocchini, an Italian subject born in 1874 and the daughter of a laundrywoman in Suez, converted to Islam in 1894. Disregarding her parents' judgment, she married local subject Ahmed Sirsar. The two had already resided conjugally for a while. Finally, Dante Baldini, an employee at the Quarantine Office in Suez who was born in Leghorn in 1841, was an Italian citizen and always expressed his clear intention to remain so, even joining the administration of the Italian school (presumably the one in Port Said). He converted to Islam in 1882, "on which occasion he was assigned the name Ahmed Sadik al Muslimin so that he could marry in front of the Kadi of Scibin el Kom the indigenous woman Kadra bent Mohammad el Ascari."²⁵⁹ Not always did aspiring converts get away with their schemes, especially when they, as transgressors or religious and ethnic taxonomies, were young and female, as in the case of a Greek girl willing to convert who was swiftly repatriated by her consular representatives in 1895.²⁶⁰

Being a Child about the Worksites

Children, whether born or going through childhood in the Suez Canal region (roughly until fifteen years of age), beckon to us from the edges of both the visual and the written record of the time, as does the little nameless girl standing in the upper right corner of a photograph appearing to be mostly occupied with another story (see figure 9 in chapter two). Yet children have the power to beget additional accounts. For example, the birth of a baby girl called Ida Gemma Emma on July 23, 1869, in Ismailia reveals the migratory trajectories of her parents, Antonietta Marmonti and Achille Ferrugio, moving about the isthmus in search of work. They had both traveled from northern Italian towns, Udine and Magenta, respectively. It was in Egypt though, specifically in Alexandria, that they had gotten married in 1868.²⁶¹ Therefore, while it is unclear whether Achille and Antonietta had undertaken their relocation to Egypt as an engaged couple or had traveled independently and eventually met on the other side of the Mediterranean, their offspring attests to their eventual encounter and relocation to the Suez Canal banks (see figure 7 later in this chapter and figure 9 in chapter two). But besides the infants who were first coming into the world on the isthmus, those children who were old enough to go to school in the first few years after the excavation's beginning in 1859 must themselves have undertaken the migration to this region of Egypt, either alone or with parents, kinfolk, or other adults. There were enough school-age children in Port Said to result in the opening of several schools. In early December 1860, forty-two pupils were enrolled at the school run by the "*iman* [*sic*]" whom de Lesseps had personally invited to Port Said.²⁶² In April 1863, just a few days after landing in town, the nuns of the Bon Pasteur Catholic order from the French town of Angers established a school in Port Said. Although the Company furnished the nuns with books, paper, ink, and all that they asked for, parents were said to be "very indifferent" and to neglect sending their children regularly. "Our little girls," the nuns lamented, only came for two or three days and missed school for stretches of fifteen days. Consequently they learned very little. The sisters had set up both an orphanage and a day school in which students paid fees.²⁶³ While the nuns welcomed girls, three Franciscan priests and one layman were, also by 1863, operating a school for both Catholic and Muslim boys.²⁶⁴ Starting in 1866, the Greek community in Port Said maintained, besides a chapel, a boys' and a girls' school.²⁶⁵ By 1873 there was also a Jewish boys' school in town, presumably a Talmud Torah, with fifteen pupils. These children alone must have amounted to about one-quarter of Port Said's Jewish population at the time, which included twenty families and numbered about seventy individuals.²⁶⁶ Overall, the state of children's instruction in Port Said by 1879 was flourishing and moved the Egyptian geographer Muhammad Amin Fikri to describe this town's numerous schools and establishments for education training (*makatib al-tarbiyya*) enthusiastically.²⁶⁷

Not only were children out and about on the canal worksites, but they also contributed to the smooth functioning of the enterprise. Children of unspecified nationality worked as "blowers" (*souffleurs*) in the atelier of ironworks in Port Said as early as 1861.²⁶⁸ In Qantara alone in 1861, around a hundred "little girls and boys" between ten and thirteen years of age

Ouvriers terrassiers du canal de Suez travaillant à la couffe. (Voy. page 20.)

FIGURE 7. Children toiling as earthwork laborers on the Suez Canal worksites, 1862. Illustration by Hurel. *Source:* P. Merruau, "Une excursion au canal de Suez (1862)," *Le Tour du monde: Nouveau journal des voyage VIII*, 2nd semester (1863): 13.

pursued their tasks with "remarkable zeal and dexterity."[269] Specifically Egyptian children under twelve years of age, even if they made up less than one-quarter of each contingent, had still been arriving at the Company's encampments under the forced-labor system that had been formalized in 1862 (see figure 7).[270] This was consistent with Egyptian children's recruitment at archaeological excavation sites elsewhere in Egypt, where children as young as five removed dirt with the aid of baskets.[271]

Children in Port Said led more precarious lives than adults. Between the summer of 1864 and the summer of 1865, fourteen infants under one year of age perished in this town alone. Seven slightly older children between the ages of one and ten, some of whom must have been born elsewhere and undertaken the journey to the isthmus, also died in the same time frame. Company health officials admitted that children in this area were particularly vulnerable to the lack of hygienic, residential, and dietary measures. At the same time, however, they held specific assumptions about the younger ones' adaptability to climate that were based on their supposedly innate ethnic characteristics (as further expounded in chapter two). They considered the offspring of Maltese and Greek parents as the healthiest, while second came Italian children and those from the French Midi. Ostensibly, southern European children needed less hygienic care than those children whose origins lay farther north and who were reputed to be more sensitive to the heat.

If interpellated, health officials would have then predicted that, for example, the five children whose birth was registered in Port Said between February and May 1863 had unequal chances of survival: three were French, one was German naturalized French, and one was Maltese.[272] But life and toil on the isthmus did not discriminate when taking their toll on the most vulnerable migrant population.

PEOPLES OF ALL SHAPES AND COSTUMES

By 1865 Port Said's beach, still deserted a mere six years before, had become a huge arsenal. Contractors were busy installing and readying the by now congested worksites.[273] The town counted seven thousand inhabitants, of which around twenty-five hundred were identified as "Egyptians" and the rest as "Europeans." Besides the five wooden huts accommodating workers, the steam-operated devices to distill and desalinate salty water, the lighthouse, and the oven for baking bread, three robust constructions had been erected to lodge supervisors.[274] There were a doctor "belonging to the encampment," a pharmacy, and a field hospital. A general store sold provisions at an exorbitant rate and deducted their price from customers' monthly pay.[275]

Various locations across town were available to the spiritually inclined who wished to tend their souls in their off-duty hours. Early on, the Company had assigned free plots of land to the Catholic, Muslim, and Greek Orthodox residents for the construction of their temples.[276] In 1862 a small Catholic church dedicated to Sainte Eugénie was inaugurated. When its bell rang for the first time a harmonium accompanied the chants. The chapel was so small that many of those attending its opening crowded around outside. A mere six years later, the church was no longer deemed sufficient to host Port Said's Christian faithful, and the town's sawmill was remodeled into a church thanks to donations. The new church of Sainte Eugénie would be consecrated and its bells blessed in 1889–1890.[277] By 1863 Port Said vaunted, besides the abovementioned Catholic chapel, a church for the Greek as well as a wooden mosque located in the "Arab village" that by 1880 was too small and dilapidated. In 1886 a new one, called Tewfiqiyya, was to be inaugurated.[278] The Muslim imam who had been active since 1860 in Port Said's "Arab village" and school had allegedly attracted a large crowd to that part of town.[279] In 1871 the Company relinquished the management of the land and buildings of the Catholic, Greek Orthodox, and Muslim congregations to, respectively, the Catholic Church, the Greek government, and the Egyptian government. In 1881 the Company conceded to Protestants a room in town for free. By 1884 it had handed free plots of land to the so-called British cult so that the "spiritual needs" of both the settled English colony and the transient population of English sailors passing through the seaport could be fulfilled. In 1884 and 1885 respectively, the Copts and Maronites in town also received free plots of land from the Company. Less munificent was the treatment of Port Said's Jewish community of 158 men, women, and children, who met in "a mere wooden shanty divided in two by a partition, for all the world like a small booth at a fair" and adorned with cobwebs, soiled Hebrew prints, a converted stand for distributing boat tickets, and a shabby curtain. By 1881 that preexisting synagogue located on the seashore had to be evacuated. The Jewish community then also solicited the Company's assistance. When offered a mere room by the Company, the community refused and requested either pecuniary aid or land similar to what the city's Catholic and Greek communities had obtained before. But this request was not granted. Eventually, two men called Delburgho and Judas Levy purchased two plots of land for the construction of the synagogue and the accommodation of the rabbi.[280] By circa 1893, Port Said boasted a synagogue located on Rue de Commerce next to Belso, a company that rented coaches, saddled horses, and funerary carriages for all budgets. Later observers related that a Jewish quarter could be found in Port Said, nestled between the "European" and the "Arab" towns. It would also be reported that at least two Jewish temples stood there: a main one, a Beaux-Arts-style synagogue called Ohel Moshé, and a smaller one.[281]

Back in the mid-1860s, batches of workers identified as "Italians," "Austrians," and "Greeks" reportedly kept arriving at the isthmus. The towns of Port Said, Ismailia, and Suez and all the camps in between began resembling big fairs, where peoples "of all shapes and costumes" hurried, played, stole, and fought. Migrants' looks and boorish manners were "no good news," and the Egyptian guards dispatched locally, the *cawas* (from *qawāss* in Arabic or *qavvās* in Turkish; "sentry soldier" or "local guard of public security" or simply "guards") were at a loss regarding how to handle this moving mass of scrappy individuals (as further analyzed in chapter three).[282]

Some scholars think that at least for the bankers, contractors, and bondholders participating in the increasingly onerous foreign loans to Egypt, the country had become a goldfield. However, others point out that Suez Canal operations did not actually generate Klondike-like revenues. Contrary to optimistic predictions, having spent the proceeds from previous loans to open the canal, the Egyptian ruler Ismaʿil later had to earmark almost every piastre of internal revenue for servicing debt obligations.²⁸³ Eventually he had to hand over Egypt's shares in the Company to Britain in November 1875. Between 1876 and 1879, the Anglo-French Dual Control or Condominium gradually took over the Egyptian government, secured a repayment of Egyptian debt, and disseminated Britons at senior levels within Egyptian ministries.²⁸⁴

Figures about migrants' movements to the Isthmus of Suez tell only part of their story. The seeming exactness of numbers does not accurately reflect the complexities posed by enumerating people and the failures of those supposed to do the counting. First, official figures did not include clandestine migrants and those who deliberately eluded consular supervision. As the Ismailia-based director general of the Company confided to the Egyptian governor of the isthmus in 1868, it was difficult to keep track of all the workers, especially after they had left the worksites or gone back to their families located elsewhere.²⁸⁵ Second, national taxonomies conceal the fact that the identification of many of those roaming the nineteenth-century Mediterranean world was far from easily achieved. While Italy was still in the making in the 1860s and beyond, for example, migrants departing from its coasts may have been more precisely singled out as Neapolitans, Lombards, Florentines, Piedmontese, Calabrians, Tuscans, or Sicilians.²⁸⁶ So-called Greeks included both citizens of independent Greece after the war of independence in 1821–1832 and subjects of the Ottoman Empire, as well as individuals who just happened to be Orthodox.²⁸⁷ To complicate things further, citizens of the Ionian Islands after 1815 were considered to be British subjects. Moving to other shores, Dalmatians could be considered coastal "Croatians," but the two groups were classified separately because the former were governed by Austria and the latter by Hungary. In turn, so-called Croatians, before 1918 and the establishment of the Kingdom of Serbs, Croats, and Slovenes, could fall under the jurisdiction of either the Austrian, Hungarian, or Ottoman regimes.²⁸⁸ Finally, categories based on supposed nationalities forcibly straightened the ragged trajectories some migrants followed. How could one neatly classify, for instance, the "Italian" man who reached Egypt from Philippeville in Algeria; the "Italian" person born in Alexandria, Egypt, and relocating to Port Said; or the "Italian" woman who fled her debts in Egypt and found refuge in Spain?²⁸⁹ As suggested by Shana Minkin, death and the archival record it engendered were one of the few ways to pin them down and box them into only apparently homogenous categories.²⁹⁰

CONCLUSION

To trace the contours of people's movements toward the Suez region in the mid-nineteenth century and onward, we must parse the concrete and seemingly mundane details of their displacement. "Volume alone is meaningless," Adam McKeown wrote, "if we do not understand the broad patterns and distinctions that shape migration" or the way in which official categories or borders themselves have come to be. Juxtaposing sheer numbers to life histories shows that neither general processes in mercantile and capitalist development nor supposedly free choices exclusively influenced migrant trajectories.[291] Far from being "free self-deciding individuals," migrants were subject to macrolevel constraints, such as larger transnational economic cycles, regional fluctuations in supply and demand, transportation advances, and information flows. They were affected by mesolevel actors, such as labor contractors, representatives of navigation companies, and border agents. But at the same time, their determination to relocate responded to mundane microlevel decisions made with their families or independently.[292] Both "structure" and "agency" were involved in their migration. Poverty at home and individual decision-making were intertwined.[293]

A few of the migrants who had relocated to the isthmus remained nameless because they drowned in the canal and their corpses could not be identified. Of other mobile lives, nothing remains but a cheap and small portrait photograph the size of a visiting card of the kind that had been invented in Marseille in 1851 and had become popular in the Ottoman Empire in the ensuing decade.[294] One example of these keepsake portraits captured an anonymous man standing still moments before rushing to the docks in Marseille to embark on an Egypt-bound vessel. It then got lost on its way across the Mediterranean and ran aground in the French consular archives (see figure 8).[295]

The need for labor all along the canal in the making both altered preexisting circuits of Mediterranean mobility and formed new ones. This chapter has highlighted the often haphazard components of migration, the contingency of mobility, the restraints, and the possibilities available to people on the move in the mid-nineteenth-century Mediterranean. It has also illustrated how moving individuals, even when unidentified, still were able to affect the imperial strategies worked out in the European and Ottoman capitals. This chapter jettisoned those metaphors, routinely used to describe the experience of migration, that rely on elements in some unthinking

FIGURE 8. Unidentified man photographed at a studio on the Quai du Port in Marseille, 1861–1870. *Source:* CADN, ARI, C1, [1861–1870].

natural world (streams, flows, waves, transplants, and roots) and that connect people to places while implicitly confining displaced people to an aberrant status.[296] Rather, it has explored "some of the realities" of their old existence, Isthmus-bound voyage, and relocation.[297] The next chapter turns to the new existence they made for themselves in Port Said and all along the developing Suez Canal.

TWO

Like a Beehive

RACE AND GENDER ON THE SUEZ WORKSITES

> Just as a land barrier could be transmuted into a liquid artery, so too the Orient was transubstantiated from resistant hostility into obliging and submissive partnership.
>
> —EDWARD SAID, 1978

BY THE SECOND HALF OF THE 1860S, the Isthmus of Suez had become a vast and bustling worksite stretching along the 160 kilometer strip of land between the Mediterranean and the Red Sea that soon would make room for water. Spirits were high. The forced Egyptian contingents were no more. They were partially replaced by the European workers and machines that came in droves to the isthmus area.[1] Employment opportunities mushroomed on and around the excavation sites. The canal did not just provide long-term opportunities to engineers or laborers within the Company worksites. In the isthmus's fledgling encampments and towns, employment could be found for minor functionaries, accountants, shopkeepers, clerks, doctors, carpenters, surveyors, topographers, barge captains, blacksmiths, masons, stonecutters, train conductors, cooks, and telegraph operators. And that was not all. There also were, among others, porters and boatmen. Some worked for the consulates as translators or guards. Others found employment as doctors and midwives. Female teachers and nurses also toiled on the isthmus. So did water carriers and errand boys.[2]

Official Company narratives spoke of an idyllic isthmus society in which healthy workers zealously pursued their assigned tasks, the excavation proceeded flawlessly, and the public enjoyed a pervasive commercial freedom. One such account read: "Workers dig the sand along the furrow. Masons build houses. Carpenters and stonecutters and other workers pursue energetically their tasks. Women wash their laundry. This place resembles a beehive, where each holds on to his [or her] occupation."³ Company narratives neatly divided up labor, casually hinted at women's presence in and around the worksites, and sustained the fiction that all workers were tireless and content with working conditions.

Such buoyant reports must have reassured shareholders and appeased skeptics. However, the hierarchies based on nationality, race, and gender that Company officials constantly strove to impose on the isthmus workforce produced arguably mixed results. Male and female migrant workers, dead set on serving their individual and collective ends, met management's impositions with ambivalence, resorting to forms of defiance and compliance as needed. In the canal-related bonanza, which lasted through the second half of the 1860s, men and women on the isthmus enjoyed a heightened degree of mobility, changing occupations and residences frequently or at least whenever opportunities arose. They shaped their existence on the isthmus for better or for worse, for richer or for poorer, in health and in sickness, till misfortunes or even death did them part from their belongings and beloved ones. Once they set foot on the Suez Isthmus, they willy-nilly became members of the young and growing canal force and targets of the Company's familial strategies.

THE DIVISION OF LABOR ON THE CANAL WORKSITES

At the destination, employment opportunities and stints in the canal works varied. On the one hand were the employees or clerks. On the other stood the laborers. The two categories were kept distinct, because the former enjoyed more advantageous economic conditions, such as an indemnity for the relatively high cost of living. Laborers either worked at the worksites (*ateliers et chantiers*) or in services related to navigation (*marine*). Moreover, the laborers were generally divided into two groups: day laborers and the much more numerous pieceworkers. While day laborers were promised 1 franc a day, pieceworkers were cajoled by the prospect of making 1.5 to 2 francs daily (corresponding to 6 to 8 Egyptian piastres of the time) or even more if they were "active and smart." Company representatives emphasized how especially Arab workers voluntarily adhered to the piecework system and absolved themselves by insisting that workers alone determined their earnings according to the output they managed to yield.⁴ Piecework was seen as especially advantageous because it guaranteed that the worker, once he had begun the work, would complete it lest he lost his pay. Work that was paid daily instead gave workers the chance to take breaks in order to pray or smoke a cigarette. Pieceworkers were given by cadres the necessary pickaxes, shovels, and wheelbarrows, as well as two pieces of cloth that they used to enwrap themselves during meals and at nighttime. They toiled six days a week on shifts that lasted ten hours, from 6:00 to 11:00 a.m. and from 2:00 to 7:00 p.m.[5]

Reportedly some workers, such as a group of Calabrians, could not get used to the piecework system (*lavoro a cottimo*) adopted by the Company and returned to Alexandria to sail back home. Yet pieceworkers earned more than double what day laborers made. By the second half of the 1860s, workers were lodged for free by the Company in wooden huts. Food cost them 90 francs a month. Some preferred to do their own cooking and put money aside. In 1866 a group of workers from the northern Italian region of Piedmont declared to the Italian vice-consul in Suez that, by living frugally and setting money aside, they managed to go home after six months of working with around 1,000 francs in savings, minus travel expenses. Therefore, these Piedmontese workers must have made at least 166 francs a month or roughly 5 francs per day. They may have been some of the skilled Piedmontese stonecutters, whose expertise with granite was sought after in Egypt at large and who then went on to jobs at the hydraulic works in Asyut and Aswan between 1898 and 1902.[6]

At first canal workers were officially "free to provision themselves as they [saw] fit." They could do so at the shops that the Company had set up, with the traders who came to the campsites, or in nearby towns: Bulbeis, Zagazig, Mansura, and Damietta.[7] After April 1861 the Company no longer managed and operated the supply shops for its workers throughout the isthmus as it had done in the preceding two years. It had opened its encampments to free trade; around fifteen hundred private traders, of whom eight hundred were Europeans, took over.[8] Greek shopkeepers in particular dominated the commerce in groceries, in either shops or street stands.[9] The Company no longer had to worry about providing the isthmus population with flour, meat, or clothes and could now dismiss the four hundred employees who had been in charge of provisioning. By the "simple facts of freedom and competition," the allegedly self-sufficient isthmus was bursting with everything at no cost to the Company.[10]

By 1864 privately run commerce in Port Said was growing day by day. More and more hotels, cafés, bakeries, and tailors' workshops were up and running, and merchants of wine, "colonial goods," and dry goods were setting up shop in Port Said, Ismailia, and at other points all along the canal.[11] The arrival of a growing number of immigrants at the isthmus around 1865 coincided with a spike in the price of food.[12] By 1866 goods such as olives, candles, cigars and cigarettes, ankle boots, and trousers could be found at the worksite of Serapeum. People in this encampment could also purchase chocolate, jam, cigars, beer, Médoc wine, Fleury Champagne, and vermouth. Incidentally, the "Arab villages" that developed piecemeal in the proximity of the "European" campsites also constituted a "precious resource" for the camp dwellers, who could find well-furnished markets there. Those settled in Al Guisr, for example, could purchase in the Arab part of town goods such as *tumbak* (Persian tobacco, especially for use in the narghile), soap, onion, lentils, oils, and so-called Arab sugar and Arab cheese.[13]

The goods available on the isthmus had to cater to the tastes of a diverse multitude. Manifold scripts and signs in disparate languages embraced spelling liberties to publicize tobacco, cigars, coffee, and wine along Ismailia's Rue François-Joseph, that town's bustling artery of trade.[14] Besides Italian, French, and Greek, signboards on Ismailia's booths also appeared in Turkish, Spanish, and "American."[15] Multilingual signs also surmounted Port Said's shops, sometimes resorting to multiple tongues and alphabets. In one of Port Said's streets, a sign in front of a shop established by Simcha Nowowolsky in 1890 promised novelties, millinery, lingerie, and haberdashery to passersby in both French and Yiddish. It must have done well because the venue was in operation at least until 1916.[16] Most signs in Port Said entertained viewers with a blend of idioms, revealing just how many nationalities either called it home or passed through.[17] French examples included "Fabrique de limonade gazeuse et eau de Seltz" or, on a sky-blue background, "Dizard, coiffeur." Sometimes wax figures helped the illiterate fathom what the business was about. Statuettes of a gentleman with disheveled hair and a coquettish lady, for instance, helped Dizard promote his barber services.[18]

RULES AND EXCEPTIONS IN THE ENGAGEMENT OF CANAL WORKERS

On the job, groups and categories of laborers were organized hierarchically. The heterogeneity of the labor force was maintained by continuously producing and recreating supposedly inherent "national" or "racial"

FIGURE 9. Widening of the canal: work on the protection of the banks, 1869–1885. A little girl is standing on the right edge of the banks, held by a man. Photo by Hippolyte Arnoux. *Source:* ANMT, CUCMS, 1995 060 1489, Association du Souvenir de Ferdinand de Lesseps et du Canal de Suez.

distinctions.[19] As in other circumstances studied by historian John T. Chalcraft, "attributions of collective identity summarized, stabilized, and reproduced occupational specializations."[20] Accordingly, documents of the time claim that each group of workers would be assigned the specialty that suited them most and that workers' gangs were, whenever possible, constituted of men of the same provenance.[21] The available sources reveal a relatively consistent pattern of grouping workers together and assigning them specific tasks according to their shared origins (see figure 9, wherein men on the sides in European dress are supervising navvies, who are clad in *galabiyyat*), all the while exposing exceptions that trouble the supposedly universal rules of workers' engagement.

The story goes that bosses were exclusively French and the upper echelons were mostly filled with European engineers.[22] But work arrangements on the ground were more complex than that. The French also controlled commerce; found employment as engineers, foremen, and supervisors; maneuvered wagons; and worked as mechanics, supervisors, and stonecutters. Only a few French became masons and earthwork laborers (*terrassiers*). Some French workers, however, specialized in carpentry alongside Austrian workers. Austrians and Italians, according to some observers, were almost all drivers, miners, and mechanics. But Italians also found employment as earthwork laborers and masons. Moreover, Austrians, Italians, and Greeks toiled as unskilled laborers and boatmen; ran small trade; and managed most of Port Said's cafés, music halls, and drinking stalls (more on this in chapter four). Italians and Greeks, however, did not account for all traders, whose fold included Arabs and Armenians. The Greeks also became mariners and hodmen (charged with the removal of materials from the place of delivery to the spot where they are used) and covered "all those tasks requiring only occasional labor." At those worksites where blowing up rock formations was necessary, "Europeans" handled the mines while "Arabs" removed the rubble. Reportedly, most Arab workers became earthwork and unskilled laborers. Some toiled away in the hellishly noisy boilerworks. Most Egyptians worked in the coal-heaving business for very low wages.[23] Nonetheless, at least a few Arabs or Egyptians did cover mostly unacknowledged clerical positions. Around 1866 one Ibrahim Nagib reported to the Company as the locally dispatched doctor in the Tell El Kebir area, while Saad Mustapha worked as an employee at the Company's Bureau of Transportation.[24] Finally, Jews in Port Said reportedly lived by money changing; tailoring; and selling small articles, fancy goods, and curiosities.[25] On the whole, while "Europeans" and French in particular tended to occupy the highest end of the isthmus labor ladder, "Arabs" and Egyptians sat uncomfortably on the opposite rung. In the divided labor market that quickly emerged around the Company's worksites, national and racial categories relegated certain populations to given segments of that market.[26]

Company officials and contemporary observers liberally employed national and racial taxonomies to divide up groups of workers and order them hierarchically. To explain and sustain such hierarchies, they evoked workers' presumedly self-evident differences in resistance to temperature, tolerance to sun exposure, or even fertility. Three main sets of ideas circulating at the time conjured up this worldview. First, in an epoch of emerging nationalisms, Company officials and canal commentators adopted ideas of "ethnic" and "national" belongings that postulated some unambiguous prior community of territory, language, or culture.[27] Second, they fell in line with most nineteenth-century evolutionists, including Charles Darwin, postulating the existence of distinct human races and the struggle among individuals, tribes, nations, and races as the chief engine of progress in human history.[28] A tendency to create a simultaneously physiological and moral classification of human types had already been developing in eighteenth-century European thought, but the nineteenth century witnessed its coupling with ideas of genetic differentiation.[29] Throughout the latter century, biology, sociology, geopolitics, and Orientalism joined forces in producing increasingly rigid theories of racial hierarchy and developing an overt discourse of Muslim inferiority and quintessential Muslim difference.[30] In Egypt, both the idea of the nation's ethnological uniqueness (and its exclusionary logic that defines what the nation was not) and the discourse of race being conducted in western Europe would become popular among Egyptian anthropologists. Moreover, Arab readers of Darwin appropriated the multifaceted notion of "evolution" and subjected it to endless debate.[31] The third group of principles dictating the differentiation of workers in Suez were the theories of environmental, biological, or geographic determinism that constituted the scientific orthodoxy of the day. According to morally determined notions of climatology, geographic circumstance determined people's physical constitution and moral build. In other words, the climatic or geographic features of a certain place dictated whether the persons dwelling in it belonged to savagery or civilization. Climatology hence became a moralizing vehicle for conveying racial judgments. While temperate "European" climates supposedly retained their normalcy, inauspicious climates were seen as occupied by inferior racial groups, who alone could bear the conditions.[32]

The environment thus persisted as a site of difference.[33] In particular, those contexts characterized by high temperatures were approached as places where nonwhite bodies could stand extreme heat before their pain limits were reached.[34] It was reported that on the Isthmus of Suez, "Arabs were the only ones to be employed as stokers, because they alone could easily bear the elevated temperatures of the heating chambers." So-called natives were assumed to be naturally able to withstand the harsh desert climate. Ostensibly, the *fellah* could tolerate the local heat and dryness, while the European farmer had no resistance to them.[35] On the whole, non-Egyptians were believed to have one relative physical advantage: they could stand the cold more easily. While Europeans perceived the winter of 1862–1863 as mild, for example, so-called blacks reportedly suffered from the temperature drop that was recorded in that conjuncture.[36] This racially inflected physiology also applied to fertility. Egyptians were seen as prolific, while second-generation European residents of Egypt were considered to have become completely sterile or to bear sickly children who died at a young age.[37] Among the "Europeans" who could neither get used to the climate nor produce lasting offspring, only Jews were said to constitute a likely exception. But this was simply because most of them, it was reasoned, were really of African descent.[38] In a word, nineteenth-century evolutionary thought often conceived biologically successful peoples as social inferiors.[39]

Not all Europeans, however, were the same under the fiercely shining Egyptian sun. Peoples from the Mediterranean littoral were said to have much better chances in high temperatures. Immediately after the Egyptians, the Greeks were rated second for their resistance to weather while performing earthwork as long as they stayed sober and ate sufficiently. They resisted the heat equally well and stood the cold even better. After the stout Egyptians and Greeks, the workers from the South of Italy came third. Calabrians stood out for their demonstrated resistance to climate. Finally, Montenegrins and Dalmatians proved their worth as sailors and crew members on the dredgers in Port Said. Workers hailing from the eastern Adriatic coast were said to be robust, sober, and docile and to make excellent earthworks laborers. The French and others from Europe's temperate regions, on the other hand, could only momentarily expose themselves to the sun's rays, and there was no chance they could tolerate continued manual work out in the open. They were better suited to toil under a shelter or in workshops and offices, or to perform surveillance tasks. In a word, supposedly delicate men from "the North" could hardly bear employment on the earthworks and therefore had to perform less bodily challenging chores. The least "intelligent and active" Egyptians, instead, could not aspire to anything other than earthwork.[40] The Suez Canal strip of worksites thus became one of the many peripheries of Europe in which, as theorized by On Barak, those racial theories deeming white-skinned bodies unsuited to hot climates and dark-skinned bodies as heat-resistant justified brutal labor arrangements.[41]

Stratification on the isthmus also emerged from the interaction among various immigrant groups, whose identities were at least in part ascribed: stereotyping and labeling contributed to reinforcing such identities.[42] Supposedly fixed features, awkwardly mixing virtues and flaws, were summoned to justify the isthmus's hierarchical society. Fittingly, the Egyptian *fellah* was said to be superficial, covetous, and passive after centuries of oppression. Laborers from the lacustrine area north of the canal were reputed to be hardy and patient. Workers from Upper Egypt were seen as vigorous but lazy and undisciplined. Neither *shaykhs* nor officials, or so the received interpretation went, succeeded in making laborers work at a suitable pace. So-called Barbarins (individuals from Upper Egypt or Nubia) were deemed to make good, skilled servers who nonetheless had a predisposition for debauchery and unruliness.[43] Among Mediterranean workers, Bretons, Calabrians, and Greeks had a reputation for causing murders and conflicts on the canal. Bretons had become, in the nineteenth-century French imaginary, picturesque country bumpkins with a penchant for trouble.[44] Calabrians were said to be hot-blooded, prompt to reply with their fists, and generally easy to provoke. Italians in general were reputed to stick to the job only briefly and to cause "very serious embarrassment" especially in the surroundings of the town of Suez.[45] Greek workers on the canal worksites made a name for themselves as particularly turbulent and in need of especially close surveillance. Greeks in Egypt, a loosely defined social group, were portrayed negatively for many reasons, ranging from their strong demographic presence to their participation in moneylending. Along with Jewish money changers, Greek ones were reputed to steal from those foreigners in Egypt who relied on their services.[46] Finally, the Maltese in Egypt were criminalized by other communities, the government, and their own diplomatic representatives. In Julia Clancy-Smith's words, the Maltese represented "the scapegoats of the Mediterranean expatriate underworld."[47]

One the one hand, the highest positions being held by Europeans may have been justified by the specific skills required to perform certain technical services that for a time they alone possessed.[48] However, donning the title of "engineer," for example, did not necessarily imply that the person held an academic degree, as "simple masons or mechanics" could avail themselves of that title.[49] On the other hand, men of different origins often performed the same job, such as the task of earth removal. The director general of the works himself, writing in 1860 from Damietta, claimed that the Egyptian *fellah* workers charged with earthwork had promptly familiarized themselves with the pick and the shovel, which "they handled equally [as] well as the European earthwork laborers." This latter group comprised, among others, Austrians, Italians, Greeks, and British subjects, as well as Georgians and Armenians.[50] The earthwork laborer Florio Choupéro, for example, was Austrian and twenty-two years old when he died in 1866 from gastroenteritis complicated by a chronic lung disease. The laborers Blassignone and Ellena were Italian subjects. The former died from meningitis allegedly contracted during a night spent outdoors. The latter, afflicted by acute sciatica, went to the hospital but refused to undergo treatment. The seventeen-year-old earthwork laborer Jean Dracoti, a "Hellenic Greek," got into a fight at the Nefiche campsite west of Lake Timsah and suffered fatal knife wounds. Finally, a certain number of British earthwork laborers were also employed in the Suez works; their names, however, remain unspecified, hinting at the possibility they had come from Malta.[51]

Company officials attempted to sustain the idea that workers' national and racial classification and specific occupation on the isthmus were directly correlated. They reported that those individuals belonging to the "white race" worked as employees, laborers, and traders. They also wrote that members of the "indigenous race" toiled as laborers, traders, and servants.[52] However, it remains unclear whether one's supposedly classifiable "race" determined his or her job prospects or whether, conversely, it was one's occupation that determined his or her "racial" identity. Moreover, this taxonomy neglected all those who fell into its interstices. People along the canal busied themselves with occupations that were not necessarily associated with their alleged "race." For instance, persons who may have been classified as "Europeans," such as a certain Pierre in the hire of a man called Mamoud in Ismailia in 1869, actually toiled as servants.[53] Finally, though oddly precise, this classificatory system glossed over the ambiguous national or racial identities found in late nineteenth-century Egypt. Claims to "Europeanness" by people roaming the Mediterranean at this time were unstable and contested.[54] "Greeks," for instance, who belonged to a highly ambiguous category to begin with (as discussed in chapter one), were intermittently embraced and excluded from the European population of Egypt. Louis-Rémy Aubert-Roche himself, the highest medical authority on the isthmus, considered them as Europeans on one occasion and as non-Europeans on another.[55] Other Company sources found it expedient to categorize "Greeks" with "Turks" as members of the "white race" on the isthmus and to subsume "Arabs," "Barbarins," and "blacks" under the label of the "indigenous" race.[56] Yet the three latter apparently distinct categories confusingly overlapped each other. For example, nineteenth-century sources defined "Barbarins" as "Arabs from Upper Egypt" or "Nubians."[57] *Barbari* could also be a derisory term translatable to "Black African."[58] Some included both "Arabs" and "Barbarins" as members of the Egyptian population alongside the *fellahin*, while others argued that individuals from the Nubian part of the Nile valley shared their origins with the "Berbers" of North Africa.[59] Yet other sources drew lines between "Arabs," "Egyptians," and "Syrians" as if they belonged to distinct groups.[60] This may have reflected common Ottoman usage, in which "Arabs" invariably and colloquially referred to black Africans while ethnic Arabs were termed "white Arabs."[61]

Endeavors to categorize people on the isthmus were thus impossibly precise. As argued by Julie Greene regarding Spaniards in the similarly heterogenous Panama Canal Zone in the early twentieth century, workers' ascribed identities on the Isthmus of Suez were "awkward, complex, and highly mutable. [...] How they would be considered depended a great deal on the exact circumstances."[62] Stereotypes continued to color the European imaginary of the Suez Canal. On the one hand stood fellow industrious "Europeans" with their shirt sleeves rolled up. On the other hand slouched the "Arabs," naked and plain-living, happy to live simply on water and biscuits. Ephemeral and yet concrete, these ideas fueled the division of labor on the isthmus and yielded very tangible effects. Since indigenous workers lived in their country of origin, had allegedly always been used to a frugal existence, and could endure its hot climate, they were then deemed "infinitely" less valuable than the European workers doing the same job. They had to meet, it was claimed, only half of the latter's expenses. Hence, they received only half of the latter's pay. The fact that once trained they could perform the same work their "European" peers pursued did not spare "Arabs" the inconvenience of being paid much lower wages.[63] Elsewhere on the globe and roughly at this time, nonwhites were conveniently thought to be able to survive on less than white workers.[64] As for Egypt's labor history, as argued by Joel Beinin and Zachary Lockman, the "ethnic" divisions and wage differentials on the canal worksites would be key to the development of the early Egyptian working class; ignite resentment against foreign workers, managers, and owners; and fuel a nationalism wherein class and national consciousness blended.[65]

The isthmus environment and climate thus came to both explain the labor hierarchy and validate variations in salary among different people often performing the same job. This refrain was so convincing that it lingered on in later literature. It would be argued, in fact, that most coal-heavers in Port Said came from Upper Egypt because they were "able to tolerate the hardships of this type of work," that the lagoon fishermen "were a vigorous race accustomed to conditions which would have been intolerable to outsiders," and that "the interminable tropical sun had precluded the possibility of employing European labor extensively."[66] An argument could be made that differences in the stamina of, for instance, the reputedly stronger Upper Egyptians and weaker delta *fellahin* actually derived from the strength-sapping parasitic diseases that thrived in the latter region's perennial irrigation system (and that later would be introduced into the former as well).[67] However, the racialization of workers on the isthmus appears to have rather been a cultural construction grounded in everyday interactions and cloaked in scientific terms that were themselves the product of contingent worldviews.[68]

Hypotheses of "friendly collaboration" among Egyptian and foreign workers seem far-fetched.[69] The national and racial taxonomies, stereotypes, and hierarchies analyzed earlier were hard at work keeping groups apart, whenever they were not actually persuading them to report on each other. In 1861 at the Al Firdan worksite, for example, Italian, Spanish, "Arab," and indigenous *fellah* workers did toil together on the same site. But Italian, Spanish, and "Arab" workers actively foiled the evasions and protestations of their *fellah* fellows.[70] Nonetheless, hints in the sources suggest that individuals rubbed elbows with others more often than has been acknowledged. In June 1862, for example, when the corvée was still in force, coerced laborers worked side by side with free workers of disparate provenance. In fact, while Egyptians, "Arabs," and Syrians may have mostly toiled in earthwork, Greeks and Maltese were nonetheless also recruited for that task whenever subcontractors could not procure enough drafted workers.[71] In another instance, between 1864 and 1869 the yards established by the Dussaud Brothers company in Port Said employed at least 350 Egyptian, Italian, and Greek workers simultaneously.[72]

At the very least, workers resorted to speaking, with sometimes bizarre outcomes, a common language that apparently resembled the lingua franca employed by Mediterranean sailors or the *petit sabir*, the "international and barbarian idiom" in use among French soldiers in Algeria, peppered with the technical terms pervasive on the Suez worksites.[73] In Port Said the *sabir* would apparently still be in use in the mid-1890s, when a visitor recorded hearing Arab porters, hotel employees, *drogmans,* and boatmen shouting "in all languages, whose natural blend forms this Mediterranean volapük called *sabir*" (the Volapük was Esperanto's less known sibling in the family of nineteenth-century "world languages").[74] To be hired as a surveillant on the worksites, knowing Italian was apparently an asset, but possessing mastery of several other languages was regarded as necessary. Exceptional candidates could handle multiple languages at once, such as the 1819-born Hungarian "refugee" Charles Megyessi, who boasted Arabic, Turkish, German, Greek, Hungarian, Italian, English, and French in his repertoire (he actually did not know how to read or write in Arabic, but he spoke it, understood it well, and translated easily when another person read it out loud).[75] In sum, finding employment as a *drogman* or "translator" on the canal worksites meant knowing at least French, Italian, and Arabic and throwing some Greek and Turkish to the mix as bonuses.[76] Since written Arabic eluded even the most eclectic of polyglots, hired translators could often speak it but were at a loss when it came to deciphering written sources. At any rate, knowing Arabic well was no longer considered sufficient on the Isthmus of Suez by the second half of the 1860s. Both French and English were seen as necessary, as "people from all over" kept coming. In both Suez and Port Said the number of Europeans was growing incessantly, and the local police lacked officials who could master the many languages that were in use.[77]

EXTRAVAGANT HOPE FOR PROFIT

Months or even years later the Suez worksites rarely yielded the economic security that migrants were yearning for. Early on, those who had flocked to the isthmus were said to have had their "extravagant hope for profit" frustrated, their enthusiasm dissipated, and their weariness made acute by the imposed solitude.[78] How little some of these workers owned is apparent from the paper trails spun by their deaths and the auctioning of their belongings, when their lives shrank to the contents of their pockets. The Austrian stonecutter Andrea Ewich, for example, left absolutely nothing behind, not even a passport or identity papers, when he succumbed to dysentery. Many did not even possess more than a change of clothes. The Dalmatian laborer (*manoeuvre*) Jean Manovich left two stockings, one pair of shoes, one shirt, one pair of trousers, one jacket, one *tarbush* (a felted red hat with a flat circular top also known as a fez) that he may have bought as a souvenir, and one leather belt. All were in very bad condition and estimated at the meager sum of 50 centimes (or half a franc), slightly more than half of an unskilled laborer's day's pay in the mid-1860s. Similarly, the belongings of the Italian subject Giuseppe Alcoré, a thirty-year-old laborer who perished from the effects of diarrhea, were in very bad condition and were sold off for 50 centimes: one pair of stockings, one shirt, one pair of trousers, one jacket, one gilet, and one hat. The Swiss mason Jean Martin died from dysentery at age twenty-nine, leaving behind a comparable array of things that were also sold for 50 centimes, but he also owned a pair of ankle boots. Doing slightly better was the French subject Jacques Ledilly, of undisclosed occupation, who was killed by chronic dysentery at age forty-seven; his single pair of shoes, one pair of trousers, one shirt, one gilet, one jacket, and two hats were all in bad condition but were sold off for 1 franc and 50 centimes. Indicating perhaps that coats were prized possessions, the belongings of the Austrian laborer Milan Marcovich, victim of an algid pernicious fever (perhaps due to malaria), yielded 5 francs despite their bad state; they included one pair of stockings, one shirt, one pair of shoes, one pair of trousers, one beany, and one coat.[79]

Among the less impecunious was a Pierre Cortari, who owned one pair of shoes, two trousers, two shirts, one flannel gilet, two coats, one hat, and one pair of glasses. Upon his passing, all was sold for 5 francs and 50 centimes. The French subject Réné Rogel owned a similar range of things but also held onto one tie and one wrought iron plate, for a paltry total of 75 centimes. In contrast, the relatively better-off Frenchman Louis Edouard Remy, a forty-one-year-old working as a medical aide, kept a shale-oil lamp, a kettle, three pairs of shoes, five shirts, six trousers, and three pairs of socks. All his stuff was estimated at 92 francs and 50 centimes. Rosina Balcowich, a woman who died in Chalouf and whose belongings were auctioned off for 128 francs, had ferried to the isthmus a chamber pot, a coffee roaster, and a fan and a sunshade to fend off the heat. The French subject Joseph Carlod, surveillant of the works, left more than 800 francs' worth of stuff. The heap of things he left behind included, among others, glasses, candles, a sponge, some soap, files, scissors, a screwdriver, an inkpot, a magnifying glass, razors, a brush, books, a mosquito net, and, again, a *tarbush*.[80] The poorest and the wealthiest alike collected specimens of the *tarbush*, the official headgear of the Ottoman Empire, perhaps as mementos of their Isthmus experience. But their possessions differentiated them in death as they did in life. In sum, individual misfortunes created snapshots of the belongings and resources in the lives of various kinds of workers who helped build the canal.[81]

Some of the migrants living on the isthmus stressed what they perceived as their isolation, the harshness of the surrounding environment, and their sacrifices. "I find myself in a desert, far from everyone," wrote one man to his father. Desert sands figured conspicuously in the description of their suffering, as highlighted in the writings of another, the father of a family who had just been fired. "I am an honest man," penned a carpenter who had been on the canal for fourteen months and had similarly been dismissed, "who has come to the desert with the intention not to enjoy himself but earn a piece of bread."[82] These migrants' troubles also survived in the memories of observers, who recounted their "apostolic sweats on the southern Mediterranean shore." A Franciscan friar claimed that in the first years of the canal works migrant workers had to withstand the blazing sun, inflamed sands, poor shelters, and scarce water.[83] Some, moreover, felt unsafe. On the one hand, they often perceived the intervention of the local police as forceful and arbitrary. On the other hand, they saw themselves as surrounded by untrustworthy countrymen who were either army deserters or jailbirds.[84]

Canal workers initially intended to stay only one year, or eighteen months at most, but their sojourn at times added up to several years. Originally meant to be short term, the migration of Egypt-bound women also could last decades.[85] Migrants were staying longer than initially planned and still not making enough money to either go back or satisfy their families' requests. Nicola Castello, for example, in a heart-wrenching letter he addressed to his father in the Alpine village of Rabbi San Bernardo, lamented that for the fourteen months of his stay he had not received any news and felt forsaken by his family back in the Italian peninsula. Maria Baraga's husband was gone from their home in the Hungary-administered Croatian port city of Fiume (Rijeka) for sixteen months. He was still not sending back enough money when she menacingly wrote to him in 1865: "If you got it into your head that you want to come back to Fiume, come, because the way is open, but think hard before you do so."[86]

TROUBLE IN THE SUEZ PARADISE

Workers could undertake a variety of actions, ranging from the refusal of labor, physical departure from the workplace, and intentional slowing down of the work pace to compliance, and even collaboration, with the authorities. No group of workers had a monopoly on any one cluster of actions. It was not, as has been suggested, just the workers on dredgers, who landed on the isthmus well into the canal project, who most frequently went on strike or rebelled.[87] Those Egyptian forced workers who engaged in earthwork in the early 1860s could rely on a long-standing script that had been tried out before in the government's public works.[88] For example, while digging at those archaeological sites that dragooned local laborers, children and adult laborers ostensibly sang the praise of the masters while perhaps tapping into a shared repertoire of ironic subversion to which the Western recorder was tone-deaf.[89]

Corvée laborers often resorted to deserting the worksites and intentionally damaging the work tools. On January 13, 1862, for example, 1,643 men arrived at Al Firdan; 500 of them fled a mere fifteen days later with their task still incomplete. The comings and goings were even more tumultuous at worksite no. 6, where 3,465 *fellahs* from Upper Egypt arrived between January 13 and 17 and as many as 2,900 absconded on the evenings of the 17, 18, and 19. Another 1,242 men arrived there on January 20, bringing the total number to 1,808, but more than one-third or 620 deserted just three days later, on January 23. What is more, before leaving the deserters smashed and burned a great number of wheelbarrows.[90] The historical record reveals what may have prompted their actions at least in Al Firdan, where workers resented the ill treatment they had received at the hands of their French foreman. They allegedly had been instigated by Riz El Abbed and Abdalla El Abbed, two "*nègres*" who were "influential" over the laborers. The fugitives were reconducted to the workplace. And to add insult to injury, "the Italian, Spanish, and Arab workers" toiling on the same site had declared "that the indigenous *fellah* have never lacked anything and have only received good treatment" and thus helped exonerate the French foreman. The *shaykhs* were accused in his stead of mistreating the workers and causing their acrimony. Whoever held ultimate responsibility for ill-treating the *fellahin*, the fact of their abuse on the worksites continued, and other fellow workers were partly responsible for it.[91] Again under the forced-labor system in 1863, workers at different worksites, including European agents and laborers, deserted the works without authorization. In August 1863 forced workers reportedly "created pretexts" and obstinately refused to keep digging the main furrow. They succeeded in "turn[ing] everything upside down." By September 15, 1863, the Company's worksite was nothing but a knee-deep "hideous swamp." Egyptians, whether *fellah* or not, were also reputed to be keen saboteurs. Ostensibly, whenever they found a stray wheelbarrow, tool, or beam, they broke it and hid it under the sand after checking that no one could see them.[92]

By 1865, after the end of forced labor and upon the intensification of international recruiting efforts, numerous desertions were recorded at all worksites. They paralyzed the worksites right at a time when entrepreneurs ought to be fulfilling the contracts they had signed with the Company. This scarcity of manpower distressed those in charge, as they risked being considered personally responsible for the deferral in the works.[93] A situation that entrepreneurs experienced as frustrating because of the widespread dearth of labor nonetheless enabled workers to relocate frequently and change employers at will. The fact that different firms operating on the isthmus occasionally poached each other's workers attests to these two sides of the same coin. In January 1865, for instance, the representative of the Société des Forges et Chantiers in Port Said complained to his higher-ups in Marseille about agents of the Suez Canal Company, to which the Société was providing dredgers, cranes, and barges. He accused Company agents of having lured away with higher salaries some of those workers in his hire who were in charge of assembling dredgers and cranes. When the need to activate a dredger arose he lacked the manpower to operate it. The Company's engineer-in-chief quartered in Port Said, however, responded that only three of the thirteen workers who had abandoned the Société had actually relocated to Company-related worksites: Mikaël Zenzafilé had moved to the Montalti workshop, Jouanin Mikaël was now employed at the Perrusson workshop, and Giorgi Elias had relocated to the Heidegger workshop. Workers on the move were not simply chasing higher wages. In fact, both Zenzafilé and Mikaël still made 4 francs daily as they had done previously. Only one in thirteen, Giorgi Elias, now making 4.25 instead of 4 francs per day, apparently improved his chances through relocation on the isthmus. The other ten workers had vanished into thin air.[94] Therefore, workers must have had incentives other than moneymaking to head elsewhere.

The newly recruited Moroccans, Greeks, Dalmatians, and Bretons were said to be especially insubordinate and prone to abandon the encampments. The Company deemed these workers especially ungrateful and unworthy because, contract in hand, they had had some of their expenses covered and could hence exercise a special leverage. Whatever country they came from, they were expected to both return the expenses that had been covered and stay put on the job. However, even if they had been enticed by potentially elevated salaries, they now came to dislike the task ahead and frequently left. In May 1866, for instance, a coppersmith called Junot left Port Said without collecting his pay or notifying anyone which way he was going. Nothing else was known of him, except that he was thought to have died near Al Guisr later that year. Workers often left the Company workshops to go work for less at the Company's contractors, toil at other firms, or even seek their fortunes in the Delta. Ostensibly they preferred to lead a "miserable life" in Egypt rather than honor their commitments.[95]

Some of the workers who did stay put intentionally delayed or interfered with the planned schedule of advancement of the works. One contemporary observer commented that the "Arabs" who had freely shown up for recruitment no longer felt "the authority of their traditional chiefs and, considering themselves as masters of the situation, advanced ridiculous demands and worked half-heartedly and intermittently." He spared French workers no criticism either. He wrote that they were notorious "for their constant and ridiculous requests, their rebellious spirit, their misbehavior." They had to be dispatched home as soon as they could be dispensed with and as eagerly as they had been initially summoned to serve the canal's enterprise.[96]

Workers may have resented what even diplomatic officials, such as the Italian vice-consul in Suez, denounced: the frequent cases of violent death, fatalities from wounds, or mutilations in the works. The vice-consul also deplored the scarcity of safety measures undertaken by Company agents to prevent such occurrences and the untimely fashion with which they notified the relevant consular authorities of workers' deaths. Finally, he accused the Company of having failed to compensate the families of the wounded or the deceased for their losses. In late 1868, for example, the Italian worker Pasquale Spinelli, a father of seven, was wounded in the leg in a mine explosion and had to be repatriated after six months of unsuccessful treatment at the hospital in Chalouf. In 1869 two workers lost their lives in the same section and by the same means: five weeks apart from each other, Nicola Lebrizzi and Macario Ponesso died in rockslides in the Company worksites on the Suez plain. Both had been the sole breadwinners of large families. While in Lebrizzi's case the Italian vice-consul was timely informed and could verify the death and notify the family, the Italian representative would have been unaware of Ponesso's passing had it not been for one of the victim's compatriots, who shared the news. The vice-consul must have still been unable to collect the money owed to Ponesso, withdraw his due salary from the Company office, and pass it on to his family, because six months after that worker's death his widow, Maria Minervine, wrote from the Tuscan town of Gimigliano that she still had received nothing. At least dead workers, such as Lebrizzi and Ponesso, but also those like Raffaello Adami, Pietro Mariutti, and Alessandro Valiandi, who perished in Chalouf in 1866, had a name to cling to. Many more died in Chalouf alone before a priest or a consular representative could record their passing.[97]

In the mid-1860s, one more incentive to flee the isthmus worksites made its macabre debut: cholera. In June 1865, attempting to contain the spread of the disease, the Company management reduced daily work schedules by one hour, then by two hours. It sheltered laborers during the hottest time of the day. It circulated coffee and other doctor-recommended beverages among crews. It extolled its own precautionary measures while blaming the spread of the epidemics on the Italian and Greek individuals who had fled Zagazig (where 105 people died of cholera on June 27 alone) and accusing them of having brought the scourge with them into the isthmus. In terror, the fugitives were trying to reach Ismailia, gain Port Said, and set sail for Europe. The Company was helpless against the mass desertion of workers triggered by the disease or by rumors surrounding its course. Many, especially among the Greeks, reportedly set off on foot from Ismailia through the blazing desert for the eighty kilometers separating them from the Mediterranean. They were loaded with heavy luggage and unconcerned with the lack of available food in crammed Port Said. "No human force would have been capable to stop the movement, reaching one encampment after the other, of this undisciplined gang" noted an author at the time to describe the cholera-triggered flight. Shedding corpses along the way, this crowd finally made it to Port Said.[98]

The choleric epidemic that broke out in June 1865 continued to make workers feel unsafe. In March 1866 Company sources claimed that, at least in Chalouf, there was no trace of cholera and yet some Italian workers were using the specter of the disease as a pretext to depart the worksite. Canal workers, nonetheless, may have legitimately still feared to end their days in dire conditions. In June 1866, for example, in the French hospital in Suez, an Italian worker from the Chalouf campsite struck by cholera lay in the same room with a young French lady and shared her symptoms. Both lay in darkness, with shut door and windows, with no lights for the night, no fire to heat their frozen bodies, and none of the water they desperately begged for. They vomited and defecated in their own beds. They were dying a desperate death.[99]

Even when accidents and sickness were averted, worksites offered plenty of opportunities to grouse. As was the case with the Egyptian forced laborers discussed earlier, workers may have suffered at the hands of abusive foremen. A Serbian worker at Al Firdan, for example, who dared to claim "some money that was legally owed to him," was hit with a dagger by a gang head (*chef de brigade*) called Cazejus, culpable for many other similar acts.[100] Paydays may have been fraught times. Reportedly, workers nervously besieged the Company offices and clashed with the Egyptian police out of fear that Company bureaucrats could flee without settling what was owed to them. Delays in the distribution of wages could trigger strikes, such as the one that erupted in the quarry of Hyènes, on the eastern edge of Lake Timsah, in August 1869, which prompted overseers to demand that troublemakers be harshly reprimanded so as to avoid any future trouble.[101] One of the Company's enterprises, in 1866, tried to pay workers' salaries with a "standard currency" (*monnaie de convention*), which must have been quite an unpopular measure. First, having the right change was often a challenge. Second, this type of money was an impediment to dealings between workers and merchants or others. Finally, even if all European currencies could be found, at least in Port Said, exchange rates would drain pockets dry. Having English or French cash on hand was strongly recommended instead.[102]

On March 1, 1866, about one hundred workers went on strike at the worksite of Chalouf, north of Suez. They lamented that the local company-run Bazin store was lacking staples that they considered essential: cheese, lard, pasta, and potatoes. This store had perhaps become, as in comparable contexts elsewhere, a "potent symbol of employer control."[103] Even if desired goods were eventually delivered, panic spread anyway following the death of a worker for reasons that sources do not specify. Some more men had fallen ill but refused to enter the hospital. A Company official had some of the "camp men" forcibly bring the sick laborers to the medical facility. Twelve sick Piedmontese workers lay in the hospital. Even if many were reportedly doing better than the previous day and only one was still in very serious condition, the locally dispatched doctor, a man known by his last name, Salemi, still expected that at least a dozen coffins would soon be needed and made the camp head prepare a few in advance. At the same time, more than 350 men had gone missing from work. Around 60 of them loitered in the campsites, singing. Others had left for Suez without even waiting for the pay that was owed them and would be disbursed the following day. They said: "If my poor mother sees my comrades arrive to the country (*pays*) without me, she will believe me dead, so I want to do as the others do and leave before having received what is owed to me."[104]

Overall, remuneration was not a sufficient incentive to stay. Ill treatment, accidents, the fear of disease and death, the aversion towards forced hospitalization, the lack of staples, and thoughts of home may have all been valid motives for workers to drag their feet, protest, rebel, or depart. The forms of canal workers' "agency" subsumed a spectrum of disparate options.[105] These exemplify that different groups of migrant workers resorted to similar tactics. The historical record offers no explicit evidence of cooperation, but toiling in proximity to one another may have led to mutual observation and joint refusal, shared pantomime, or collective flight. Even apparently individual acts of desertion may have depended on collective networks.[106]

AROUND THE CANAL: NO WORK, OTHER WORK, AND "FEMALE WORK"

As predicted by a clairvoyant observer in 1858, the army-like throng of "20,000 partly Egyptian and partly European workers descending upon the Isthmus of Suez would be accompanied by 1. Male and female innkeepers 2. Laundrymen 3. Merchants of all kinds 4. Domestic laborers 5. Vagabonds and people out of work." The agglomeration of a heterogenous population, he continued, would likely attract to the worksites a bunch of people "disinclined towards honorable behavior."[107]

Since the early days of the canal excavation, "a floating population moving constantly in search of any kind of deal" inhabited the isthmus side by side with the "Arab" and "European" workers. By 1862 Company officials claimed that "unemployed hence useless" individuals had been allegedly arriving in Port Said with "unknown goals." They accused them of taking accommodations away from the Company workers, those truly "entitled to these advantages," and drinking their expensive water. For the local economy as well as for security, it was argued, the situation had to end. Only theoretically, access to Port Said ought to remain easy and open to anyone. In actuality, "special reasons," including the swelling of Port Said's population especially during the summer months, justified the idea that no one but the "necessary individuals," defined as either the workers or the merchants, should be let in.[108] Company representatives claimed that vagrants of dubious morality were plentiful on the worksites and that they refused all types of work. Among these unattached individuals, officials identified the culprits in brawls and other violent episodes that occurred daily, such as homicides in Ismailia and Chalouf or disorder and thefts in Port Said. By the mid-1860s, they reasoned, the "situation created by vagabonds showing up in the desert" could no longer be tolerated.[109] Around the same time, newspapers reified the idea that Egypt was being invaded by "evil-doers, the refuse of all Europe, unable to live in their own countries," who could there "revel in almost perfect lawlessness, the scourge and terror of honest men."[110] Even if being off work may have been only a temporary status, unemployed migrants on the isthmus or those with no steady jobs were seen as unpredictable and unmanageable. Their presence on the worksites, where everything supposedly revolved around assigned occupations, was equated with a constant threat to public order.

For its part, the Egyptian government declared its willingness to receive those industrious foreigners who were equipped with a trade they could live on. But in 1866 it also announced it was no longer able to welcome those masses of laborers (*braccianti*) who had no craft, were forced into vagrancy by the lack of jobs, and were prone to disrupt the safety of the public. Unemployed Calabrians were one such worrisome lot. Once they landed in Alexandria, which was expensive and offered no jobs, they had nothing to eat and nowhere to shelter. They assailed the consulate, yelling and shouting that they were hungry. Others were stranded in the streets, begging for charity. Ostensibly the Company was willing to employ them for the canal works, but it refused to pay or lend them the money for their railroad transfer to the isthmus. The Egyptian Police wanted no vagabonds in the streets. Public health authorities worried about poor people gathering in the city's squares, concerned about the daytime heat and the nighttime cold. Meanwhile, the Italian consul wondered helplessly: "How can I abandon them in the streets, starving?" In only two days over the course of January 1866, 250 Calabrian migrants were repatriated to the peninsula almost for free.[111]

Disparate measures were attempted to tackle the drifters allegedly roaming the isthmus. Foreign consulates and the Egyptian government agreed that governments in the migrants' countries of origin ought to keep a closer watch on departures. Those individuals unable to meet their first expenses upon arrival ought not to be allowed to leave in the first place. Contractors should have been forced, before obtaining migrants' passports, to make a deposit to cover both the migrants' relocation as far as Suez and their return trips. Upon landing, those with no regular passports ought not to be allowed to disembark.[112] Company representatives agreed that as for the human cargo of ships coming from Europe or elsewhere, the Canal Company or the contracting companies were to admit only those individuals with regular papers or certificates of good conduct. They added that in order to avoid local trouble, the points of embarkation for Port Said should be kept in check: Damietta, for example, from where undocumented Greeks apparently used to embark. A local police force should be established, and boats ought to be surveilled, so that those with no papers would not be permitted to disembark. A regulation proposed in 1862 prescribed that no one was to be admitted to the Port Said worksites, unless they prove to be in possession of engagement contracts or identity documents. Everyone else was to remain confined on board and be sent back to their respective places of embarkation. Individuals who were not working and could not justify their means of subsistence had to immediately leave the town.[113]

Nonetheless, the undocumented still found ways to reach the isthmus and take up residence, especially in Port Said. The fledgling port town thus came to witness, at the same time, the efforts by officials to block the unemployed and the homeless and the latter's success at circumnavigating official regulations. Most undocumented migrants entered Port Said mostly from Damietta and mainly via land. The surveillance that had been established in 1862 on the Alexandria-Port Said maritime line turned out to be vain. That was not the route of choice for the *sans-papiers*, as only authorized merchants and workers on leave journeyed on it. Purging Port Said of what Company sources labeled "all the bad subjects residing there" appeared to be easier said than done. Still in 1862, no police personnel were available to take on that task. Boats could enter Port Said's harbor freely. No agent in either Port Said or Damietta checked newcomers' papers or barred disembarkation to those who lacked them.[114] By 1868 the Egyptian government had established a "sanitary bureau" in Port Said, whose employees were supposed to visit the incoming steamship, note the arrival of each passenger, examine passports, and send individuals to their respective consulates (or, in the case of Ottoman subjects, to the local government). But boatmen of different nationalities, thanks to their knowledge of languages, found it easy to disembark passengers into their dinghies and obstruct the bureau employees' efforts. Alleged criminals and ill-doers, especially those with no passports and those who had been exiled before "for serious reasons," could therefore sneak back into town.[115] Finally, consular authorities became resigned to the fact that many of their nationals were in the country with no papers, which hindered consulates from undertaking any administrative or judiciary measure and checking on evildoers.[116]

People's movement toward and presence on the Suez worksites engendered an array of job opportunities in fields other than technical or manual canal work. Migrant men and women promptly seized the chances at hand to scrape together a living all along the canal. At different times they held disparate kinds of jobs all along the waterway in the making. They did not necessarily stay in one occupation for long. The trajectory followed by a Mr. Maurice, among others, in a matter of just four years' time is telling. He was one of the "old" employees of one of the contracting companies, since presumably 1859. Soon thereafter he was only temporarily occupied as head of the workshops warehouse, at 7 francs per day in Port Said. He left for Damietta on sick leave. He then came back to Port Said, where he fell sick again and was hospitalized. By 1863 he was off to Cairo. Sometimes those doing odd jobs were the spouses of Company workers, such as the unnamed wife of a foundry worker in Ismailia who topped up the family's income with the refectory she was running. Madame Geroli, the wife of a Company laborer in Ismailia, strove to do the same by washing laundry.[117] At other times occasional migrant workers were singletons. While internal migration within Egypt was usually a "family business," as whole peasant families tended to transfer from the countryside to the city, many foreign women moved to Egypt as sole breadwinners.[118]

The Canal Domestic Labor Market

Foreign women and "Arab" men seem to have dominated the market for domestic labor on the isthmus. Domestic service constituted at once a temporary occupation and a way to ensure board and lodging. Foreign women may have lacked knowledge about or access to employment opportunities other than personal and domestic service.[119] Lucia Luciani, for instance, born in the Italian Piedmont's town of Lesegno, had been working as a housemaid on the isthmus when she died at the Suez hospital at age fifty-one, in 1865, from an abdominal edema. The fact that her suitcase and objects were to be sold suggests that no next-of-kin was around to acquire her belongings.[120] Research has shown that by the nineteenth century domestic work was undergoing a process of "feminization" and becoming almost totally a woman's domain. Moreover, with their homebound tasks, domestic workers in particular may have reified patriarchal family ideas and a gendered division of labor.[121]

On the isthmus this sector became accessible also to indigenous men, enterprising individuals who had no prior training or experience in the field, who mostly received belittling treatment. In nascent Port Said specifically, Egyptians or "Barbarins" found employment in the service of foreigners early on. They were commonly employed as doorkeepers, house servants, grooms, and runners, as well as coachmen and cooks. They enjoyed a reputation for cleanliness, honesty, and subordination, but at the same time they were not spared allegations of lasciviousness, either.[122] Seldom could these individuals' places of origin be identified precisely. A rare example is provided by a character in the pages of a semifictional memoir; called Kemsé-Abdel-Kérim, he was a young "Barbarin" servant from Khartoum, then in southern Egyptian territory.[123] In the archival record, however, a boy called Hamit was simply identified as "Arab." Another just called himself Khalil. One Hassan Mohammad was referred to as the last of a series of servants within the same household with the same name. Yet he must have been memorable. He was apparently adding potatoes to soup without peeling them and plucking rabbits instead of skinning them. He unsurprisingly remained employed for only about two weeks.[124] Among other chores, servants were also supposed to water the household's vegetable garden, if there was one, and wash their masters' laundry in the canal.[125]

To female immigrants, domestic service offered more security and respectability than the occupations that exposed them to the rough and tumble of the street. Living in proximity to one's employers presented advantages as well as risks.[126] When the Italian Luigia Frigieri entered service in around 1867 as a waitress in a brewery at the encampment of Serapeum, she also went to live with her employer. What made her even more vulnerable to his demands is that he advanced her a considerable sum of money, which she needed to pay back a debt (she allegedly had returned the money after a few days). Terminated after only one month on the job, not only did she receive no salary for her work, but her former employer also retaliated by demanding from her the sum of 250 francs, withholding her clothes and other belongings, and spreading insinuations that she sold herself for money.[127] Working women were also at risk of accidental pregnancies, "the biological manifestation of their economic and social vulnerability."[128] Marie Angelique Guerin, an unmarried French woman who hung herself in 1863 on the Hyènes Plateau, left a note hinting at the reason for her suicide: she was pregnant. She owned one trunk and two boxes with clothing items and had twenty-one letters among her belongings, suggesting that she could read and write.[129] During one single month in 1869, five women were registered as giving birth in Ismailia. Four were French and one was Austrian. Only a minority was married: while two out of five were in a marital relationship, three were "unaccompanied," which indicates they had born children out of wedlock.[130] The rate of illegitimate births within foreign immigrant communities in Egypt appeared to be higher than in their respective home countries. Consular authorities explained this by adducing that most Egypt-bound immigrants were either married already or were highly mobile singletons who had no intention of marrying.[131]

The Canal Prostitution Trade

Prostitution also offered women options for paid work.[132] As early as 1861, de Lesseps recounted the presence of some "almées" on the Seuil worksites.[133] *Almées* (from the Arabic ʿawalim, "learned women") initially were respected professional female entertainers, far above any hint of impropriety.[134] But those Westerners venturing into nineteenth-century Egypt were often unable to distinguish *almées* from prostitutes.[135] What they tentatively knew was that dancing in public, as a Company cadre remarked, was a job reserved to this female group.[136] Western visitors portrayed *almées* as skillful and handsomely dressed but lascivious and overall irreconcilable with truly modest Western ladies (who were nonetheless interested enough to attend their shows).[137] De Lesseps had known about *almées* since at least 1854, when he attended one of their performances under the auspices of his host, the Egyptian viceroy. In 1855, in Luxor, he again enjoyed a show organized for him by the local governor with five *almées*, musicians, and a clown. He commented that "it had been recommended to them that things be kept within the limits of decency," thus manifesting his understanding that such limits could well be violated on other occasions.[138] De Lesseps's forays into Upper Egypt occurred in the wake of the decision taken by the Egyptian ruler 'Abbas (1849–1854) to lift a ban on prostitution issued by his predecessor, Muhammad 'Ali, who had expelled prostitutes from big urban centers and military camps in 1834. With the ban lifted, those women deemed to be prostitutes were allowed back into Cairo but could no longer practice their occupations publicly.[139] Beginning in the late 1850s, even though there was still no official recognition of their trade, "prostitutes" appear in official documents without being indicated as being outside the law.[140] At least some of them must have headed to the Isthmus of Suez, where significant labor migration and new urban centers created a demand for commercial sex.

The Company attempted early on to regulate prostitution practice in the canal area through a French model of "regulationism," which entailed the licensing of brothels, their confinement within designated urban quarters, the police registration of prostitutes, and their subjection to mandatory and periodic medical inspections.[141] This system had been in operation in France for most of the nineteenth century and in Algeria since the 1830s; it would be implemented in Tunisia in the 1880s and in Morocco in the 1920s.[142] The earliest record of regulation in matters of commercial sex on the Isthmus of Suez dates from the early months of 1865. That is when Company doctor in chief Aubert-Roche formalized measures about the surveillance of "public girls" that de Lesseps himself approved. Involved in their implementation were Company cadres at different levels: from the director general of the works to the engineers in charge of a certain portion of the excavation, the chiefs of the Bureaux Detachés below them, and the municipal superintendents (*commissaires*) further down. The system of prostitution control rested on the presumed collaboration at the local level of municipal superintendents, brothel keepers (assumed to be female), and medical personnel. Prostitutes, immediately upon their arrival at a campsite, had to present themselves to the locally dispatched doctor for inspection. If they were healthy, he would issue to them cards carrying their name, description, and date of last visit. The doctor was to inspect each girl once a week. He would fix the day and the time of the visit, after which he would stamp her card. Every now and then municipal superintendents had to visit the public houses and make sure that the girls had their cards with them, with a valid date. If brothel keepers failed to enforce these measures in their houses, they would first be warned; a second infraction would trigger their expulsion. All prostitutes who fell sick, failed to produce a card, or produced one with an invalid date had to visit the doctor immediately, under threat of expulsion.[143]

Brothels were available for men's leisure as soon as encampments were up and running. All along the length of the isthmus, in the bigger centers that were sprouting along the canal banks and in the more-or-less temporary encampments, women could toil as prostitutes but also as brothel servants or managers, as exemplified by a lady called Albertini who ran the house of public women at the Serapeum worksite. Women's employment in this sector could be either permanent or just temporary and occasional. An Italian woman called Emilia, for example, first toiled at a brothel in Ismailia, where she had ostensibly been led by a lying seducer. After having been ill-treated and ignominiously exploited by the brothel's mistress, she fled and found work as a waitress at the drink stall of a certain César, who moonlighted as a carpenter.[144] In Port Said, where brothels welcomed clients past midnight in defiance of regulations and inhabitants' laments, commercial sex was destined to become a thriving business.[145] By the second half of the 1860s venues that were identified as establishments for prostitution were found in the "Arab villages" of Qantara, Al Guisr, and Serapeum. Brothels were also in place in the so-called Greek village of Ismailia. In Chalouf, a brothel operated at least until 1868, when some coals and a cigarette that had rolled over against the mats set it on fire.[146] Around the same time, Suez boasted no less than four brothels (*magazzini di donnacce pubbliche*) in the surroundings of its Catholic church, and both European and native prostitutes reportedly resided in the very heart of that town's "European quarter."[147]

Migrant Women's Scattered Livelihoods

Other kinds of jobs, also available on the isthmus in the 1860s, defy the idea that migrant women came only as either domestic servants or prostitutes. They also flout the assumption that nineteenth-century chain migration was necessarily initiated by men. Women on the Isthmus of Suez ran their own businesses. A Miss Elisa in Ismailia, for instance, borrowed the accommodation of a Y. Cohen to do laundry while he was away. She consumed so much water that she was presumably trying to make a living washing other people's dirty linens.[148] Women were also storekeepers, such as Madame Serre of the Boucherie Française or French butchery located on Commerce Street in Port Said. Another, Madame Chai Gouthman, entered into a contract for renting a one-room coffee hall with attendant kitchen and paraphernalia in Ismailia. Women owned and ran hotels and taverns. Maddalena Giuffrida, for instance, owned the Hôtel d'Europe in Port Said.[149] One woman by the last name of Soldaini, originally from Leghorn, ran "more than one tavern" (*osteria*) on the isthmus and had done so well for herself that she "even had servants" working for her; her brother, back in Italy, begged her repeatedly around 1870 to fund his Egypt-bound travel and let him work for her.[150]

Women also rented out property or lent money. Al Hadja Zarifa Bent Abdelrahman Hezzi, for example, owned an oven in Port Said's so-called Arab quarter. Estranged from her Tunisian husband, who had built the oven in the first place, she resided in Damietta. In the same area of Port Said, specifically in the "fourth section of the Arab village," Oum El Keir owned and rented out her property.[151] Like their male counterparts, women became enmeshed in disputes over debts. For example, Françoise Töller, a destitute Austrian widow in Port Said, forcefully requested the consular intercession to claim from a Company worker a debt of 272 francs for board and lodging. Antonia Campi, an illiterate Italian migrant, decided to keep the trunk of her debtor in exchange for the board and lodging that he had failed to pay. The French woman Claudine Lapeyrouse, who provided lodging for defaulting workers at Al Guisr, first requested the intercession of their boss to hold what was owed her from their pay. Then she talked to the municipal superintendent. When both refused to help her recover her losses (including liquor and broken glasses), she threatened the municipal superintendent that she would "ask from him all her credits in order to settle her affairs."[152] Finally, women explored even more drastic options. On an August day of 1866, a lady called Blanchet, who ran a "kind of café" in the Arab village of

Ismailia, "abandoned her home and the country" rather than face a remarkable debt of 1,243 francs. Her creditors could not do much, but wondered: "Under what circumstances did she leave? What dispositions did she take before her departure and is it possible to be paid either in real estate or in furniture?"[153]

Notwithstanding the spate of new jobs created by the Suez Canal Company on the isthmus, women migrants on this strip of land still had a relatively limited range of occupations available to them. In many cases, they had no alternative to seeking employment in areas considered "women's work," such as personal and domestic service.[154] However, even if they had to bear often exploitative conditions, women made use of all means accessible to them and demonstrated an ability to control their circumstances.[155] Recruitment and placement agencies targeting foreign women for work within Egypt appear to have been established only at a later time (see chapter four).

In the 1860s, what female migrants on the Isthmus of Suez could do was to relocate from one encampment or town to another, relying on a novel circuit for mobility that the Isthmus of Suez now had to offer, and in turn contributing to it with their movements. Miss Marie Dutruy, for example, decided in 1863 to transfer the store she was running at Al Guisr to Ismailia, where she rented the portion of a house located on "rue Negrelly [sic]." Claudine Lapeyrouse, at first employed as a housemaid in Al Guisr, later opened an establishment in that same location, where she provided workers with board and lodging. But by 1866 she wanted to leave Egypt for good.[156]

Whether standing still or on the go, unmarried women had an especially hard time keeping professional and familial arrangements separate. Different fortunes awaited them at the end of their paths, crisscrossing through different jobs, men's houses, and locations. This is exemplified by the trajectory of Miss Marie Servat, a twenty-two-year-old French woman also known by the name Antoinette, who had moved from Alexandria to Port Said at the beginning of 1867. She had been working there as a cook at the house of Mr. Quiquizot, a wine merchant, for four or five months. From Port Said, she moved to a spot known simply as "Kilometer 34" of the canal, where she found another job as a cook at a Mr. Legroud's house for another few months. She then moved again to Chalouf, where she was "taken in" by a French man called Marius Delbergues, a mechanic. Servat and Delbergues were said to live as a couple (*maritalement*). Chalouf was her last stop, since she died there of pneumonia in 1868. She was completely destitute, and the clothes she was wearing were neither inventoried nor auctioned off; they were deemed of no value.¹⁵⁷ An arguably happy ending awaited Marie Thomassin, a "lost and abandoned girl" who first worked as a servant in Port Said. Her French master, who had already been condemned for blows and lesions inflicted on his legitimate wife, subjected Marie to "all sorts of ill treatment." Marie was then "taken away" by another man, an innkeeper in Serapeum, who made her his lover (*maîtresse*). Reportedly they lived happily thereafter at Bitter Lakes, where he had a job as a stationmaster (*chef de gare*).¹⁵⁸

In both Servat's and Thomassin's cases, sources employ euphemisms to describe their circumstances (both were said to have been passively "taken"), conceal the hardship these women experienced, and downplay the resilience they found to move on time after time. It is necessary to read between the lines to reconstruct the skein of these migrant women's paths across the Isthmus of Suez and create a fuller picture of their "scattered livelihoods."¹⁵⁹

WORKERS AS CHILDREN OF A BIG FAMILY

The Company was purportedly not content with just handing out salaries. Company workers became, for their employers, like children of a big family.¹⁶⁰ Laborers were suffering from the heat, drinking brackish water, eating biscuits, and sleeping on the blazing sand. Ostensibly they did all this for the sake of the endeavor's success. In exchange, the Company was allegedly taking care of its workers during and after their employment in the name of mutuality, their participation in benefits, and the peaceful association of capital and labor. In theory, this found expression in the fact that when Company workers died, their widows and children could receive their pensions in proportion to the length of the deceased's employment.¹⁶¹ The Company adopted a managerial style that provided for workers' needs while depriving them of independence and responsibility. While Company officials might have sincerely committed to workers' well-being, they were at the same time exercising a form of paternalism and implementing yet another instrument of company control. They simultaneously emphasized the ideals of family and cohesiveness and chastised those who did not comply with such principles.¹⁶²

Some of the workers admiringly thought that the Company's founder, Ferdinand de Lesseps, appeared simpler in his manners than all the rest of the Company engineers and employees. They claimed he commanded so much respect that "Arabs" used to lift their hands to their forehead and then kiss them.[163] Others recounted that de Lesseps was in the habit of asking laborers about their families and impressed them with the feeling that he knew their uncles, aunts, and cousins personally. He seemingly "cared about and loved everyone." After his visits, morale was boosted and workers toiled with renewed enthusiasm, to the point that "some demented ones (*enragés*) refused to leave the worksites" even under the midday sun and fell down crying, "Long live the Canal, long live France." That is how they died, uneventfully and yet heroically, dripping blood from their mouths and noses.[164] May this have been the destiny of Jean Valentino, Italian subject and mason of twenty-five years, who succumbed to sunstroke-induced "cerebral meningitis"?[165] According to one of the Company historians, de Lesseps was a great philanthropist, who "loved the humble, knew how to talk to them and inspire confidence in them." He cherished thinking of "his" personnel as constituting a "big family."[166]

De Lesseps himself therefore supposedly headed the isthmus family and accordingly sported a manly and father-like demeanor. Not only did the Company founder and his admirers have a penchant for gendered metaphors—the canal would "pry open the land to the peoples" and "inundate the desert bosom with life-instilling waters"—but de Lesseps himself was presented as the embodiment of a masculine ideal.[167] According to Western written and visual accounts, the Company founder easily overcame the illness-inducing Egyptian weather and came to the canal region twice or three times a year, which bolstered his health greatly. His "rich constitution, iron health, and youthlike ardor" set him apart from the average man, to begin with.[168] A caricature by Etienne Carjat published in June 1862 in *Le Boulevard* (1861–1863) suggested that De Lesseps had bravely defied the horrors of the isthmus desert and confronted its exotic and dangerous looking creatures (including a crocodile, perhaps symbolizing Lake Timsah, meaning "crocodile" in Arabic, or ancient Egypt itself).[169] In the guise of a modern Hercules, his benevolent and fatherly smile gleamed above a mass of rock-hard muscles. Thanks to the powerful steam-operated technology sprinting between his legs, he had conquered the land of the pyramids peeking in from the misty horizon (see figure 10). A hypersexualized and repulsively grinning Sphynx with female features, arguably representing Egypt's pharaonic glory, passively bore testimony to the European and male undertaking. This 1862 caricature expressed the ideal of the male prowess expected in the canal enterprise and hinted at the pervasive power of Company management. Whether it had been realized with a satirical intention or not, the masculine

FIGURE 10. Caricature of Ferdinand de Lesseps by Etienne Carjat, 1862. *Source: Le Boulevard*, 29 June 1862, Bibliothèque nationale de France, Paris.

ideal and Orientalist setting it conveyed remained intact and immune to laughter.[170] The caricature's very subject evoked its absent antihero: the weak and unmanly worker unwilling or unable to take up de Lesseps's challenge.

The Canal's Infantilized Male Workers

Abroad, male workers on the Suez worksites were triumphantly portrayed in pioneer or soldier-like fashion. The Académie française sanctioned this hyperbolic portrayal through its selection of a poem by Henri de Bornier as the winner of a prize for an epic on the canal in 1862, around the time Carjat published his caricature. The poem instigated (French) workers to be as brave as their fathers had been (presumably during Napoleon's 1798 foray into Egypt) and fight their own kind of battle "at the bottom of the pyramids." It urged them to form a "battalion of workers" and follow war chief Ferdinand de Lesseps into the subjugation of the Isthmus of Suez.[171] Yet male workers on the canal worksites apparently failed to prove their valor at the destination. The Company infantilized the men on its payroll by considering them as essentially unable to resist the lure of gambling, liquor, and women. It castigated them for wasting time and energy at the gambling joints, brothels, canteens, and music halls (*café-concerts* or *café-chantants*) that were opening up all the way from Port Said to Suez.[172] It considered those frequenting these places as weaker individuals and fretted that the workers "who spent part of the night playing" would "not be in the condition, the following day, to complete the assigned task."[173] In a word, the Company saw them as constantly at risk of sapping their vitality and, most important, their productivity.

In particular, the male workers who liked to visit gambling halls were paternalistically treated as prey to illusionary gain. Even if gambling in Egypt appears to have been legally prohibited only in 1891, when activities within "public establishments" were regulated (see chapter four), in the mid-1860s and beyond it still seemed to have occurred mostly under cover.[174] The venues where workers seemingly squandered their savings over the roulette and card games, often simple houses in the daytime, were surreptitiously transformed into gambling dens at night. In one of these establishments in Port Said, a table covered with a green rug was found in the middle of a room. It was especially at nighttime, the French vice-consul pleaded with the governor, that government officials ought to visit the suspected houses to catch the culprits red-handed. In Port Said around 1866 four different gambling halls could be found; they were run by an Armenian Ottoman subject called Vincent, a subject of Bavarian origin called Valentin, a French man called Bousquet, and a Greek called Demetrios Caralambos. In Al Guisr there was one gambling hall located in the Arab village next to the European encampment.[175] To the knowledge of consular authorities, three more had been established in Suez by Greeks and Italians. In Suez the French consul, local employees, and sometimes the governor himself attended the joints (*bettole*) where gambling occurred. Even if they were apparently more interested in watching the shows than in gambling, their presence still raised eyebrows.[176] Even some agents were caught frequenting gambling houses, a fact that was met with disapproval by Company echelons. But in the eyes of Company representatives, it was for the working-class population of either workers or employees who made up the bulk of Port Said that the gambling habit appeared to be truly dangerous.[177]

Company managers and foreign consuls framed gambling as a gateway to profligacy, debauchery, crime, and misery. It acted as a catalyst for all the "pernicious tendencies" that had seemingly emerged in Port Said and elsewhere on the isthmus. In regard to the worker with a penchant for playing for money and reveling, it was said: "[I]f he has lost, he is exposed to dishonest temptations that occur when he has no money and needs to discharge his debts. If he has won, it is likely that he will spend his day in debauchery and will lose, the same night, more than he has gained the night before."[178] Those laborers who spent the whole night gambling were also said to indulge in drunkenness and get into bloody brawls. Liquor and hashish were available. Petros Ribao in Al Guisr, for example, complained in 1862 that his roommate was consuming the psychotropic substance and bringing women into their shared tent. Reveling workers were also blamed for contracting and spreading syphilis among the population (even if the ultimate responsibility for venereal diseases rested squarely upon female shoulders).[179] Finally, the inebriated reportedly screamed and sang so loudly as to keep everyone awake all night long. Company and consular authorities, perhaps exasperated by the lack of repose, connected workers' nighttime activities to their alleged criminal tendencies. Gamblers, it was argued, could only fulfill their needs by resorting to theft. Said to be "by nature" lacking in foresight, they always ended up forfeiting huge sums. Losses on the gambling table therefore ostensibly forced them to commit acts that they would not have perpetrated otherwise.[180]

To counter the "plague" of nocturnal gambling and revelry, it was recommended that workers be led back to "a strong regularity in life and of conduct" so that they could "fulfill their duties and provide for their needs through their work."[181] By 1862 Company representatives attempted to have establishments close at 10:00 p.m. The measure produced its malcontents, such as Mr. Bornet in Ismailia, who at first insisted in vain that he be authorized to close his café at 10:30 or 11:00 p.m. and later was closed down altogether for disregarding the general rule.[182] In order to guarantee a stricter respect for rules, more police measures were invoked and the appointment of an old guard from the French vice-consulate in Damietta was sought. In 1863 the French, English, Austrian, Italian, and Greek consuls subscribed to a proposal crafted by the Egyptian government. This regulation forbade the carrying of arms except for hunting weapons; prescribed the carrying of a lantern at nighttime; and proposed that investigations and arrests be carried out in music halls, breweries, and taverns (an ambitious attempt, given that capitulatory agreements granted foreign establishment managers the right to refuse entry to local law enforcement, as further explored in chapter four). Finally, it called for the closing of such venues at midnight.[183]

Efforts to make workers bed down early continued, suggesting that previous attempts had failed. In 1866 Company functionaries on the isthmus were still helplessly trying to set a closing time for taverns in Chalouf and

Ismailia, which remained open for part of the night and were said to fuel several brawls. In 1867 the guards patrolling the Greek village of Ismailia, for example, found that cafés, taverns, and brothels were routinely open at midnight and beyond. The Egyptian government decreed that all venues had to close at 11:00 p.m., and all those who ventured outside of their homes at night had to carry a lamp. Company functionaries had to deal with the fact that only the local government could handle such matters. Yet they asserted that, Ismailia, for one, was a worksite that belonged to "their" Company, where they had "the right" to ask for intervention to stop the gambling, which "[by] its nature" produced "the biggest disruptions in the execution of works."[184] The Italian consul in Suez relied on the local government when he threatened the Italian owner of one of the town's music halls where gambling took place. However, the consul's warning that he would "tell the local authority to intervene as much as possible, destroying the games and denouncing the players" was met with the businessman's mocking reply that "he had nothing to fear from the local authorities."[185] Finally, the situation reached a climax near Al Firdan in 1869: cafés and taverns, of which there was a disproportionate number, were open all night and kept workers and employees up till very late. Gambling houses had been established side by side and even in the midst of workers' houses. They attracted "vagabonds" who allegedly compromised the security of the encampments and encouraged bad behavior. Despite Company requests to stop authorizing the opening of such establishments and to move some of them away from residential areas, venues for workers' leisure continued to waste time, generate trouble, and damage the work pace in ways "greater than one could think."[186]

The Canal Wives

While Company officials often treated male workers as if they were rambunctious children, they may have proved receptive to the idea that women could bring "civilization" and order to the land they were trying to settle.[187] They frequently approached female migrants as wives rather than women. By doing so, they acknowledged women's presence on the isthmus but envisioned, managed, and then recorded their presence in instrumental ways. Initially, in order to "destroy the prejudice of the alleged (*prétendu*) forced-labor system," they emphasized that "some veiled women working with their quaint costumes" could be sighted among earth movers on the sites. Later they adduced their presence to prove that the Company cared about the stability and the wellness of the workforce and insisted that the Company's founder himself had favored the establishment of wives and families on the canal worksites (see chapter one). Accordingly, those Company agents, but not the workers, who had been recruited in France could bring their wives and children and be reimbursed for the related expenses.[188] Of Port Said, Company sources commented that "several workers, well-paid and confident in a long-term job with no displacement, brought over their families and thus contribute to the development of the city."[189] This optimistic portrayal did indeed capture the presence of wives and households, but it omitted the hardships of family life on the worksites. Jusepha Rubena, for example, lived in a tent in Ismailia with her mason husband Raphael Sciva and was seriously sick with dysentery. The spouse of a *cawas* or guard called Ahmet, a guard in the employ of the Transit department (Service), similarly dwelled in a tent in Al Guisr with her spouse. Not much is known about the circumstances of a couple of *raya* (*ra'aya*, literally "flock") Ottoman subjects from Izmir whose baby boy was born in Qantara in 1863.[190] As in company towns elsewhere, the Company management fostered on the Suez worksites a clear relationship between marital status, employee stability, and the solidity of the canal undertaking.[191]

Despite the Company management's ambitions to foster domesticity to promote stability, not all women on the isthmus became the embodiment of idyllic home life. Many were not married, settled, or even steadily employed. Contemporary narratives would insist that these wayward females needed guidance, as supposedly did a pitiful "girl" living by herself at the campsite of Al Seuil. A semifictional feuilleton, published in Paris in 1877, described another young woman by the name of Georgette, toiling as a servant in a rundown tavern that catered to canal workers, as similarly deficient and needing rescue.[192] As illustrated earlier, women worked as servants, laundrywomen, waitresses, innkeepers, or prostitutes. They may have toiled on the worksites themselves, carrying pitchers filled with water to the laborers (see figure 11).[193] They also found employment as artists in music halls, such as Rosa Calamita's daughters, who were unmarried (*signorine*) and whose mother negotiated an engagement contract at an establishment due to open in Ismailia from September to November 1869. One would sing and the other one would dance, six days a week, "agreeing on one evening each to their benefit, excluding though the cost of the ticket."[194]

Wives and husbands may have parted ways upon landing. Victorine Bayard, for example, remained for at least one year by herself at her job

FIGURE 11. Hut of *fellah* workers in Ferdane, 1869. *Source:* Marius Etienne Fontane and Edouard Riou, *Le canal maritime de Suez illustré: Histoire du canal et des travaux* (Paris: Aux bureaux de l'Illustration A. Marc et Cie., 1869), 129.

at the Hôtel des Voyageurs in Ismailia after her husband had returned to the French town of Carcassonne. He had been beaten up with a stick by the manager of the hotel, who had convinced them to come from France (they got married in Marseille) but never reimbursed their travel expenses of 550 francs. Moreover, their superior allegedly treated them "like beasts." Consular authorities finally decided to look for Victorine. By the time they did so, however, it was unclear whether she was still alive or simply estranged and unwilling to communicate, as she had never replied to her husband's missives.[195] Widows also may have decided to remain and take advantage of opportunities that were not available elsewhere. Madame Toussaint, for instance, witnessed her spouse's death from a fever. She was still in Ismailia one week after his passing. Would she stay on after her husband's death? Perhaps she was still figuring it out. But others did stay put. A woman who ran an inn in Ismailia and survived in the consular records for hurling pool balls at an abusive patron, for example, was known as "widow Gauthier." Adele, the widow of deceased carpenter Alexandre Dumas, stayed on in Ismailia despite having a sick child residing in the French town of Bordeaux. She may have eventually remarried and relocated within Egypt, because she declared that "Mr. Lecler entrepreneur of Suez Canal in Alexandria was interested in her."[196] Her destiny may have pushed her away from the isthmus, but not far enough to turn back across the Mediterranean.

Company and consular officials had an opportunity to pronounce on what was morally acceptable in familial life on the isthmus, especially when summoned to resolve conjugal disputes. When Mr. Bialkowski, an Austrian innkeeper, died in Al Guisr in 1866, a woman called Buoncore declared she was his legitimate wife. When she turned out to be unable or unwilling to present evidence that would qualify her as the veritable widow Bialkowska, Company officials grew suspicious. In their eyes, she "voluntarily muddled up the situation by refusing to give proof." When the "real" widow showed up complaining about the way her husband's assets had been disposed of, therefore, they replied that she should have been grateful instead, as they had done everything within their power to safeguard Bialkowski's *legitimate* succession.[197] In another example, when Domenico Giorgini wrote to the Italian vice-consul in Suez to beseech the expulsion of another man, he claimed that his wife and the latter "were inconsiderately soiling the legitimate nuptial bed [. . .] taking away the domestic peace from an honest father [. . .] committed to provide a chunk of bread to his family." He declared their trysts offended public morals and derided his honor. Given that he signed with a cross, we might surmise that the pompous rhetoric was not his own. Yet the clichés conveyed in his letter must have proven persuasive, as his competitor was indeed expelled, even if proof was lacking, for "reasons of public scandal and danger for the interested parties."[198]

Male and female workers consciously traded benefits for autonomy and exploited the Company's paternalistic rhetoric for their own purposes. Even if the underlying structure of power remained unchanged, they learned the skills to negotiate paternalism.[199] The Italian Gaspare Lucia, for example, "in Ismelia [sic] with no work, no money, and no one else to ask," pleaded with the French consular authorities sometime in 1869 for "guidance" and support (*assostenimento*) so that he "would not incur in any shameful occasion." A woman called Jeanne Felicite, residing in Port Said since 1864, similarly relied on the Company officials' attitudes, protective and intrusive at the same time, for her own interests. She wrote to the general director that the man with whom she had "shared her domicile," a mechanic working at the service of the contractor Lavalley, had left her "to start again" with a woman of ill repute. Because of this other woman, he was on the verge of plunging again into the "common life" from which his employer had supposedly "saved" him some time prior "when he was still in Damietta." Felicite declared that she was not asking that he come back to her. Ostensibly, she would be "satisfied to just see him avoid a second danger" that she deemed serious and imminent. Felicite and the man were not married either, but that did not prevent her from hinting to his employer about the doubtful morality of his new relationship. She also appealed to the employer's fatherly mission toward his employees: saved once, they might as well be saved again and rebuked about "their duties." Mr. Déré, a dredgeman in Port Said, resorted to a comparable strategy and attempted to top it up with extradition. His wife, Thérèse Stinger, had left him to follow to the Al Firdan worksite an Italian man, Mr. Clément Conti, who had previously worked as an interpreter at one of the Company's enterprises. The husband begged consular authorities to remind her of "her duties" and send her back home to France.[200]

Debates raged about women's work in nineteenth-century France, contrasting productive and upright male workers with dissolute and improvident female workers.[201] Hence it is unsurprising that echoes of those debates should find their way to the isthmus worksites and affect both the bourgeois thinkers close to the Company and Company personnel alike. Claimants conveyed clichés of honest men toiling for a meager wage while treacherous women, behind a façade of honesty, seriously endangered the domestic and public peace. An Italian man in Ismailia, for example, threatened violence when his former French mistress took to insulting him in public "loudly and with the most shameful words." He explicitly admitted that "his blood" was about to force him "to an evil action."[202] If, on the other hand, females tried to react, they were likened to furies. According to a newspaper article of the time, women were essentially unable to contain their verbal insolence. They were seen as slyly dangerous. Ultimately they deserved to be slapped and disarmed by the authorities, who had given shelter to the "poor" men who had in vain tried to calm them down.[203] Wayward women on the isthmus had a chance to become "good, dignified, and one would even dare to say honest" individuals, such as the orphan Marie Thomassin. But male intervention was key to their redemption. Thomassin, for example, was identified as a "lost and abandoned girl" who had found refuge from an abusive employer in another man's house and had, perhaps reluctantly, turned from the latter's servant into his lover. It was thanks to his male "good advice" that she had allegedly "changed route" and now conducted an irreproachable lifestyle. On another occasion, Ahmad Effendia, a man whom some wrote off as just a drunkard and a womanizer, was given the chance to redeem himself by saving a "little girl" who had been terrorized and starved by her former master at one of the Company worksites. Effendia declared to a Company official:

"She has come to me so that I could feed her and give her some money." On the one hand, he claimed that he wanted to end her suffering in the name of justice.[204] But on the other hand, he also clarified the amount of money that he had already spent for her, thus quantifying how much he yearned to keep her for himself. Authorities on the isthmus enforced and reinforced such platitudes, which held sway in official narratives and pillow talk alike.

In 1884 the words of a concerned grandmother, Clemence Vial, a French midwife in Cairo, lay bare how workers themselves, including women, tapped into and contributed to collective notions of male and female respectability. Her missive addressed to the French consul also outlined the financial and emotional challenges faced by workers in Port Said and elsewhere on the Isthmus of Suez: "I think my daughter-in-law is honest and a good mother. [. . .] But I am distraught to know she works in a refreshment hall. I would prefer she made a living with ironing." Clemence Vial discounted the fact that her twenty-three-year-old daughter-in-law supported a family of three on her own by waiting tables in Port Said. Her being married to a parasitic thirty-two-year-old man who had lost his job, "had no dignity," and was prone "to create scandals" did not acquit the young woman. What mattered, to Clemence Vial and perhaps others, was the existing labor hierarchy and attendant workers' worth. Vial went on to thank her institutional interlocutor because her grandson and granddaughter, Henri and Matilde, had been accepted in the Catholic boarding schools in town.[205] Vial neglected to specify whether her daughter-in-law, the hard-at-work waitress, had actually agreed to her own children's relocation away from her, their mother.

While male workers, often moving and working collectively, and their visible acts of resistance or compliance do deserve their place in the isthmus's history, so do the women, who moved in smaller groups and were frequently employed behind closed doors. Among the most ignored in labor history, in fact, have been those who took up jobs in private houses, such as domestic servants and wet nurses.[206] By leaving them out of discussions on the labor market, official accounts have rendered women invisible or dependent and thus economically irrelevant.[207] Whenever selective attention has been paid to foreign female servants in Egypt, Orientalist fantasies about an exotic and licentious Middle Eastern world that fundamentally endangered unaccompanied women have been in the way. Yet by way of example, Austrian women in Cairo, Alexandria, and Port Said outnumbered Austrian men: nearly two-thirds of the 4,505 Austrians in Egypt in 1900 were women.[208] Presented as either prostitutes or wives, women and their role in the canal region, economic and otherwise, have been largely overlooked. Not only does their presence demand acknowledgment, but paying attention to their local, short-range, domestic relocations crucially reveals that they both counted on and sustained a nascent circuit of mobility along the canal in the making.

CONCLUSION

On the canal worksites, differentiating between engineer and earthwork laborer, "European" and "Arab," righteously employed and idly unemployed, well-behaving spouse and illegitimate paramour served as a powerful tool of control. The workers toiling on the canal worksites were pigeonholed in hierarchies built on preconceived ideas about their national and racial affiliation, assumptions about their tolerance to the desert environment, and gendered expectations. If they spent time or money in cafés and brothels, which the Company, the Egyptian government, and foreign consuls persistently threatened to shut down, they were treated as irresponsible children in need of supervision. As for migrant women, they were at best approached as vulnerable dependents rather than the self-confident and enterprising individuals they may have been, occasionally enabling the migration of brothers, husbands, or other males.[209] At worst, migrant women were stigmatized as fundamentally immoral, constantly lured by or falling into prostitution.

Far from being solely confined to the workplace, the hierarchies described thus far were also inscribed in the urban space. Port Said was divided from its inception into a so-called European quarter and an Arab part of town and maintained unequal sewer and water infrastructure across its two halves.[210] Because of the inflammable materials it was assembled with, the latter part of town was more prone to fire accidents, such as when a cracked clay bake oven let out flames that set on fire the fibers and planks of the hut where it was located as well as the surrounding ones; by the time the pumps were brought, the whole neighborhood was on fire. Smoking in straw-filled storerooms adjacent to stables could also become an incendiary habit.[211] Street names in the "Arab" part of town reflected the diversified provenance of much of its immigrant population, such as Minya, Sharqiyya, Aswan, al-Daqhaliyya, and Damietta.[212] Ismailia would also develop a "European" section that would remain separate from two distinct Arab "villages," a Greek quarter, and a Calabrian quarter, in all four of which the overwhelming majority of residents lacked proper sanitation.²¹³ Urban dwellers constantly crossed the lines that were drawn in the urban layout. People in Port Said persisted in setting up shop, peddling, consuming, or residing where they were not supposed to and challenged the Company's efforts to prevent their mundane transgressions.²¹⁴ Even if limited because of their spatial separation, there were still marriages between Ismailia's Egyptian men and foreign women, which "contributed to the mixing of cultures and languages and habits in the city."²¹⁵ The canal towns were not just the products of blueprints devised in Cairo or Paris, but rather became theaters for performances of power and unintended urban realities.²¹⁶

Official reports conflated the unemployed and the would-be workers who happened to have no engagement contracts or identity papers on them with criminals. Especially those migrants doing odd jobs or toiling outside the realm of the Company eschewed the mechanisms that were supposed to regulate life on the isthmus. The measures contemplated or taken throughout the 1860s by Company management, foreign consulates, and the Egyptian government revealed their shared concerns about individuals out of work and workers out of place, whose existence raised red flags precisely because they eschewed the national, racial, and gendered hierarchies and expectations in place. Nothing much could be done about the former group, composed of the allegedly numerous "unemployed outcasts" hovering around the worksites and frequenting its entertainment venues. "No means but legal means" could be applied to them, which in substance meant they dodged Company surveillance.²¹⁷ Instead, the Company could attempt to regulate the behavior of those on its payroll by imposing or threatening punitive measures. If workers fell sick due to venereal disease, intemperance, or voluntary brawls, they would receive no salary. If they left voluntarily, were frequently absent, performed poorly, or refused to work or to respect discipline, they would be dismissed without the fifteen days' severance pay that would otherwise be granted upon dismissal. The same applied if it was discovered that the worker suffered from a chronic or contagious disease that had not been declared at the time of the engagement. Conversely, productivity was encouraged and prized under the often misplaced assumption that remuneration could act as a powerful stimulant to the fatigues of labor. Workers could obtain supplements in their salaries by working at night, laboring extra hours, or taking on exceptionally challenging or dangerous tasks. Some observers commented that the laziness of a few created confusion and caused delay to the works, but that most were diligently enforcing the rules and making sure their companions also contributed to the common tasks.[218]

Company, consular, and government officials may not have cared about the stuff of workers' everyday lives. In 1863, for example, a man called Geyler, a municipal superintendent and French consular agent in Ismailia, pronounced on the reimbursement of 18 francs that Grouaz, a trader in town, demanded from a laundrywoman, Madame Geroli, for a lost jacket. Geyler said that he and other Company officials had no time or willingness to deal with such "minimal issues."[219] Profoundly disagreeing with Geyler's position, this chapter has lingered on issues that can be considered as minuscule as Grouaz's lost clothing item and Geroli's missed gain. Paying heed to mundane everyday life on the isthmus reveals that migrant workers could indeed be at once indolent and watchful, adopting actions that ranged from the trivial to the theatrical. Workers reacted to unpopular measures and circumvented often daunting realities in manifold ways and for disparate reasons. Men and women, single or married, abandoned the worksites, went on strike, fled the country, switched jobs, or had one drink or one bet too many. They changed occupations and residences frequently and resorted to mobility as an effective strategy whenever their circumstances changed for the worse. When opportunity arose, male and female workers conveniently utilized officials' paternalistic notions to maximize benefits or to seek out new opportunities, even at the expense of their peers.

THREE

A Semilawless Borderland

THE PRESENCE OF THESE PEOPLE
COULD BRING EVIL

> There is a rather sizeable number of workers in Ismailia and in the other camps who do not have jobs or resources. [...] [M]ost of them are now without a home and sleep in the streets. Here it is, their situation: how will they make do in the future, where there will be no jobs left? At this time, there have been accidents like theft or others; it is to be believed that the people who commit such crimes, whenever left with no money or resources for their daily living, are set on a path to commit evil.
>
> —ISMAʿIL BEY, 1869

IN 1869 TALK OF THE SUEZ CANAL'S official opening was on everybody's lips. The waterway's inauguration had become a heated topic of conversation especially in Port Said, at once the canal's first encampment, northern harbor, and Mediterranean outlet. Some exalted the certain success of the undertaking, while others voiced skepticism.[1] A few cherished the festive atmosphere, but those whose livelihood was threatened by the end of the canal works must have felt less jovial. In the months prior to mid-November 1869, contradictory rumors had been circulating about the completion of the enterprise. By the preceding summer even the most incredulous had to acknowledge that the goal was nearing achievement. However, despite the extraordinary efforts of "thousands of workers" and "very powerful machines," a few points still needed intervention.[2] In October 1869, one month before the inauguration, the quay of Port Said's Grand Basin, right in front of Customs, was still encumbered by a great deal of materials. These needed to be removed to enable the circulation of the crowds who would attend the ceremonies. While wood and rocks could be easily carried away, rubble posed a more insurmountable problem, as its owners were missing. The Company had everything removed by the beginning of November.[3] A mere ten days before the official inauguration date, fourteen dredges were still operating to prepare the passage through the four-kilometer-long section near Port Said. Rocks along the canal bed were to be destroyed with explosives. Strong transverse winds, curves, sand, and opposing currents raised concerns. On the eve of the inauguration, if a self-laudatory anecdote recounted by de Lesseps himself is to be believed, a huge stone had slipped into the main basin and threatened to impede the passage of vessels.[4] Still, the show had to go on.

After 1869 Port Said fleetingly appeared destined to thrive as a toll collection point, a purveyor of goods and services to vessels, a point of transshipment, and especially a coaling station (until 1887, ships were restricted to transiting the canal exclusively in the daytime and spent time anchored at Port Said and taking on coal).[5] But the slowing down and the completion of the canal works left thousands of people destitute. Throughout the 1870s instability remained a feature of isthmus life. By the mid-1870s the population of Port Said was described as generally poor and made up of workers of all kinds. Whether transient or sedentary, inhabitants were described as "nothing but auxiliary personnel to the boats flowing by, to or from the Canal." Italians and French in Port Said and Ismailia were said to be living there only "precariously."[6] British subjects may have not fared much better. In 1875, for example, after losing his business in Port Said, the British subject James Testaferrata arranged for his entire family to leave town. As elsewhere in Egypt, British "remittance men" in dirty clothes occasionally knocked on the doors of their consulates asking for relief.[7]

The immigration (see chapter one) and employment (see chapter two) of a large-scale, unskilled workforce of disparate origins enabled the massive canal undertaking to reach completion. As the project neared achievement, however, most of the isthmus inhabitants became disposable. Before and after the canal's official inauguration, the concentration of a heterogeneous population on the Isthmus of Suez posed challenges to Company representatives, as well as to Egyptian officials and foreign consuls. Company bureaucrats worried about "keeping order and discipline" within what they defined as a "veritable Babel tower."[8] For its part, the Egyptian government perceived its own potential inability to manage conflicts. It fretted that with the isthmus becoming an international crossroads, the old jurisdictional system would no longer suffice to guarantee legal interactions among the government, the indigenous population, and the foreigners. According to the Egyptian ruler, all the powers interested in fostering relations between

Europe and the "Orient" ought to negotiate the old system of the capitulations anew.[9]

The canal's opening in 1869 opened the floodgates to these and other overwhelming challenges. Egyptian, consular, and Company authorities attempted to exercise surveillance and discipline migrant workers on the isthmus and in Port Said, but the isthmus population, thanks to local and international circuits of mobility, circumvented and shaped the disciplinary regime that authorities had often haphazardly assembled. From the perspective of isthmus dwellers, both the canal's inauguration in 1869 and Egypt's occupation by the British in 1882 proved to be momentous, but in ways that challenge the conventional interpretation of these two watersheds in the history of the canal region, Egypt, and the Mediterranean.

THE CANAL INAUGURATION IN 1869: FESTIVENESS AND FRUSTRATION

November 16, 1869, was a sunny day. A "ragged crowd" was getting ready to attend the celebrations that accompanied the inauguration of the Suez Canal planned for November 17. For three days, day and night, everybody could travel for free on the steamers through the canal. An immense multitude hastened to Port Said from Aswan, Alexandria, and Cairo. Visitors also hailed from Trieste, Brindisi, and Marseille and had to be hastily lodged in the few houses in town, at the time still barely more than an encampment.[10] At night, "Chinese lanterns" of all colors gleamed throughout Port Said. The glitter of artificial night helped transform the town into a "most enchanting place." More than forty thousand people also gathered in Ismailia "to dull the pains of their miserable lives in the inebriation of the celebration."[11] While the event in Port Said may have had a more formal and exclusive character, Ismailia seems to have been more accessible. There, too, lodgings were insufficient; some guests made do by spending the night on board vessels. Countless visitors slept in tents especially provided for the occasion.[12] There were performances by those local female entertainers encountered earlier and known as *almées*, especially arranged for the occasion.[13] Ordinary folks congregated to celebrate the event, marked by reveling as well as thieving. Egyptian authorities worried about the opportunities that such a gathering might offer to ill-doers. They fretted about protecting the hereditary prince of France or the Russian czar from targeted assassination attempts, but they also voiced concerns about what the "general masses" could do. They even considered requiring all hotelkeepers and room renters to keep a registry of all the guests coming and going. It is unclear whether they followed through on this.[14]

In considering whom to invite, the Egyptian authorities had made sure nobody's sensibilities would be hurt. They extended invitations to kings, counts, and beys, without forgetting their numerous retinues and journalists. Notably, however, Ottoman and North American representatives steered clear of the festivity. The British formal presence was relatively small, which did not prevent Victorian writers publishing profusely about the canal construction and inauguration.[15] In the hours leading up to inauguration day, Port Said's harbor filled with French, Italian, and Austrian vessels, carrying the elite of European monarchies and important governmental functionaries. Outside the harbor itself, five British men-of-war, dispatch boats, two Austrian ironclads, and some Italian ships waited. The Egyptian ruler's yacht, the *Mahrusa*, was anchored nearby. Cannon blasts shook earth and sky.[16] They were fired at intervals as princes, ambassadors, and other celebrities paraded by in rapid succession. At about 8:00 in the morning of November 16, the eve of the canal opening proper on November 17, the most feverish pitch of enthusiasm marked the arrival of the empress Eugénie of France on her vessel, the *Aigle*. Religious services were held at 3:00 in the afternoon on the beach in Port Said, west of the harbor. Three kiosks had been erected on the seashore for the occasion (see figure 12). The Shaykh of Al-Azhar, Monsignor Bauer (empress Eugénie's confessor and prelate of Pio IX), and Monsignor Luigi Ciurcia pronounced benedictions. Coptic, Roman Catholic, Armenian, and Greek clergy also attended. The movement of the sundry royals and nobles through the central kiosk, the parade of Egyptian troops, the rich attire of the Catholic priests, and the confused and at times inaudible litany of inaugural addresses must have offered an interesting spectacle indeed. The dull sound of the waves, lightly swelling and slowly breaking on the sands, accompanied the ceremony.[17]

On November 17, at about 8:00 a.m., the Suez Canal was opened formally. A procession of about seventy steamers of various nationalities began. The lineup of ships was a politically delicate affair, given the considerable international turmoil caused by the canal project. Vessels were headed by the French yacht *Aigle* steaming out of the harbor. Other ships followed at intervals of ten to fifteen minutes until forty vessels were afloat. Twelve hours later the French empress's yacht and Isma'il's yacht were the first to moor

FIGURE 12. Three altars at the canal's inauguration at Port Said, November 16–17, 1869. Photo by Hippolyte Arnoux. *Source:* ANMT, CUCMS, 1995 060 1489, Association du Souvenir de Ferdinand de Lesseps et du Canal de Suez.

in Ismailia. As the royal guests landed, they were conducted by the Egyptian ruler to his new palace, built especially for this occasion.[18] The festive arrangements were conducted with unimaginable prodigality and unprecedented splendor, which prompted the French empress Eugénie herself to allegedly declare that she had never witnessed anything as luxurious.[19] In the efforts to display Egypt to Western eyes, the opening celebrations replicated elements of the world's fairs.[20] Large temporary marquees were erected, where thousands dined and drank at the viceroy's expense. Champagne and other costly wines flowed like water. Thousands gathered to dance, talk, and eat together. A military exhibition of Arabs and Bedouins was arranged to please the visitors, and fireworks ended the night. It would afterward take several days, and various special trains, to clear away the fittings and furniture provided for temporary use.[21]

It was a night no eyes had ever seen before, nor would those of future generations.[22] The viceroy had even tried to convince composer Giuseppe Verdi to write an opera for the occasion. It was only later that Verdi took up the offer and wrote *Aida*, which debuted on Christmas Eve of 1871 in the newly erected Opera House in Cairo. But Verdi did participate in the ceremonies arranged for the opening of the canal and allowed the Egyptian government to perform his opera *Rigoletto*. He also wrote several new cantatas and smaller pieces to celebrate the joining of the two seas.²³ Once the cannon blasts had died away, the brilliancy of the illuminations had faded, and the floods of costly wines and the luxury of free tables had ceased, the account to be paid was a hefty one. Some claimed that the expenses reached 4 million golden Egyptian pounds (or 1 million francs).²⁴ Historians have argued that Isma'il's sybaritic celebration showed nothing but his craving for Western approval.²⁵ Others defended Isma'il's confidence as understandable given the circumstances. The recently concluded North American Civil War had enabled Egypt to corner the world's cotton market. The high expenditures incurred by the pageantry of the canal did not, for the time being, disrupt the festivities.²⁶ But immediately afterward the Company promptly sent an invoice for the recovery of its expenses in Port Said and all along the canal for the inaugural ceremonies "on the account of the Egyptian government." The bill amounted to 197,688 francs in Port Said and 304,318 francs in Ismailia.²⁷

The carousing had successfully transformed the desert, even if fleetingly, into "a fairy-land."²⁸ Yet many soon woke up from what turned out to have been a bittersweet dream. Even before November 1869, the slowing down and the completion of the canal works had brought about the dismissal of hundreds of workers previously employed at the Company's worksites. In October 1867 alone, contracting companies such as the Gorin company and the firm managing the Company's foundries and the worksites (*Forges et Chantiers*) were about to let go around twelve hundred men in Port Said: nine hundred Greeks, Italians, and Maltese and presumably three hundred "Arabs."²⁹ More dismissals occurred in the following months. Out of a sample of nine individuals who had all reached the isthmus between September 1862 and May 1866, including single males, Franciscan friars, a Greek pope, a police caporal, a man with his family, and a woman who signed on behalf of her husband, only three were still in Port Said in 1870. The other six had all left. Three had returned, in 1868, all the bedding or furniture they had temporarily received from the Company, while three rushed to do so between November and December 1869. In October 1869, in Serapeum, an Austrian man who had been out of work for eight months vented his frustration by firing gunshots in the middle of the night and was subsequently detained.³⁰ In the "Quarantine" worksite north of Suez, which had once housed as many as ten thousand workers, only two thousand were left by early November 1869. Laborers were not the only ones heading out. So were tavern owners and small shopkeepers with their wives and children. The Italian population of Suez, for example, quickly dwindled. By December 1869, the number of

Italian workers in the southern town had dropped from around two thousand to less than five hundred.[31]

After 1869, acute misery was to be found all along the canal. The brand-new waterway may have given even outlying urban centers, such as Damietta, a "new impulse." But Damietta also relied on a railway that connected it to Cairo and Alexandria and steadily amplified its relations with the cities of the Egyptian interior.[32] On the opposite end of the isthmus stood Suez, which, notwithstanding its railway connection to Cairo since the 1850s, still entered what some defined a "rapid decay" as early as 1871. The canal's opening and the completion of the quays had made the workshops for boat repairs in Suez redundant and transshipment less remunerative.[33] By the early 1890s the decay of the southern town would be utterly manifest. Once the ancient Red Sea port for Egyptian and British trade, it now lost its past commercial primacy with the opening of the canal that ironically bore its name.[34]

Ismailia, for its part, based its hopes on the inauguration of a sweetwater canal that carried its name and had been in the making since the 1860s. In 1877 the Ismailia canal finally carried Nile water from Cairo's district of Bulaq into Lake Timsah, at the heart of the isthmus and near Ismailia. Nonetheless, even if the voluminous trade in goods from Upper Egypt and Middle Egypt to the Mediterranean (such as cotton, wheat, and construction materials) could now pass via the brand-new Ismailia canal, the town's aspiration to become a port regularly attended by maritime commerce was soon frustrated. The town at the center of the isthmus remained too distant from both the Red Sea and the Mediterranean. The Company-imposed tonnage fees for navigation through the Suez Canal increased the price of freight so much that they made Ismailia's competition with Alexandria untenable. Moreover, the ships transiting through the canal had a full load already and were unwilling to waste precious time for the meager profits that stopping in Ismailia would entail. Both Suez and Ismailia then, after 1869, were declining. "Much more life" could be found in the "small city" of Port Said.[35]

Port Said, presiding over the canal's Mediterranean gate, occupied a relatively advantageous position. Steamships had to stop there to replenish coal, provision stocks, and transfer their cargoes. More steamers brought coal, of which a great quantity was sold.[36] By 1868 people in Marseille, in the Sicilian cities of Palermo and Messina, and in Izmir, Rhodes, Mersin, Alexandretta, Latakia, Tripoli, Beirut, and Jaffa could reach Port Said directly on one of the ships of the Messageries maritimes françaises. Their boats left three times per month. One steamship of the "Spanish company" connected Port Said and Barcelona once a month. The Marseille Company of Steam Navigation, also known as Marc Fraissinet and Sons, connected Port Said to Marseille and Alexandria with three boats monthly, while also linking it to Malta. By January 1871 people could embark for Port Said from both Odessa and Trieste thanks to the Austrian Lloyd. Port Said–bound travelers in Odessa could also opt for the Russian Navigation Company, leaving every fortnight and welcoming further passengers in Istanbul, along the Syrian coast, as well as the then Ottoman Chios and Rhodes. For those willing to depart from or reach other Ottoman ports, a ship of the Egyptian Khedivial Company left every Thursday from Alexandria and stopped in Port Said on its way to Jaffa, Akko, Beirut, Latakia, Alexandretta, and Mersin. Every Sunday, it headed back and stopped once again in Port Said.[37]

Port Said became a vital node in central and eastern Mediterranean networks of mobility of both goods and passengers. At the same time, it pulsated as the heart of the isthmus circulatory system. By 1865, around the time when its population reached seven thousand, boats brought their loads from other Egyptian areas such as Manzala, Matariyya, Damietta, and Rashid. Goods sold easily and profitably to both residents and those passing through. More vessels came from Europe, carrying either the iron, copper, wood, and food that the Company demanded, or loads of commercial goods, like food or clothes, to be sold to workers.[38] After Port Said's internal harbor was opened in 1866, the town became the port of arrival and revictualling for the whole canal area down to Suez. Boats could now set their anchors right in the middle of the city, and provisioning became easier. A range of products, including fresh fruits and vegetables to be transported further into the isthmus, found their way from Syria, the southern Anatolian province of Karaman, the Greek islands, "Turkey," Russia, France, and, via Alexandria, Austria and Italy. According to some, Port Said was a pleasant and lively town, whose bazaars were stocked up with all things necessary to life. Its population was mostly made up of Greeks, and especially those hailing from the Candia (Heraklion) region of the island of Crete.[39] By the end of 1869, some ten thousand people were reportedly dwelling there. Of these, fifteen hundred were said to be Egyptian.[40] Greeks amounted to three thousand or thirty-five hundred; French to two thousand; and Austrians, predominantly Slavic and Dalmatian, to fifteen hundred. The rest were Italians and Maltese.[41] By the end of its first decade of life, Port Said had acquired a post office, a telegraph station, stores and warehouses, a "European" hospital and an "Arab" one, a Catholic chapel with an annexed boys' school, a convent of the Bon Pasteur nuns with a girls' school, a Greek chapel, a mosque, public bathrooms, a theater, and two markets. With its elegantly built wooden constructions, regular streets, tree-shaded squares, coffeehouses, and shops brimming with all that could be desired, it was favorably compared to a small French port (minus the individuals in "Oriental" apparel). Optimistically, it was expected to further swell its population and fulfill "a great destiny."[42]

After 1869 and the completion of the canal works, however, Port Said also began to suffer. Trade opportunities waned. Its population dropped from 10,000 to 8,671. It also changed in composition. It was recorded that more Egyptians (4,461) and fewer foreigners (4,210) dwelled there.[43] Relatively more Egyptians had moved to the port city, while fewer "foreigners" decided to settle there than in times past. By 1874 Port Said was said to produce nothing, to have no periphery (*banlieue*), and to lack a sufficiently easy and quick way to communicate with the rest of Egypt. The prospect of a railway rapidly and regularly linking Port Said to Damietta was touted as early as 1872 but would never come to fruition. Port Said consumed dearly taxed goods. The merchandise produced in the rest of Egypt came with "transportation fees," such as the one needed to ship it across Lake Manzala. The goods coming from Europe had an 8 percent importation tax collected by the Egyptian customs. And those commodities hailing from Ottoman ports were made pricier by a fee for internal export imposed by the customs at the place of embarkation.[44] Everything had to be imported to this northern isthmus town; the prices of staples were exorbitant. Rents, too, were incredibly high. Notably, in 1879 a lady called Giuseppina Stanich obtained from her landlord a remarkable one-quarter reduction of 50 francs off her monthly rent of 200 francs for her house in Port Said. Her argument that the bad state of business in town did not allow her to pay rent that high must have been persuasive.[45] Port Said had become the stopover port par excellence, where businesses from all over the world came to provision themselves. Meanwhile, its population, who owed their living to the perpetual passage of ships, could hardly scrape together a living.[46]

SUSPICIOUS, CLANDESTINE, AND CULPABLE

Suspicion met the droves of stranded individuals roaming the isthmus around 1869. The Egyptian governor claimed that with no money or resources for daily living, they would unavoidably resort to thieving and other delinquency to survive. As the inauguration loomed near, it had already become clear that the isthmus works were no longer what they used to be. The number of men throughout the camps along the canal left with no jobs or means of earning a living was mounting. For as long as they had been employed, the governor continued, they had spent their money in restaurants, wineshops, and elsewhere (see chapter two). Now, most of them had no homes and turned to sleeping in the streets. Some killed time playing dice. How would these people, the isthmus governor wondered, make do in the future, when there would be no jobs left and they would scatter throughout the campsites? The Egyptian governor, along with the consular agents of Egypt's allied powers, whom he had summoned, agreed that those foreign migrants with no employment or resources were to be forcibly sent back to their own countries. They would be able to return only when and if work started up again. These high-ranking officials feared that after the definitive completion of the canal work, "the presence of these people could bring evil."[47]

As the canal work slowed down in the second half of the 1860s, Port Said became notorious. It had apparently turned into a semi-lawless borderland, where impunity was granted and dangerous outcasts thrived.[48] Those in positions of authority were convinced that the robberies, the attempted thefts, and the attacks of all kinds in Port Said were increasing day after day. They insisted on the confidence and audacity of ill-doers, who dared to commit armed holdups in plain daylight. They described how, in the evenings, gangs roamed the city and attacked the lingering passersby. Fights were frequent and deteriorated quickly. Private citizens of means were obliged to hire guards to protect their homes because of the recurrent burglaries and the frequent desertions of seamen.[49] Storekeepers submitted complaints about nocturnal attacks on their property to the French authorities in town, who in turn protested to the Egyptian governor. Consuls lamented that Port Said's Quay Eugénie was not safe at night and argued that a police post should be established in that area and patrols ought to make more frequent rounds.[50]

Thieves in town were no naïve first timers. In 1878 they unlocked Mr. Colomb's door with false keys and locked it right away behind their backs, so that they could proceed unnoticed. They then pried open the wooden closet where Colomb kept his clothes and money with a carpenter's chisel. On another occasion, George Bedos, a French subject living in a room located on Port Said's shore, had his door pried open with an iron instrument without the neighbors realizing what was going on. One blanket, one cotton covering, and five white shirts were stolen from him. In Mr. Baun's case, robbers acted even more boldly. Two or three of them climbed into Mr. Baun's apartment during the night and succeeded in stealing a birdcage with thirty birds, including six pairs of canaries and specimens from China and India.[51]

In Port Said, coveted merchandise quickly changed hands and easily crossed urban boundaries, weaving together different neighborhoods in the same geography of illicit exchanges. Even if stolen goods' ultimate landing is unknown, following their trail still opens a unique window into the "social life of things" and the relationships they engendered.[52] A golden watch chain, for example, passed from hand to hand until it disappeared into thin air. In July 1888 the Italian subject Marco Boesmi went to the police station in Port Said's "European quarter" with Abdalla Filele, a local subject, holding a golden watch chain. Boesmi declared that the chain had been stolen from him six months earlier, together with a golden watch, clothes, and money. After the theft, an unidentified Greek whose whereabouts were unknown had sold the chain to a grocer called Mikail Nicolas, another Greek subject who had already been handed over to the Greek consulate. Nicolas had in turn sold the chain to a clockmaker in Port Said's "Arab village" called Selim Abdel Nour, as became clear when the latter was summoned to the governorate. Finally, Abdel Nour had sold it to Filele, who had acquired it "in good faith" and was therefore allowed to keep it. After the visit at the police station and before the inquiry had come to an end, Filele eventually gave the chain to an unknown Englishman through the mediation of a man called Syra. Boesmi never saw his chain again.[53]

Besides offering ground for crime of various types, Port Said also emerged as both an inlet and an egress for unregulated streams of things and people. Individuals from Mediterranean and Red Sea ports resurfaced in Port Said, their occasionally accidental landing, after traveling clandestinely in steamship steerage hideouts. In 1871, on two different occasions, captains of the Messageries françaises company apprehended stowaways who had introduced themselves aboard their Indochina-bound steamers in Marseille. The concealed travelers must have taken advantage of the lack of active surveillance in the French port city, either in the harbor or on the vessels themselves. Whether the stowaways wanted to jump ship at the first stopover or continue their voyage farther east was unclear. They had no money or documents to prove their identity. The French consular authorities who had a say in their affairs could not conceive of these stowaways as anything but runaways from judiciary pursuit.[54] Irregular travel could occur within Egypt itself. In 1880 a Syrian called Ibrahim Kherges embarked clandestinely in Alexandria on board a French ship. Unfortunately for him, his trip was stopped short in Port Said, where he was curtly invited to leave with all his luggage.[55]

In a few years' time the uncontrolled movement of people through Port Said had taken on such proportions that consular authorities in town began investigating the existence of an organization for "clandestine migration." Their suspicions fell on an Italian man, Vincenzo Calò, a native of Brindisi and resident of Port Said, whom they suspected of arranging the expatriation of suspects or convicts from the former port city to the latter. Reportedly he could rely on a network of seamen employed on the Peninsular & Oriental steamers connecting Brindisi and Port Said. He could also rely on the fact that docks in Brindisi were vast and unsupervised. Previously a storekeeper and a dockworker, he declared he was now just a middleman for fellow Brindisi sellers, on whose behalf he traded in fruit and vegetables. Yet he embodied the very accusation he was trying to refute: he had himself come to Port Said from Brindisi clandestinely, he claimed, "to flee hunger."[56]

Through Port Said people came and went seemingly unhindered. In the northern port city, for example, a burglar had "found refuge" after having struck a household in Ismailia. Even though the search for him was in vain, as "individuals with no free conscience wander from one place to another without ever stopping anywhere," Port Said had sheltered him at least for a while.[57] Disparate rumors about another man on the run from the Port Said police had him sailing away to Greece, enrolling in the Salvation Army and leaving for India, becoming a monk in Jerusalem, departing for America, or becoming an Ottoman subject and converting to Islam in Akko.[58] Most of this gossip must have been unsubstantiated but still evoked the multiple potential paths that branched off Port Said. The fact that a man called Harvey left Aden for Port Said, where he wandered for some time, and then was taken on board a Saigon-bound steamer "out of pity as passenger paying his passage with services rendered on board" suggests that Harvey's trajectory was indeed another viable option.[59] Other transiting visitors included Russian pilgrims (perhaps on their way to Jerusalem).[60] More pilgrims from Syria and North Africa had to wait in Port Said for the vessels that would carry them to Jedda or Yanbo. They slept in the sandy streets, exposed to heat and cold. Their presence sparked the apparently ineffectual demands that they be housed in a hospice built for them, receive bread subsidized by the *Waqf* (the religious endowment in town), and be forcibly evacuated to

Ismailia and Suez with the midnight ferry of the Poste Egyptienne. Yet they continued to crowd Port Said's streets and perhaps look for clandestine ways to continue their trek toward Mecca.[61]

Stuff on the Move in and through Port Said

Stuff also circulated in and out of the canal area unsupervised. In 1866, for example, one boat, on behalf of one of the Company's contracting firms (Dussaud), reportedly entered the port to download its freight without authorization or preliminary notice.[62] That same year, the Egyptian government had frustrated the Company's hopes to make a free port out of Port Said. It had established an Egyptian Customs post in town.[63] The French consul thought this decision was aimed at establishing a "harmful fiscal system." It deemed it nothing but a vain desire that damaged "a city that, not asking and not owing anything except for its own toil and having never known the advantages of governmental favors, has the right to at least some restraints." Quite simply, though, the Customs post in Port Said was intended to prevent smuggling and issued orders to other bureaus and to the city's vice-consuls.[64] In the 1860s the items that were destined for the provisioning of workers throughout the isthmus, in fact, were exempt from custom rights and ought to be accompanied by the Company's franchise certificate or by a receipt from Customs showing that rights had been paid. But a certain number of individuals smuggled materials; food and beverages; and prohibited items such as foreign salt, tobacco, and powder outside the canal area. They bootlegged them toward Salahieh (east of Ballah Lake on the road from Qantara to Cairo) through Qantara, toward Zagazig and Suez through Nefiche, and to Damietta through the Manzala lake. At the destination, they attempted to sell the smuggled wares without disbursing custom rights. If caught, smugglers had to pay double customs fees; their wares would be confiscated only if they belonged to a prohibited category.[65] All in all, the relatively light punishment may have made smuggling throughout the canal area an enticing field for profit. Potentially, just about any kind of moving object could become contraband as long as the ruling administration decided it was in their interest to label it as such.[66]

Within Port Said itself, pilfering became a thriving business that by the early 1880s apparently employed four or five hundred people, mainly Greeks.[67] Greeks in Egypt bore the blame for all kinds of misconduct and corruption. In particular, they developed a reputation for bringing illegal goods into the country.[68] Apparently coveted goods included hashish, which could not be legally cultivated in Egypt or imported into the country after 1884.[69] Greece functioned as Egypt's foremost source of this substance into the late nineteenth century, only to be replaced by Syria in the 1910s.[70] Gin, French and English tobacco, "Turkish" tobacco, and tobacco from Morea (Peloponnese) also found their way into Port Said. This happened through orchestrated unloading at night, when a handful of men would elude the moonlight and the Port Office lights and take the freight from a boat that had quickly approached the shore when they whistled. Alternatively, in plain view and yet unnoticed, accomplices would transfer the prohibited load, little by little, into town.[71] The revenue that Customs lost by losing track of this stream of tobacco pouring in through Egypt's coastline, especially at Port Said, Damietta, and Suez, was enormous.[72] Only occasionally did informing ease the work of Customs agents, such as in the instance when a "native" informer enabled boxes filled with hashish to be seized from an English steamer. Sometimes smuggled goods found their way into the country under the very eyes of Customs agents in Port Said. Cognac bottles coming from the Greek port of Piraeus and offloaded in Port Said from a Russian steamer, for example, contained little tin pipes in which around 350 grams of hashish were hidden, for a remarkable total of 48 kilograms. Their secret contents had been given away to Customs agents when either the bottles' darker-than-usual color aroused suspicion or said bottles started leaking through the cracks of the sealing wax that was supposed to hold them together.[73] On another occasion, a large quantity of gold, silver objects, and diamonds was found on travelers' own bodies. In 1900 three Syrian men and two women, just disembarked from a steamer of the Messageries françaises hailing from Syria, were trying to smuggle them into Port Said.[74]

By 1883 Customs had functioning branches in Port Said, Suez, Ismailia, and Damietta. Guards were also appointed along the coast in Rashid and Alexandria. Camel riders were to patrol the coast west of Port Said up to the border between its *Muhafaza*, "administrative district," and that of Damietta.[75] However, the very low coast to the westward of Port Said and the sandy dunes that divided the adjacent Lake Manzala from the sea must have made smugglers' jobs relatively easy. They landed on desert spots where they unloaded their cargoes in no hurry.[76] Further, the canal itself conveniently supported new smuggling opportunities and functioned as a channel for objects in flow. People as varied as hotel managers, boys, sailors, and mechanics threw goods into the sea after sealing them in rubber bags to protect them from the water and make them float. They did so at agreed-upon points where a boat would collect the stuff. Sometimes the sailors among them pretended that what they were throwing into the sea was nothing but rubbish and casually let go of boxes off board.[77] As historian Eric Tagliacozzo suggests, much of the history of smuggling was a game of knowledge of local conditions on the part of both traffickers and authorities on the lookout.[78]

Authorities swiftly pointed fingers at culprits for the vice that had apparently taken on notable proportions in Port Said, a maritime city embracing "the most different elements." They blamed its dense, undisciplined, working-class population.[79] Indeed, especially after the completion of the excavation works, unemployed laborers may have indeed behaved untowardly.[80] Whether the motley crew of unemployed men took to crime more than others, however, is difficult to assess through the available sources. Authorities in charge, in fact, perceived migrant labor as intrinsically disorderly, especially but not exclusively when not harnessed by the yoke of steady employment. Unemployment and unruliness often confusingly coincided in the minds of those who attempted to rule the improvised and yet labor-regimented isthmus society, where stereotypes were rife (as already expounded in chapter two). As elsewhere, the mobile poor were viewed as potential criminals and labeled accordingly.[81] Not surprisingly, the suspicions of the abovementioned Mr. Baun about his vanished birds fell "on a couple of young bad Greek subjects living off thefts" and were shared by his neighbors. In other cases, "some Italians" were fingered as suspects for trying to sneak into houses on the Quay Eugénie and for committing the many burglaries that were recorded in the area.[82] Yet less conspicuous others bore as much responsibility. None other than an elegantly dressed Frenchman, for instance, tricked the jeweler Minas Folaitis in Rue du Commerce into purchasing some supposedly golden powder that turned into worthless metal upon smelting.[83] Most of the surviving written traces were left by officials or the people who sanctimoniously appealed to them, who tried hard to harness and denounce what they perceived as a suffused delinquency peculiar to the city.

AN INHABITED CITY WHERE JUDGMENT FOLLOWS THE LAWS

A police bureau was created in Port Said as early as 1862 (before those in Timsah/Ismailia or Suez). Already counting many thousand inhabitants, both Port Said and Ismailia were deemed too distant from the preexisting authorities of Zagazig and Damietta, which made "a local and direct authority" sorely necessary.[84] The Port Said Police Bureau comprised a *chef de la division*, who was in charge especially of Europeans and was aided by a *cawas*, a "guard." The chef de la division and the *cawas* were paid by the contracting firm (*entreprise*), which later charged the Company for the expense. The Police Bureau also included a *shaykh* of so-called Barbarins and a *shaykh* of Arabs, also called "Cheik-el-beled."[85] The Egyptian ruler Isma'il, who had succeeded Sa'id in January 1863, required several provincial cities by law to organize or restore their police forces.[86] In March 1863 the ruler created the governorate of the canal. He appointed Isma'il Hamdi Bey as the governor (*muhafiz*), who would reside in Port Said.[87] Isma'il Bey, previously in charge of the surveillance of the drafted workers' contingents (circa 1861–1863), now supervised the local police, among other tasks.[88] The guards' number went up to one hundred in 1866. Reportedly, most of them were "Turks" (many of whom hailed from Crete), unruly, and prone to flee on board Greek ships. By May 1870 a supposedly stable Egyptian police force had been instituted in Port Said.[89] The men on duty were to inspect all parts of town, transmit information to the governorate, perform night rounds, make sure shops were closed at night, ensure that passersby carried lanterns with them, capture wrongdoers, quell brawls, retrieve stolen goods, and ferret out fleeing debtors. They also had to research the causes of fires, make an inventory of losses, relieve the wretched, and identify the corpses of the drowned. Finally, they were expected to investigate suspects; seize their weapons; and drive away those to be expelled and the undesirables, including thieves, criminals, and the unemployed. At least initially, the Canal Company covered the costs of bread and other staples needed at the police headquarters, such as lamp oil.[90]

Theoretically, as elsewhere in Egypt, cases of general policing on the isthmus empowered the local police to arrest anyone. Company agents should never interfere. They could intervene only if the police wanted them to mediate amicably among the "Europeans" involved in individual disputes. It was the nationality of the defendant, often difficult if not impossible to establish, that determined jurisdiction. In case of arrest, if the persons in custody were "indigenous," then the judgment followed its course with local authorities. If the arrested were "European," then the individual was to be handed over to the pertinent consular authority, who was responsible for investigating and judging the person in custody. In cases of simple contraventions such as the violation of a police order, Europeans would be fined by their respective consulates.[91] In Port Said, contraveners could be temporarily held in custody within their own homes and guarded by consulate "Janissaries" (acting as messengers and guards and assisting the local police whenever required), from whose oversight, however, they could easily abscond.[92] In Port Said as in the rest of Egypt, Europeans had become immune from the implementation of Ottoman law to the extent that they eluded even simple policing regulations. The consuls had assumed jurisdiction in all criminal matters involving foreigners even when these simply transgressed municipal regulations.[93] Custom and the capitulations merged to provide foreign consuls with virtually "absolute jurisdiction" over their nationals. The ill-doers among them took advantage of the confusion caused by the simultaneous functioning of multiple laws and eluded adjudication with ease.[94]

To make law enforcement even more troubled in the isthmus territory, both the Canal Company and the Egyptian state vied for the monopoly on police functions over this uniquely unruly place. In 1861, for instance, there was talk of appointing a Bavarian Company engineer to head the Austrian consulate in Port Said, given the sizable number of Dalmatians in the city. Even if the plan did not work out eventually, the idea that the engineer could simultaneously serve his employer and foreign interests was at least contemplated.[95] In 1863 the Port Said police were said to be initially "somehow" under the direction of Laroche, then vice-consul of France, previously acting as the Port Said–based director of works of the Company. In 1866 the Company official who supervised one of the central portions of the canal lamented that there were two thousand workers of all nations on the sixteen kilometers of works under his supervision. He claimed that "the small number of *cawas*" provided by the governor was insufficient and "forced" Company personnel "to act as semi-police."[96] Company representatives hence presented themselves as super-consular powers eager to replace state-mandated officials. They argued that if a nascent settlement was made up entirely of workers, then the Company was free to proclaim those internal regulations it deemed useful in the interests of the work. Egyptian authorities, they claimed, had no right to intervene in these internal regulations. They could only tackle crimes and misdemeanors.[97]

Company officials both treated the canal area as their own factory floor and expected the Egyptian government to fulfill certain management tasks. Egyptian authorities, for their part, were piqued by the leeway that the Company had obtained from Sa'id and later expanded. They claimed the Egyptian state's sovereignty over the isthmus, all the while expecting the Company and the population to cooperate in running public affairs.⁹⁸ Accordingly, the governor of the isthmus, Isma'il Hamdi, was annoyed when one of the Company representatives failed to sign a joint inquiry into a violent dispute involving an Italian and a Frenchman in Ismailia. On the one hand, the governor expected the bureaucrats at the designated office, the Company's Bureaux détachés, to intervene as soon as the local government called for them in all affairs involving Europeans on the sites. On the other hand, however, he assumed they would do so unofficially. He also wondered what kind of assistance they would eventually be able to offer to the local government. Finally, he subtly questioned whether their presence would prove useful.⁹⁹ Each side strove to maintain its own sway while curtailing the other's power to intervene effectively. In 1864, when people, especially Greeks, began target shooting in the heart of Ismailia, the Company engineer-in-chief in town argued that "the Company should be able to prohibit this to its employees and workers while the Egyptian authority should take the same measure of prohibiting to Turkish subjects and other Europeans as well." Exactly four years later, the Egyptian governor protested about that still apparently popular pastime. He argued that the Company ought to warn those Europeans who loved to aimlessly shoot their guns day and night either in Ismailia or in campsites and that this was a dangerous diversion that sooner or later would get someone hurt. By doing so, he implicitly put the responsibility for preventing the acts in question on the Company. Residents continued to be terrorized and small birds, "the joy" of the campsites, to be killed mercilessly.¹⁰⁰

Consular officials dispatched to the isthmus, for their part, mostly felt helpless. On the one hand, they rehashed the notion that the canal worksites, albeit managed by a substantially French company, were under Egyptian authority. They stressed that when seeking information about the whereabouts of foreign workers in the Company's hire, it was local authorities who responded and not French ones.¹⁰¹ They insinuated that the Company had an interest in manipulating the Egyptian authorities and presenting itself as the exclusive representative of French interests, much to the French consul's chagrin. They also complained that the Company rarely acknowledged the existence of consular authorities in Port Said and did so only when it served the Company's interests. Consuls felt that they were usually summoned only when it was too late. Understaffed and overworked, they felt they had no allies in the Egyptian administration, either. They saw it as unstable and unreliable and claimed that Egyptian police agents were insufficient and incapable.[102] They reviled Egyptian authorities for allegedly "not knowing how to deal with Europeans." They slandered Governor Soliman Bey, in power since the end of June 1870, claiming he had no command of the administration, only knew Arabic in a place where Europeans were the great majority of the population, and "fell easy prey to the small passions of a bad entourage." In other words, Soliman Bey was maligned as an unsophisticated, monolingual, and easily manipulated bureaucrat.[103] In 1867, when storekeepers in Port Said petitioned Laroche, the French vice-consul in Port Said, about the nighttime burglaries and attacks they were reportedly suffering, he informed the governor of their grievances. The governor claimed that the best solution was for storekeepers to hire night guards. Laroche instead pointed the finger at Port Said's ineffective "municipal services" in areas such as, among others, policing, public health, and street lighting. Meanwhile, private residents of means, such as the entrepreneurs attached to the Company, threatened to organize a police service or hire armed guards themselves.[104]

In spite of the barrage of criticism from all sides, the Egyptian state did strive to assert its control over this territory by monopolizing the right to regulate people's movements in and out of it and the authority to exclude unwanted outsiders.[105] In March 1869, in order to ensure the "general quiet," the government instituted the Passports Bureau in Port Said, similar to the one it had already established in Alexandria. This institution was to work in unison with the harbor master or *maître de port*. The Passports Bureau was meant to ensure that all those passing through the harbor, including captains as well as owners of stationary boats, conformed themselves to the instructions issued by government officials. According to the statutes of the Passports Bureau, all boatmen ought to preliminarily inform the government agent before taking anybody from whatever ship on board, including those individuals coming from abroad.[106] By 1872 an official in white dress and crimson sash appeared from the Egyptian corvette in its six-oared gig and welcomed incoming vessels with some ceremony. Upon being received by the captain, the official "presented a roll containing the signals, given as a guide to all ships going through. This done, with salaams he departed."[107] Passports Bureau agents oversaw visiting incoming ships, asking captains for the lists of passengers and taking urgent police measures if necessary. In a word, the Passports Bureau took on the functions of a local authority "in charge of the public order." The Egyptian government considered worthless all passports, patents, and tickets of residency that bore no signature by an

Egyptian official and that fell under no preliminary agreement with representatives of foreign powers.[108] After 1884, when Ottoman authorities began regulating international passports, they stipulated that "foreigners who intended to enter the Ottoman Empire had to carry a passport with an Ottoman visa" and that authorities could still "forbid entrance if someone was marked as a security risk."[109] By 1895 Passports Bureau agents were to fine all "local" subjects coming from abroad or even from "Ottoman countries" who lacked either a passport or a "transit permit."[110]

Despite the increasingly firmer stance of the Egyptian and the Ottoman administrations, the standing of the Passports Bureau remained precarious. Consular representatives in Port Said, dead set on subordinating agents' visits to their preliminary authorization, constantly challenged its authority. Moreover, some navigation companies actively opposed the agents' visits by making it difficult for the bureau's boat to pull over the incoming ships. The Russian Company, the Fraissinet Company, and the Austrian Lloyd were said to be particularly uncooperative.[111] Later moved to the quay, the control of passports continued to be a fraught procedure, targeting men and women of varied provenance and garb and requiring both a fence and surveillants to keep up appearances of the state's effective monopoly over its borders and surveillance over border-crossers (see figure 13).

The job of other government officials operating in Port Said's harbor was hardly easier. Those working for the Service of Docks and Entrepôts in Port Said, at first temporarily entrusted to the Maison Bazin and retaken by the Egyptian government in 1874, were ineffective in enforcing the regulations vigorously in all ports of Egypt. Similarly futile, apparently, was the role of the controller of Ports and Lighthouses, active in his position since at least 1876. In 1888 Port Said's harbor master still lacked a guard post. The surveillance that he exercised from a barge in very bad condition anchored at kilometer 1 of the canal was reputed to be insufficient.[112] Also unsuccessful were the measures taken by Health Bureau officials, active since at least 1875. Mr. Joice-Bey, controller of Port and Lighthouse of Port Said, complained in 1880 that ship chandlers constantly hindered the Health Bureau officials in their functions by approaching incoming boats before the sanitary formalities had been accomplished.[113] The health inspection and dealings of a different nature occurred simultaneously, as on one day in February 1875, when both health officer Eugenio Malia (a British subject working alongside an "Arab" coxswain called Mohamed Nassar) and a ship chandler happened to be on board an incoming English ship at the same time.[114]

FIGURE 13. Examination of passports in Port Said, 1914. *Source:* EST EI-13 (411), Agence Rol, November 1914, Bibliothèque nationale de France.

The presence of smaller vessels engaging in various kinds of business (see figure 14) fomented the chaos in the port waters and complicated police operations. Dinghies embarked and disembarked passengers and their luggage. Tankers brought coal or water to the ships. Small Greek or Maltese boats swept up the coal fragments that had fallen into the water during the loading of steamers using a special net they dragged behind them. These and yet more barges, fishing boats, and war boats animated the gleaming surface of the water, shining as if it were an immense mirror.[115] A visitor passing through Port Said's harbor in 1885 counted more than twenty steamships, several sailing ships, and hundreds of smaller fishing boats. Cameleers and donkeys, a few moving wagons, porters, flaneurs, and other busy individuals scurried around Port Said's quays.[116]

At every incoming ship, there was a struggle among provision boats and boats for hire. On the abovementioned day in February 1875, for example, at least five boats crowded the space around an English steamship. The five boats came so close to each other that the boatmen got into a fight, jumping into each others' vessels, exchanging blows, and grabbing each others' oars.[117]

FIGURE 14. Port Said, Quay François-Joseph, 1854–1901. Photo by Hippolyte Arnoux. *Source:* ANMT, CUCMS, 1995 060 1490, Association du Souvenir de Ferdinand de Lesseps et du Canal de Suez.

Several indigenous and foreign individuals routinely approached the ships stopping in Port Said's harbor to sell food, alcoholic beverages, and more. Sometimes they exchanged their wares for objects that the crew had stolen from the ship's passengers. These enterprising ship chandlers, together with those fishing coal out of the port's waters, were blamed for thefts on board ships and for contraband.[118] Interpreters (*drogmans*) of all nationalities, some of whom worked for the hotels in town, also visited incoming ships and created disorder on board. By April 1882 they were forbidden to approach vessels and were urged to wait for passengers at the embarcadero.[119] Drivers of *bumboats* (boats for provisions) and those fishing for coal were also barred from pursuing their businesses starting in January 1884. However, since there still was no law prescribing what punishments ought to be inflicted, violators remained unprosecuted and presumably continued in their transactions.[120] Fathalla Darviche, for example, kept selling rosaries and other goods from Jerusalem on board French ships passing through Port

Said's harbor. The governor authorized him to do so in 1885, while forbidding him to barter or peddle "obscene photographs" and liquors.[121]

Still in 1887, however, the governor in Port Said was annoyed by those boatmen who embarked or disembarked people and goods at places other than the pier in front of the Passports Bureau and who thus eluded surveillance. As was their habit a few years earlier, ship chandlers reportedly assaulted approaching boats, eluded their watch, had no regard for sanitary officials or captains, caused constant delays, and placed officials at risk of drowning or hurting themselves by falling. They endangered the whole port. The governor issued a regulation, in both Arabic and French, declaring that all boatmen needed a governorate-issued authorization and had to provide their supposedly solid and clean boats with a serial number, in both Arabic and European figures. They were forbidden to conduct the passengers of those ships to whom the Passports Bureau official had issued no permit (laissez-passer). And they were prohibited from disembarking people and goods from the ships in quarantine or under surveillance.[122] As late as 1888, when the governor announced the need to suppress unlawful traffic in the port by means of a formal decree, the Ministry of the Interior reassured him from Cairo that the measures his interlocutor had taken that far were sufficient.[123] Evidently, however, the local official experienced limitations on his authority that were not readily apparent in the capital.

AUGUST 1882 AND BUSINESS AS USUAL

Even though local authorities continued to focus their attention mainly on port hustlers, Port Said's waters and quays did undergo a seemingly momentary business interruption in August 1882. The Suez Canal area had become the first theater of a conflict that quickly overflowed into the rest of Egypt and tied this waterway's history into national history.[124] The country's massive indebtedness, French and British encroachment, and Egyptian military defeats during the Ethiopian campaign in 1875 and 1876 had been fomenting resentment of European influence and unhappiness in the Egyptian army. Tensions had come to a head when Egyptian officer Ahmad 'Urabi seized power and established his own Council of Ministers. Khedive Tawfiq fled to Alexandria and sought foreign intervention. By the beginning of 1882 the revolt was in full swing, and the Ottoman government was aware of Britain's intention to invade the country in order to protect the Suez Canal.[125] On

FIGURE 15. Queen Victoria statue on Port Said's Quay François-Joseph, 1903. *Source:* Archives CeAlex/CNRS, fonds J.-Y. Empereur.

August 20, six hundred Englishmen occupied Port Said. The British moved into the offices of the Canal Company and blocked navigation in the canal until August 24.[126] The administration of the Suez Canal, however, was not divested, and traffic was interrupted only for the time required by military operations. By August 24 the situation was "under control" and "maximum safety" was guaranteed.[127] On September 13, 1882, British troops defeated 'Urabi's Egyptian forces at Tell al-Kabir, west of Ismailia, and advanced on Cairo, marking the beginning of the British occupation of the country. Khedive Tawfiq was soon reinstated in power.[128] The occupiers would settle the headquarters of their navy at the so-called Navy House in Port Said, formerly a Dutch establishment.[129] Even though the occupation was supposed to be temporary and Egypt remained formally a part of the Ottoman Empire, never becoming a full-fledged colonial possession, the British were to visibly inscribe their presence in the town's landscape by placing a statue of Queen Victoria at the beginning of the Quay François Joseph (see figure 15).[130]

Many of the Egyptian residents had at least temporarily left Port Said and found refuge in Damietta. Western testimonies ridiculed them by alleging that they were "crying like babies" and "losing their mind[s]" out of fear. Yet fright may have been justified. Apparently the British had directed the mouths of their cannons toward the "Arab village" and intermittently inundated it with artificial light at nighttime. To one of the French nuns remaining in town, a domestic laborer expressed in broken French the rationale for leaving: "*inglesse faire boum! Boum! Arabes tout casses! Nous partit!*" (The English will shoot! All the Arabs will be hurt! We are leaving!).[131] Numerous Egyptians living in Port Said resented the foreigners' privileges and invoked governmental intervention. Many responded to 'Urabi's revolutionary call. The *Qadi* of Port Said, Shaykh Muhammad Ibrahim al-Shahir Ba'ba 'Aisha, for example, supported 'Urabi and withdrew his loyalty from the khedive. The dismissal of the second supervisor of the coast guard in Port Said, Hasan Efendi Ibrahim, formally for "not having fulfilled his duties" but actually for supporting 'Urabi, also exacerbated nationalist sentiments among Egyptian residents and government employees. So did the flight of the governor of the isthmus, Isma'il Hamdi, who sought refuge with his deputy on board a British ship. Finally, the sight of warships also contributed to residents' agitation and anger.[132] When four large French naval frigates appeared on the horizon, the whole "Arab village" of Port Said was set in motion. Women and children ran to the beach, and men blocked the quays. Yet despite panic and cries of "Death to the Christians," ostensibly heard from the dormitory of the orphan girls at the Bon Pasteur, no anti-European demonstrations took place.[133] Overall, Port Said appeared quiet and orderly. A certain captain 'Abdallah, loyal to 'Urabi and heading a police force of some 150 men, was appointed as the revolutionary leader's representative on the isthmus.[134]

After 1882, the then Anglo-Egyptian authorities continued to approach Port Said as a place that required especially rigorous surveillance. By October 1883 the Port Said police had the numerically strongest personnel of all canal cities. In the northernmost harbor there were as many as thirty noncommissioned "officers and men," forty-eight reserves, and two officers. Six police stations (*caracols*) were operational, as many as in Suez and twice as many as in Ismailia. In both Port Said and Ismailia, stable pickets could be found. There were horse-riding town patrols in Port Said, Qantara, and Suez, while Ismailia had none. Some of the individuals who had been previously deployed on the isthmus in the realm of law and order continued to exercise their functions. All the native policemen in Port Said, consisting of both "old-town police" and "*mustaphizin* [*sic*]" or reserves, were deemed fit to continue their service. The latter force was especially praised for having remained loyal during the 1882 "disturbances."[135] There were also thirty-one "Turks" in town, but without arms or uniforms, many of them were pronounced "quite unfit." The six "Turks" at Qantara were to be dispensed with altogether. Similarly, the Turkish artillerymen at Suez ought to be disbanded. Instead, a certain Cipollaro, calling himself "chief of the Civil Police," got to maintain his functions as police officer in Port Said thanks to his experience and knowledge of Arabic. He also covered the position of civil inspector or head of detectives in town. Both Cipollaro and Gasparini, the chief of police in Suez, whom both the governor and the British consul held in esteem, had served in the Italian gendarmes (*Carabinieri*) and held prior civil police appointments in Italy. They both got to keep their pre-1882 positions.[136]

After 1882, a "very essential if not absolutely necessary" stronger English element in the Port Said police was repeatedly called for, but in vain. "Most experienced officers," in fact, leaned toward the recruitment of specifically English personnel to handle the Australia-bound emigrants and British sailors who, with two-thirds of the ships passing through the canal being English, had begun making a name for themselves as a "rather rowdy class of people" in town (for their drinking habits, see chapter four). But the appeals for their recruitment fell on deaf ears. Ironically, the one and only Englishman who had been enlisted in 1886 among the city's constables, much to the satisfaction of his superiors, had to be dismissed on "account of drink."[137] According to British sources, one single young lieutenant of the English army could have performed the work done by as many as three of the European officials employed in town. This newly appointed official would have enjoyed both the status deriving from his membership in the English service and the sympathy of men and local authorities alike, thanks to his supposedly "kinder" attitude. Nonetheless, by 1886 no Englishmen could be persuaded to serve in Port Said despite the British consul's appeals, the inspector general's awareness of the city's needs, and the personal support given by Lord Cromer, the British consul general in Egypt at the time.[138] In 1891 an Englishman with the rank of colonel eventually moved in as the head of the police of Port Said. However, he reportedly lived outside the city and at such a distance that, when summoned, it took him no less than about an hour (*un'oretta*) to reach the scene of the infraction or crime.[139]

Finally, a lack of communication and coordination plagued the police bodies operating in Port Said. On any given day in 1883, for instance, fewer than twenty men from the reportedly ineffective gendarmerie performed police duty in the "Arab town." A different assemblage of thirty men operated in the "European quarter." In the latter, six presided at the central police station, four guarded the Arab Police barracks, and six were stationed at the

European police barracks and prison. There were two more at the slaughterhouse, one at the post office, and two at the Passports Bureau. Of the remaining, two were orderlies, two were odd-job men, and one carried water. Five brigadiers superintended beats and other duty, and one native officer oversaw the prison.[140] The Swiss men who were detailed in Port Said in 1883 reportedly displayed an "impoliceman-like" appearance. They did not belong to "a good class" of people and only spoke an unintelligible French patois, creating serious communication issues with their superiors. The Austrian captain Monham spoke Italian and German but not a word of French and could not communicate with his men. Sublieutenant Blondel, over forty years of age and of Scottish appearance, could speak French but was ostensibly "in the habit of drinking and keeping doubtful company." None of them seemed to understand the necessity of learning Arabic. Finally, not only did European policemen lack experience in drilling, but most of them could not understand the Turkish words of command.[141]

Port Said's segregated urban setting complicated police matters. The Italian consul in town claimed that Port Said was indeed "divided in two distinct parts, the Arab and the European," the latter of which was "inhabited and frequented mostly by Europeans of all nationalities who either live in the country or are passing by." Rather than having a force composed simply of Englishmen, he optimistically recommended that a "good police service staffed by European guards" would master different languages and possess the bravery that Port Said's peculiar situation required.[142] However, Europeans in Port Said's police corps must have relied on the cooperation of Egyptian policemen, as they presumably did in the search for an Italian merchant in food staples, known by his initials "N.A.," who was under arrest by his consulate in May 1895 for abusing "in the most depraved way" the seven-year-old girl "N.C." He had managed to hide overnight in the basement of a house on the edge of the "Arab town." From there he had hoped to flee at dawn, but he was eventually seized by a police unit headed by the "valiant" Captain Clementi before the sun rose.[143]

Overall, while sundry surveillance measures came to be established at Port Said, people and goods, whether licitly or not, could still enter and exit the isthmus region with ease. "We are not in the desert, but in an inhabited city where judgment follows the laws," declared a frustrated French vice-consul in Port Said in 1867.[144] Nonetheless, different authorities in town seldom agreed on which laws ought to be followed and in what ways they were to be implemented. Scrutiny all along the Egyptian shoreline remained irregular and unreliable. Port authorities only requested that foreigners establish their identity, but they seldom took the documents of those disembarking and passed them on to the respective consular authorities, as was expected of them. Governmental and consular authorities scarcely had a sense of people's comings and goings through the Egyptian borders.[145] And Company sources had to reluctantly admit that not even its employees or those of its contracting firms strictly observed the issued regulations.[146] Port Said had become a site where a constant tension, registered in other coeval ports, surfaced between the need of authorities to exploit sea-borne networks of trade and mobility and the threat to their control that openness to these flows implied.[147]

CHECKPOINT AT PORT SAID

Even if the Company, the Egyptian government, and consular powers skirmished over what ought to be done about the rowdy isthmus population, they did share one ambition: to gather information about "all personnel, not just the workers, but about all the people who came to settle" in a place they respectively perceived as their own (*chez nous*).[148] When it came to actually identifying the recipients of consular or local justice, however, the undertaking could only proceed in fits and starts, as demonstrated by what befell someone called "N. Caruana" in 1893; puzzled Italian consular personnel in Port Said first classified him as "Maltese," then as "Greek," and finally as "English" (see figure 16).[149] Moreover, consulates had access to only rough estimates of the population under their administration, which was unsurprising given the unreliability of census data generally.[150] By 1877 the presence of numerous European colonies was indeed recorded throughout Lower Egypt.[151] But the precise-looking number of 13,260 provided by the Ministry of the Interior for the foreigners living in the delta (as well as the isthmus) in 1873 was so only in theory. The figures provided by consuls to the Egyptian government in 1871–1872, in fact, were admittedly inaccurate. In Alexandria alone, for example, they represented only "half of the real number or supposedly real number." One reason is that some of their subjects swelled the ranks of the temporary or passing population. And others, like the Swiss, enjoyed the protection of different foreign powers at the same time.[152]

Counting Egypt's foreign residents was imprecise also because many avoided registering themselves with the authorities out of negligence, ignorance, or a

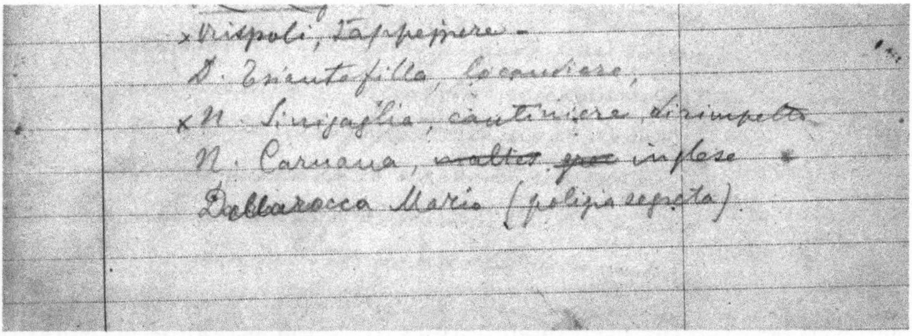
x Vrispoli, tappezziere -
D. Triantafilla, locandiere,
x N. Sinigaglia, cantiniere dirimpetto
N. Caruana, maltese greco inglese
Dellarocca Mario (polizia segreta)

FIGURE 16. "N. Caruana" variously classified as "Maltese," "Greek," and "English," 1893. *Source:* ASMAECI, ARDCE, RC, B46, folder 33, Port Said, 9 March 1893, Italian Consul to Italian Consul General in Cairo.

purposeful desire to elude hassles. While foreigners almost always reported deaths because doing so was necessary to obtain a burial permit, many failed to register themselves upon arrival, report religious weddings, and account for their children's births.[153] The extent to which they may have been cautious when dealing with their own respective consulates is suggested by a rumor that spread during the turbulent summer of 1882. According to the gossip, consulates, when handing over passports, adopted some unspecified "conventional signs," known by their respective governments, with the goal of preventing their bearers, of which there were thousands, from going back to Egypt after what was supposed to be only a temporary absence.[154] A stark example of not just wariness toward consular authorities but outright disrespect is provided by one James O'Connell. While the British consular court was in session to hear a case, he was seen standing at the door of the court and "making water" into said court. In his defense, he simply declared he "could not help it."[155] As has been argued before, the loyalty of many foreign passport holders to their respective consuls stemmed from a mix of rational choice, calculation of a means to an end, and the weighing of economic and social standing. It could be reversed if the benefits of a certain nationality faded.[156] By the same token, as local Catholic priests used to lament, European adults had a tendency to show up in church only when taken by the whim of regularizing "old unions" and legitimizing their grown-up children. Seldom could religious authorities verify whether adult self-alleged Catholics had been married before or not. Nominally belonging to various rites and sects of Christianity, they were denounced as impious for the most part: the "scum of Europe."[157]

The Isthmus of Suez provided countless hustling opportunities for those willing and able to don multiple identities, overcome social boundaries, and cross language barriers. An Austrian man found dead in the canal with wounds from of a cutting tool near the Chalouf worksite was known as both Martin Barnolich and Andrea Juretich. The latter name was corroborated by a passport, issued by the Austrian consulate in Alexandria, that was retrieved from his belongings.[158] Comparatively more successful was an enterprising man by the name of Soleiman el Mamluck or Francesco Salemi. In 1869 he was arrested for vagrancy as well as "simulation of nationality and religion, self-professing Italian, Greek, Arab, Catholic, and Muslim." Apparently he had spent months in Port Said dressed up as a Bedouin "under the pretext of waiting for a caravan of blacks with objects from the Sudan" that he was planning to transport to "Turkey." He had reportedly been arrested in the past because of his frequent contacts with Greeks known for suspicious conduct and reputation. The governor of Port Said, "having recognized his lies," had him expelled from his district. But the trickster reappeared at the Italian consulate in Alexandria, where he introduced himself as Francesco Salemi. He was then given expulsion papers (*foglio di via*). But after a while he showed up in Suez and claimed that he was seeking traces of certain "blacks" who had stolen some of his goods. Once in Suez, he conducted "a lazy life" and only in the company of Greeks "of the lowest class," whose language he spoke "with the utmost easiness." In all the time he was in that town, he never engaged in any relationships with Italians. When he was apprehended, five Greeks created a melee around the police horses with the explicit intention of preventing his arrest. Once in prison, Soleiman-Francesco declared he was a native of Istanbul. But a Tripolitan guard who had spent some time in his company realized that the captive was well informed about the city of Tripoli and suspected, also because of his accent, that he was a member of the Jewish community of that city. At that point, when the governor interrogated the Italian vice-consul in Suez about the swindler's actual identity, the diplomatic official was at a loss.[159]

Hassan Ben Ahmed did not cross as many boundaries. His position as the head of the *cawas* of the French consulate kept him in Port Said. The port city welcomed immigrant men like Ben Ahmed but simultaneously opened up some trajectories for them and foreclosed others. Ben Ahmed had moved to the canal town from Qena, in Upper Egypt, where he was born to parents belonging to a nomadic tribe from the Hejaz. In 1874 he submitted a request to be "admitted to residence in France." Perhaps he had been persuaded to do so upon hearing of the successful case of two *cawas*, "loyal servants" in the Izmir consulate general, who had been admitted to the benefits of naturalization. At first Ben Ahmed's request was received favorably. The French deemed potential problems with the local authority unlikely, because Ben Ahmed "was not Egyptian" and never figured in the rolls of personal levy. However, notwithstanding his intelligence, industriousness, and exemplary private conduct, his Ottoman identity got in the way. According to a law promulgated on January 19, 1869, whenever an Ottoman subject was naturalized abroad without the authorization of the Ottoman government, that change of nationality would be considered as null in "Turkey." The subject therefore needed to "regularize his situation with his government" first, which prevented Ben Ahmed from achieving his upward move.[160]

What people declared about themselves did not necessarily coincide with the stories told by the papers they carried or the officials they encountered. For example, a brewer called Amelia Vranch, who claimed to be Austrian, had arrived in Port Said with an Ottoman *tezkeré* (in Arabic "tadhkira," permit or pass). She was ultimately categorized as a "local subject." Clotilde Nava declared she was French but was revealed to be legally Italian by her estranged husband and father of their twelve children. Before landing in Egypt, they had been living in Istanbul, where at least one of their surviving daughters was born, living proof of the past Mediterranean wanderings of the couple. In yet another case, a man accused of a "a heinous attack against the propriety" of a six-year-old girl claimed to be Austrian. But neither the Austrian consul nor any other consul or vice-consul in town recognized him as one of their own subjects.[161]

An individual could change or switch affiliation multiple times. An employee of the Egyptian Telegraph in Qantara called Albino Paoletti, for example, renounced his unspecified nationality and made a request to become a local subject in 1887. The outcome of his supplication is unknown; local authorities redirected him to his own consulate and waited for the latter's decision.[162] Paoletti's desire might have been denied or fulfilled. But apparently changes in nationality could be reversed. For instance, the Port Said resident Aniello Malatesta, brother of the anarchist Errico, had given up his Italian citizenship in 1884. He tried to get it back in 1891 and again in 1899. He eventually succeeded in 1901.[163] Finally, claims of a certain affiliation or protection could yield unpredictably different outcomes. A self-declared Algerian was denied French protection because he had no passport or other means to prove his identity. The previously encountered Fathalla Darviche or Darouiche was born in Beirut and resided in Port Said and similarly had no passport, but he was nonetheless designated a French subject. He could perhaps count on being registered at the French consulate in Alexandria, where he had been transferred to complete his education in 1860, at the "time of the massacre of Syria," with other orphans who, like him, had been raised by the Lazariste Brothers in Beirut.[164]

Discipline, Expel, and Document

Ever since the first half of the 1860s, authorities on the isthmus had contemplated expulsion as a remedy for the malaise created by those "troubling the order on worksites." In 1863 lack of discipline was apparently pervading Port Said, where a certain number of employees and workers constantly challenged those in charge by disregarding measures taken in the distribution of lodging and the general policing of the city or refusing to obey orders. All workers who threatened the general discipline and public tranquility on the worksites were to be expelled with no hesitation.[165] Most individuals were scared off by intimidation and serious admonitions. If the Suez local police arrested Italians in Suez for vagrancy, for example, the Italian consul would try to persuade them to look for an occupation. He would also remind them that if they did not find a job, they could be held in jail at the discretion of the local authorities because maintaining public safety required their captivity. Only after all this failed would they be expelled.[166] Consular authorities and Company representatives agreed that once troublemakers were arrested by the local authority, they should be handed over to the competent authorities and shipped off. Still, when a suspect's presence alone could not be proven to be outright hazardous, it was admittedly difficult to indict such individuals and make them leave.[167] Although consuls could successfully expel many of the rebellious undesirables from the most populated Egyptian cities, such as Cairo, where they were closely monitored by the local police, they had less success in outlying centers.[168]

Upon the canal's grand opening in 1869, expulsion was also envisioned to rid the territory of those who could no longer live off canal work. The governor decided that anybody who was found "loitering in the streets or otherwise conducting an idle and vagabond life" and could not demonstrate his means of subsistence would be arrested and brought in front of the authority to which that person was subject. The owners of dives (*bettole*) and hotels in town could no longer provide lodging to strangers "whose conduct they could not be confident about." He fretted about returning Suez to its "quiet" and removing those he identified as "evil-doers."[169] The inability to demonstrate how one made a living was then associated with criminal conduct and bore comparable consequences. In December 1869, in the wake of a string of violent murders and heists along the canal, the governor and European consuls in Suez met to discuss exceptional measures for security. The Italian vice-consul in Suez wrote to the Italian consul in Cairo to ask whether he would be authorized to expel "those Italian individuals unable to demonstrate how they obtain their means for a living" from that southern port city.[170]

Yet expulsion was seldom effective.[171] First of all, consular authorities in Port Said or other provincial sites often had their hands tied because the consuls general or the chief consular courts in the capital had the final say on expatriation measures. For example, when James Testaferrata, a British subject whose character was "known," was sentenced to two months of imprisonment and a fine for the drunken use of foul language and threats, all the Port Said consulate could do was plead with the judge of the chief consular court in Egypt to deport him. Similarly, when the governor urged consular authorities to remove from Port Said the Algerian Aly Mohamad el Naggiar, a French protégé who had repeatedly caused trouble, his request went unheeded. The French consul in Port Said replied that "only the minister can take the decision to expel him."[172] Second, the locally based consul could only remove wrongdoers from his own "district." It was out of his power to force his colleagues in other provinces to expel the same individuals from their districts. This deincentivized the Egyptian government from cooperating. Local Egyptian authorities, even when notified with the expulsion decree, argued that they could not enforce the measure unless it had been ordered for the whole of Egypt.[173] The third and last main reason for the ineffectiveness of expulsion was that deportees often did not even need to bother relocating in another Egyptian province. After having been expelled multiple times from a city, they could freely come back to their starting point.[174]

Canal towns and Port Said in particular acquired a conspicuous reputation as runaways' sanctuaries. Reportedly many of those who had been chased away from Alexandria or Cairo settled in Port Said and transformed it into a "haven for all the worst elements expelled from other cities of Egypt." Ostensibly, ill-doers gathered in Port Said because they were certain they would not be molested.[175] Port Said thereby proved that, as suggested earlier, the partial expulsion of subversive elements from one province did not bring any advantage to Egypt as a whole, given that they could freely come back to the country either via land or sea and take up residence in a different province. For instance, the trickster encountered previously, Francesco Salemi or Soleiman el Mamluck, received expulsion papers (*fogli di via*) from the Italian consulate in Alexandria. However, he resurfaced shortly thereafter in Suez, where he sojourned for more than a month without the Italian vice-consul in town being able to do anything about it. As he exemplifies, the same individuals who had been previously deported from Port Said and the canal area used to simply return, notwithstanding the fact that they had been repeatedly expelled before from those same places.[176] Throughout the 1890s, Port Said continued to house a great number of vagrants of seemingly reproachable conduct. To "liberate" the city, their respective consuls strove to chase them away via expulsion decrees. Unimpressed by these and seemingly unbothered by the local police, they circulated freely in the city, attended gambling halls (*bische*), and persisted in arousing scandal.[177]

Consular officials and Egyptian government representatives disagreed on what tools ought to be employed to monitor the return of unwanted visitors. It appears that as early as the 1870s the Egyptian Ministry of the Interior had decreed that photographs of Europeans found guilty of theft or armed violence be placed at passport control bureaus in Port Said, Suez, Cairo, and Alexandria in order to prevent their return.[178] But unlike Cairo and Alexandria, Port Said had no anthropometric bureau.[179] First introduced in Egypt in 1895, this institution applied an "anthropometric system" based on the Bertillon method of identification of reoffenders. This method involved the taking of fingerprints and measuring of the head, the extremities, and the rest of the body of the offender. The bureau recorded and classified such information. Starting in late January 1895, each police station in Alexandria was to be provided with a measuring apparatus and plates for making fingerprints. In 1897 one such office was created in Cairo by the commandant of police, Colonel Harvey Pasha. In the office 243 drawers held records bearing the measurements, fingerprints, and brief accounts of the main bodily markings of roughly twenty thousand previously convicted individuals. The anthropometric system seems to have slowly extended to other parts of Egypt, but only a few local bureaus followed through.[180] British-controlled Egypt (as did coeval British India) may have functioned as a testing ground for foundational surveillance practices.[181] Still, the anthropometric system of identification proved to be only partially effective even in France, where

Alphonse Bertillon had first devised it and authorities would extend it in 1888 to the entire resident foreign population.[182]

In Port Said, the governor general was keen on the idea of taking and distributing the photographs of the exiled to its various police stations so as to intensify surveillance and prevent those individuals' return. However, he was not willing to take on the associated costs.[183] Neither was he amenable to taking responsibility for the re-expulsion of the individuals who had reappeared. The foreign consuls in town then had to repeatedly repatriate undesirables at the expense of their institutions. The Italian consul, for example, lamented that either the local governorate or the Ministry of the Interior should have been charged for such costs. He also claimed that even if Port Said still lacked an anthropometric bureau of its own, the police had found a way to photograph and measure at least three criminals in the past. Persuaded of the value of such methods, the consul declared he would take such pictures himself, if the local authority would promise to deal with the unwanted returnees by expelling them across the sea or preventing their disembarkation altogether.[184] The case of Ulisse Marcantonio might have lent support to the consul's appeals. In 1884, after having been exiled from Egypt multiple times already for thieving, Marcantonio had once again made his return to Alexandria. Upon his arrest, an empty purse and a golden Remontier watch coated in dark blue enamel and adorned with diamonds, which he had stolen from his own sister, were found on him. The photograph that had been previously taken of Marcantonio may have facilitated his umpteenth capture.[185]

Nowhere was the need for increased surveillance, it was argued, as dire as in Port Said, a port through which more than one hundred thousand passengers passed every year and whose "special situation" required no less control than Cairo or Alexandria.[186] The governorate, the police, and consulates in Port Said kept an eye on what was going on in other Egyptian cities and experimented with their own forms of registration and bureaucracy.[187] The officials who were eager to apprehend criminals operating across borders had to fine tune and diversify their methods accordingly, as was happening with anarchists on the run in coeval Alexandria.[188] Potentially, anthropometric measures could indeed be adjusted to accommodate Port Said's highly mobile and possibly criminal population. The dozen or so pieces of information vital to a positive identification could be transmitted by telephone or telegraph. Photography, three decades after the invention of the camera in 1839, was to be zealously put to use for police records. In 1868 the Ottoman sultan issued an edict ordering logbooks be kept that should contain the photographs and the names of convicted pickpockets and thieves. Copies of such logbooks were also to reach administrative centers throughout the empire, Alexandria included. It is unclear whether the practice was consistently followed up.[189] Nonetheless, the acceptance of anthropometric technologies on the part of the different authorities reflects their common desire to identify criminals and shows their preoccupation with the classification and taxonomical organization of different social groups, one of the dominant scientific trends of late-nineteenth-century European thought.[190] The apparently unbounded possibilities opened up by anthropometric measures had become the talk of the town. Around 1895 newspapers in Port Said explained to their readers what the measures meant and how they were to be applied in police investigations.[191] Thus, for once successfully overlooking their divergences, the various powers based in this Egyptian port city and the canal area all shared the ultimately fleeting goal to quantify and capture the "unavoidably elusive nature of the individual," especially the transient one.[192]

LEADING A NOMADIC LIFE AFTER 1869

After 1869 the hopes of those residing in Port Said, moving to the isthmus, or roaming through it looking for work were often frustrated. In 1876 a twenty-one-year-old French man based in Port Said, unable to support his parents and siblings after six months out of work, begged to be enrolled in the French army. In 1877 another French man testified that he could find absolutely no work in Port Said. He had been, he wrote, sleeping in the streets, out of food for three days, and had not changed his shirt for twenty-eight days. He ill-advisedly decided to walk to Ismailia, some eighty kilometers south of Port Said, which he reached in a state of fatigue, hunger, and cold. Scarcity of employment opportunities was not, apparently, limited to the isthmus. In 1878 Jourdan Marcelin asked to be repatriated. He had come to Port Said from Alexandria, but in neither city could he make ends meet.[193] Those who looked to Cairo and Alexandria to fulfill their aspirations did not fare much better. A self-ascribed "artist" called Gava, who had left Trieste and its commercial languor, had his wine-selling plans in Alexandria thwarted. Luigi Isella, also writing from Alexandria in 1878, reportedly had no connections or means and begged for a job in the Egyptian administration. Cairo and its surroundings were not faring any better. A Jerusalem-born Ottoman subject had to leave Helwan in 1891 because there was no work left there for him to support his family. In 1892, her hopes to find a job in Cairo unfulfilled, an Ottoman woman called Sylva Amira found herself evicted from her home and begging for money to pay her debts.[194]

Nonetheless, Egypt and the isthmus region continued to offer options that were not available elsewhere. Even after 1869, many foreigners continued to head toward the country after having wandered in and out of their native lands and led a nomadic life (*vita girovaga*). They hurried from one village to the next "seeking bread and fortune."[195] In the 1870s adventurers of all nations were still flocking to the "Egyptian Tom Tiddler's ground picking up gold and silver."[196] In 1879 the pianist Leoncavallo, in his willingness to try anything to leave his native Italian hometown Potenza and demonstrate his musical abilities, set off for Egypt, where a well-connected uncle already lived.[197] Another pianist, called Bianchi, was in Egypt and jobless but lived off the earnings of his streetwise mistress and promptly spent pocket money on brandy (*eau-de-vie*) rounds for his acquaintances. A third prototypical representative of the contemporaneous mobile population of artists was a female singer (*canzonettista*) from Naples called Ida Ritelli, who performed at the Ba-Ta-Clan club in Cairo for a small but enthusiastic audience.[198]

For political haven seekers as well, Egypt continued to be a destination as it had been in previous decades. Antonio Rizzo, for example, fled Leghorn as a political suspect after the revolutionary period at the end of the 1840s. Rizzo had found refuge in Egypt, where he traded in leather and pearls.[199] A few years later a man called Jules Deslandes, who had taken part in the insurrectional movement of the Paris Commune in 1871, also chose Egypt. After his arrest, deportation to New Caledonia, and escape from the penal colony, he found refuge in Port Said. He arrived in this port city in 1874, where he lived off his job as a turner and apparently regretted his tumultuous past.[200] In 1877 anarchists established a section in Port Said, while plans were under way to create a nucleus in Ismailia (and a female branch was about to be formed in Cairo).[201]

After the British occupation, foreigners' access to jobs in the public administration came under increasing regulation. Education and professionalization opportunities for Egyptians multiplied, and lucrative options for newly arrived foreigners decreased. But the abovementioned performers, artisans, and unskilled laborers and service workers, whether of Italian or Maltese or other provenance, continued to try their luck in Egyptian centers and deploy all means at their disposal.[202] Italians in Port Said, among others, still found employment there in the public administration of the post, telegraph, saltworks, and health bureaus through the mid-1890s and beyond. In 1899 an Italian engineer called Michele Guastalla, who had created the Port Said saltworks, obtained a fifty-year-long monopoly.[203]

The arrival, transit, and permanence of a highly mobile and often unemployed population raised anxieties among Port Said's wealthy residents and ruling authorities alike, as was happening at the same time in Istanbul and countries around the Mediterranean rim. The stuffed shirts within the foreign communities regarded occurrences of criminal activity as irrefutable signs of the Egyptian state's weakness. Newspapers became sounding boards for such apprehensions, fueling sensationalistic news about Port Said's lack of security in particular.[204] Some contemporary observers appreciated the job of the press precisely because newspaper articles troubled the reassuring but false official reports claiming that Egypt was generally safe.[205] Nonetheless, as recounted earlier, Isma'il had already been expanding the police and the military in order to exert greater control over society before 1882 and the British occupation. The ruler even attempted to include Europeans on the force to be better able to deal with the growingly powerful foreigners in the country. Egyptian rulers had understood that building a modern state required the active production of institutions and ideologies that could improve their ability to monitor the population. Accordingly, they had supported a modernizing and centralizing state and favored the development of novel means of counting and identifying people as early as Muhammad Ali's era.[206] As in European centers and in the Ottoman capital itself, the Egyptian state progressively tightened its control over society by qualifying, quantifying, and archiving criminal acts and offences while honing criminological practices that sought to secure the identity of the individual within the boundaries of the nation-state.[207]

By the mid-1880s, Port Said may well have ceased to be "a den of thieves and assassins."[208] In 1891 "Arab" policemen in European-style uniforms stood at every road junction.[209] Nonetheless, by the onset of the 1890s some people still perceived Port Said as unsafe. Newspapers demanded more energetic, active, and diligent police. In 1892 the governor demanded from the Ministry of the Interior the dispatch of twenty additional soldiers to keep the peace in town.[210] People felt they had to protect their vulnerable lives and belongings. Alarmed and indignant European residents accused their consuls of being passive in the face of the countless thefts, aggressions, and assaults still occurring in public establishments. Consuls in turn faulted the local police for being inadequate, intimidated, and incapable of finding and arresting culprits. They perceived the guards in town as too few and inexperienced, consisting of "almost all Arabs, poor recruits," who were fired daily and unable to get their bearings. There were no horse-riding guards, reputed to be necessary in a port city surrounded by sand and desert (by 1895, police horses were being replaced, at least in Cairo, with European-made bicycles). According to the Italian consul in town, Port Said had the potential of being "like a trap" for criminals: a little and well-organized police force would have been enough to avoid disorder and arrest ill-doers. But, the consul sallied, Port Said was instead a (legal) "desert" that seemingly lacked any constituted authority.[211]

CONCLUSION

On the Isthmus of Suez, Egyptian authorities, consular representatives, Company officials, and after 1882, British bureaucrats awkwardly joined efforts to assemble efficient police and expulsion systems that would cleanse the particularly troublesome Port Said of its unruliest elements. However, they created no perfectly centralized institution, well-oiled state apparatus, or omnipresent observer.[212] They swore by the principle that "the rules to follow on the Isthmus were the same regulating the rest of Egypt."[213] Nonetheless, Port Said quickly emerged as exceptional because of its permeable waterfront and elusive population. By the last years of the century, Port Said outwardly displayed customs and coastguard stations, a quarantine, and border control sites. It became the site of multiple forms of policing and regulation as well as the locale of intense interaction and conflict.[214] But authorities often failed to count, monitor, and even selectively expel undesirables. After the canal's official inauguration in 1869 and the demise of labor all along the length of the canal, law and order on the isthmus became the site of local negotiation, experimentation, and failure that would continue beyond the British occupation in 1882 and into the 1890s. Contemplated from the streets and the docks, both 1869 and 1882 become reconfigured: while the former appears less like a festive carnival and more like a frustrating occurrence, the latter emerges as a moment of sheer transition within the extant geography of illegal exchanges and movements that continued to pivot on Port Said and the Suez Canal.

FOUR

Entertainment in Port Said, a Sink of Immoral Filth

> I do not know what Port Said may be in its working clothes, but in its holiday dress I bear testimony to its being a most enchanting place.
>
> —WILLIAM GEORGE HAMLEY, 1871

AFTER THE CANAL'S OFFICIAL INAUGURATION IN 1869, a large portion of the workmen were discharged from the then complete canal works congregated in Port Said. The most prudent of them had apparently gone back home with their earnings. But the ruck of laborers and mechanics loitered in Port Said in hope of finding fresh jobs. They reportedly spent their money at the cafés, saloons, dramshops, and houses of ill repute with which the port town teemed. Brawls and drunken riots were a constant occurrence.[1] By 1879 Port Said had recouped the population loss recorded after 1869 and had grown to 12,300. By 1882 its populace had swelled to 15,000 and surpassed that of Suez. By 1896 about 37,000 people resided in Port Said, almost four times as many as in Suez and ten times those in Ismailia.[2] Contemporaries identified the sources of Port Said's moral blemishes in its numerous gambling venues and houses of ill repute. They blamed the port city's apparently "hopeless abandonment to immorality and debauchery" on the many migrants or "refugees" who hailed from "Turkey," Greece, Italy, France, and England, as well as the rest of Egypt and Sudan (with pride of place assigned to "low-class" individuals among the Greeks, Italians, and French).[3] Port Said's reputation for debauchery and suspicion of its mobile population thus became intertwined, persisting into the turn of the century, when Western visitors still whined about the "immorality" of the place, noted the "dissolute looks of many of the men and women" they met in the streets, and likened Port Said to a "sink of immoral filth."[4]

Turn-of-the-century Port Said elicited mixed feelings but seemingly left no one indifferent. In 1881 a visitor commended Port Said for enabling an "absolute repose of mind" but observed that "to the active spirit its monotony must be unendurable."[5] In the 1890s the press similarly associated Port Said with tedium, which late-Ottoman official discourse often equated to harmful idleness and unproductivity.[6] Articles claimed that boredom affected both the local youth and the half million transiting passengers, who would "come spend a few hours in this part of the World called Port Said, and [...] get bored after half hour."[7] Allegedly, well-heeled visitors felt compelled to go ashore only because "the dust from coaling operations made it very unpleasant for those who did not do so."[8] For those who set foot on land, there may not have been much to do in the port city, just an "odious little town" displaying a few shops selling canned goods.[9] Others stressed that there was "nothing so sad and so unsightly" as its crossroads, which only came alive with the traffic of entertainment-seeking visitors. Reportedly, nothing was there "but a little European scum jettisoned upon the edge of the desert," where all streets came oddly to an end.[10]

Nonetheless, according to other accounts, Port Said did offer recreation opportunities with its vista of the canal, blazing sand, and procession of ships. Its gaming tables, saloons, and dance halls prompted "strong excitement" among those trudging its streets in search of amusement. Most observers could not help but notice the port city's large quantity of casinos, where patrons could enjoy live music and forsake respectability after 11:00 p.m.[11] They also were struck by its "far from reputable cafés," small theaters, and gambling shops, where "the dregs of all European countries" seemingly congregated. Reportedly, by 1899 nearly every second establishment in Port Said was either a cheap drink shop, a gambling den, or a brothel.[12] Music halls, shops of photographs, and stores of novelties were legion. In sum, the city was abundantly provided with all the entertainment that weary workers or sailors might have craved after a long crossing, ranging from "wicked places" to opportunities to purchase "obscene photographs" and excellent tobacco from Latakia.[13] "All things are for sale in Port Said," Rudyard Kipling had one of his characters proclaim after an orgy of drinks and lascivious dances in one of Port Said's late nineteenth-century shady taverns.[14] An assortment of desired commodities was immediately available to disembarking passengers and residents alike. A motley gang of "Arab" porters, hotel employees, *drogmans*, and boatmen peddled them forcefully and loudly within reach of approaching steamers (see figure 17).

Colossal signs in French, English, and Italian advertised the wares in Port Said's shops.[15] The town capitalized on the canal's incoming and outgoing

FIGURE 17. Peddler of various objects in Port Said, 1880–1890. Photographer unknown. *Source:* Alinari Archives-Favrod Collection, Florence, FCC-F-020006-0000.

traffic to boost its sales. Its busy Commerce Street, Shariʿ al tijara (see figure 18), epitomized the intensity of the exchanges taking place in town. A shop located on this thoroughfare advertised its vast gamut of European, Chinese, and Japanese objects, as well as Egyptian and "Turkish" wares, and enticed its clientele with fans, silks, ostrich feathers, lace, and corals. A late nineteenth-century snapshot of this street shows that a local store run by associates Attard and Vella sold Maltese lace and jewelry. It also reveals that the Simon Arzt store (before it moved to the quay in the 1930s), founded in 1869 by a Jewish New Yorker, produced and advertised its own photographs of the canal as well as Egypt and Palestine.[16] The signs of shops selling ready-made clothes and shoes peeped out. Others promised diversion at a restaurant, an "international coffeeshop" (*caffè internazionale*), and a grand casino. Meanwhile, a stream of peddlers and passersby flowed by apparently undeflected.

FIGURE 18. Commerce Street in Port Said, 1854–1901. Photo by Hippolyte Arnoux. *Source:* ANMT, CUCMS, 1995 060 1491, Association du Souvenir de Ferdinand de Lesseps et du Canal de Suez.

Workers' salaries begot the possibility for activities other than just toiling. Recreational sites created novel forms of sociability that were autonomous from the workplace.[17] By 1890 it was reported that most people in Port Said "lived haphazardly" and wasted whatever they saved with "all those vices that mother nature created."[18] Various leisure cultures came into being in late nineteenth-century Port Said. While some, like newspaper reading and attending festivals, delineated the boundaries of groups that identified themselves as national and religious, attending drinking establishments seemed to appeal to everyone. In the latter realm, there were options for all budgets. While theaters and hotels offered places and pastimes of bourgeois sociability, hole-in-the-wall cafés and gambling may have attracted a crowd of workers and sailors.[19] As in the Ottoman context, where coffeehouses were allegedly popular among "commoners" but usually held in contempt by elite authors, similarly highbrow views suggested that coffeehouses in nineteenth-century Egypt were mostly frequented by people of "the lower orders" and craftsmen. Many, however, for example during Ramadan nights, admittedly visited coffeehouses for the sake of society or attended performances of reciters or musicians.[20] Some contemporary intellectuals indiscriminately viewed coffeehouses and "places of perdition" as sites in which rich and poor mingled. They were "equal in the eyes of vice."[21]

Isthmus residents and visitors reportedly yearned for the kind of comfort they could find at a hotel's canteen, where they were warmly welcomed, heard the most recent news, listened to traveling singers, and, depending on the depth of their pockets, sipped lemonade, Bordeaux, Champagne, absinthe, or just a glass of cheaper "raki" (also known as *arak* or *rakı*, a strong spirit distilled from grapes and flavored with anisette).[22] But drinking establishments in Port Said provided more than sheer leisure. Unlike other pastimes available on the isthmus, they provided room and time to meet and blend with individuals belonging to groups that sources identified as distinct. To workers, such spaces might have disguised a subversive function by enabling them to interact independently and away from the authorities' watchful gaze.[23] At the same time, however, drinking establishments also set the stage for often alcohol-fueled claims about national belonging and gender roles by bargoers. Gender played out in the wake of men's drunken sprees as well. The young, single, or unattached males among them, perhaps uprooted from definite expectations for social behavior, yielded to exaggerated outbursts of unbridled behavior.[24] Meanwhile women, especially those working in the entertainment sector, bore the blame for igniting male violence.[25]

By focusing on consumption and leisure patterns in turn-of-the-century Port Said, this chapter brings together multiple scales at once. In the optic of entertainment, Port Said emerges as both a way station and a terminal for global flows of commodities and people. Simultaneously, it appears as a small town that can be further parceled into the individual venues where people liked to spend their free time. The chapter then zooms in further on Port Said's coffeehouses and bars. It follows the comings and goings of its workers and patrons, arguing that this microscopic scale of mobility lays bare how the isthmus society developed, patrolled, and transgressed group boundaries, letting some in while refusing entry to others.

THE ISTHMUS SOCIETY HAVING FUN AND GAMES

Port Said's surroundings sported a peculiar mix of sand, water, and large flocks of birds. Ducks, teals, gulls, herons, ibis, flamingos: all aquatic species

FIGURE 19. Beach at Port Said, 1912. *Source:* Archives CeAlex/CNRS, fonds J.-Y. Empereur.

were represented.[26] By 1865 lake baths had been established in the town. Some claimed facilities of this kind needed to be expanded to comprise sea baths, since the town lay on "one of the most salubrious [beaches] in the world." Further, they ought to be made as comfortable as possible to attract all of Europe's hypochondriacs.[27] It took at least thirty years, however, for "speculators" to realize that making cabins available to beachgoers on Port Said's shores would have been a great investment (see figure 19). By 1893 a tramway would connect the town to the newly built Minerva sea baths. These came equipped with showers and towels, as well as a restaurant, café, and verandah.[28]

On Easter day of 1866, a soon-to-be popular society of regattas of the isthmus made its debut in Port Said; it was followed by other similar events at Lake Timsah near Ismailia. But a society for sailing became defunct in November 1870, after many of its members left the isthmus upon the completion of the bulk of the excavation work. Fishing and hunting had been popular pastimes since the early days of canal work and continued to be.[29] Yet these could be divisive activities. Fishing and hunting regulations prevented Egyptians from using firearms to shoot birds from their boats, while non-Egyptians were exempt from this restriction.[30] The campsites' population was reportedly terrorized by those hunters who "mercilessly" killed the small birds that livened up life in the isthmus camps, seized the pigeons belonging to private citizens, and even caused accidents by firing stray bullets into private homes. One day in 1897, in Port Said, a European hunter and a "native" one scuffled over two ducks that each of them claimed to have shot.[31]

Other forms of recreation on the isthmus appear to have been based on exclusive membership. As early as 1861 or 1862, three medium-level Company cadres, perhaps eager to refine the tastes of a generally unrefined isthmus society, established an "artistic, literary, archaeological, and numismatic Society" that only enlisted fifty-six members.[32] It appears that by around 1860 masonic lodges existed in Port Said, Ismailia, and Suez. In Port Said, L'union des deux mers, extant since at least 1868, had a French orientation. Later, the Progresso (1907–1925), with ties to Italy, would also be established in town. In Ismailia, L'isthme de Suez, founded in 1868, was oriented toward France. Of the two lodges recorded in Suez in the 1860s, Mount Sinai was founded in 1865 and was answerable to Italy's Masons. It had comprised Italian, French, and English components but was deemed defunct by 1868, after its members had coalesced along national lines. The second masonic lodge in Suez was called The Love for Truth and obeyed the Masons' leadership in France.[33] Notwithstanding their national orientation, members of both lodges in Suez were apparently busy spreading "poisonous ideas and books against Catholicism."[34]

The Sound of Isthmus-Bound Migration

Various kinds of music developed on the isthmus thanks to the advent and circulation of migrants who brought along their own instruments and repertoires. As early as 1861, the "Arabs" in the budding Suez encampments were said to play music with the tambourines and flutes they had brought with them.[35] By 1868 the nuns of the Bon Pasteur in Port Said had managed to have a small harmonium shipped to the port town from Marseilles.[36] Isthmus-bound migrants like Saul Croce and Anna Forneroni, settled in Chalouf, had managed to bring their piano with them, although they eventually had to pawn it and then lost it in 1869 because they were unable to redeem their debt. By 1893 there were enough pianos on the isthmus to support the businesses of both Mr. Gildo Lifonti, a resident in Port Said's Rue de la Division who worked as a tuner of that and other instruments, and a lady by the name of E. C. Keyser, who offered piano lessons in Port Said, for a price to be agreed upon.[37] In the last quarter of the nineteenth century, the *simsimiyya* genre (indicating both a genre and a thousand-year-old bowl or box lyre with five strings) blossomed out of Egypt's eastern Nile Delta and especially around Port Said and Ismailia. It developed around the distinctive dancing of street peddlers awaiting ships (the *bambutiyya*, named after the "bumboats" plying the canal) and included the mimicry of work-related movements such as rowing, winding ropes, and bargaining. Key to the development of the *simsimiyya* was the performance of male musicians known as *suhbajiyya* in Port Said's coffeehouses. Other meaningful influences on the development of the *simsimiyya* may have come from Nubian migrants hailing from Upper Egypt; their instruments; and another genre of collective singing influenced by Sufi music and known as *damma*, "gathering" in Arabic, characterizing the get-together of men at the end of their workday. The *simsimiyya* thereby was brewed in the peculiar social and cultural mix of the canal region. It percolated throughout the Sinai, extending also to the Saudi Red Sea coast and southern Yemen.[38]

Several musical societies flourished in the canal cities. From time to time the Port Said–based society L'Harmonie joined forces with the Timsa music society in Ismailia, where a great number of people from Port Said occasionally partook in revelry. Members of the Timsa organized parties during which participants "promptly invaded" the excellent buffet and danced under the presumably benign stare of de Lesseps. He overlooked the merrymaking from a portrait hung in the dance hall and adorned with flower garlands and flags of all nations.[39] The isthmus also registered less ecumenical tunes. By 1895, among the musical societies operating on the isthmus was the professedly Italian Margherita. Further, the Società filodrammatica, which ran its own hall where all of Port Said society apparently gathered, was rumored to have an "Italian" character. Two-thirds of its 150 members were originally from Italy, the president was Italian, and plays were performed in the Italian idiom. Italian youth were given the exclusive opportunity to study music for free and play together as a group.[40]

Reading and Publishing in the Isthmus

Reading as a diversion was available to that relative minority of the population who could read and write. Many of the isthmus-bound Egyptian and foreign migrants, like those hailing from Calabria, were illiterate.[41] Still, in January 1871 a reading "cabinet" was created in Port Said. It displayed a library of three thousand volumes.[42] Some of those relocating to the isthmus had brought their own books along. A man in a supervisory position at the canal works, for example, had carried with him religiously themed books (*The Love of Jesus Christ, The Invitation of Jesus Christ, Catholic Instructions*) alongside a portrait of the Virgin Mary, unspecified publications in both French and Italian, a French-Arabic dictionary, treaties on arithmetic, and the *Histoire de Bayard*, a popular knight's biography dating back to the sixteenth century.[43] In 1882 a shop located on Port Said's Rue du Commerce sold publications in "French, Italian, English, etc.," in addition to school textbooks; atlases; paper; and black, blue, red, and purple ink.[44] Around 1893, at the "libreairie et papeteri [*sic*]" of J. Horn (calling itself Anglo-American Book Depot by 1896) in Port Said, devotional books in both English and French at moderate prices were available.[45] In March 1908 the Dante Alighieri Society (founded in 1899 in Rome "to protect and spread the Italian language and culture outside of the Kingdom of Italy") instituted a local committee in Port Said. It established a library complete with books, magazines, and theater plays. Visitors transiting through town could rely on the good offices of society members to borrow books for a maximum of fifteen days.[46]

Publishing activity in town provided ready-at-hand reading material. There were at least four presses in late nineteenth-century Port Said. Two were run by, respectively, Omero Barsotti and Leone Goldman. The Maltese Ferdinando Cumbo managed the third one. Finally, Jacques Serrière had a "French" printing press installed in town.[47] As for local newspapers specifically, the French-language press began operating in Port Said in 1867, when the *Journal du Canal*, sympathetic to the Company, began publication.[48] Between 1870 and 1872, *Le Journal de Port-Saïd* was available. Starting in 1871 and 1874, respectively, people on the Isthmus could read *L'Avenir Commercial de Port-Saïd* and the *Courrier de Port-Saïd* (later resuscitated as *Le Moniteur de Port-Saïd*).[49] The year 1880 saw the foundation by the above-mentioned Serriére of *Al Busfur* or *Le Bosphore égyptien*, appearing in French and Arabic, which was transferred to Cairo in 1881. The daily *Le Phare de Port-Said et du Canal de Suez* saw the light of day in January 1888. Some argue that its creator, the "Italian" Enrico Pacho, who directed it until his death in 1895, managed to maintain an Italian character despite its reliance on French. Others attribute its creation to editor Marius Jauffret and highlight its pro-Company orientation. Regardless, the latter may not have been popular among the French throughout the isthmus; in 1890 *L'Écho de Port Said: Journal politique, commercial et littéraire* began publication from Port

Said twice a week, claiming to be "the only French mouthpiece of the Isthmus of Suez."[50] If it made the French happy, *L'Écho* aroused the antipathy of others: in 1893, three "Hellenic subjects" who claimed to represent "all the inhabitants of Cassos" (Kasos) complained to French consular authorities about G. Colomb, *L'Écho*'s director and owner and a French subject, for publishing in the "news from Turkey" feature an allegedly libelous article that offended that island's navy.[51]

Even though, in 1897 alone, other comparably provincial towns vaunted their own eponymous newspapers, under the titles *Tanta*, *Mansura*, *Helwan*, and *Fayyum*, Port Said lacked an Arabic newspaper in its first decades of life.[52] But the French monopoly on newspapers would nevertheless soon be challenged. In 1891 a man called Kyriacopoulo founded *Le Progrès* in Port Said, to appear in both French and Greek (he later transferred it to Cairo). It strenuously defended the British regime.[53] In 1892 the "Italian" newspaper *L'Elettrico: Giornale politico, letterario e commerciale* was founded in the port city.[54] By 1893 a reportedly "Maltese" daily titled *Il Telegrafo* was also in operation there.[55] In 1894 or 1895 the bookseller Horn established a new small-format journal, *La Verità* ("the truth"). Appearing in Italian at first, it later switched to French.[56] In 1895 another paper written in Italian appeared in Port Said, called *Spartaco*. In 1896 the *Corriere del Canale*, vaunted as "the only Italian newspaper of the Canal area," was started but did not acquire much of a following. It was probably published only until 1901.[57] Other newspapers appearing in the 1890s in Port Said were the *Bollettino di Porto Said*, the *Corriere di Porto Said*, and the *Cairo*. In 1903 the weekly humoristic *Ficcanaso* ("busybody") began to be published in town. Three years later *Le Petit Port-Saidien* was established, headed by a man called Mazzolini.[58]

Overall, newspapers divided Port Said's readership along linguistic and political lines. Each paper often spoke to a single community. Newspapers in Italian, for example, had professedly no readers outside the so-called Italian community.[59] Moreover, these publications expressed quite clear political orientations that must have jibed with those of their readership. Both *L'Elettrico* and *Le Phare de Port-Said et du Canal de Suez*, for example, claimed to have annoyed the Company with their unfettered content and language. In 1895, for example, the former published an article on "white-glove thieves" and hinted at the activities of the Company.[60] Additionally, newspapers produced by foreigners, thanks to the capitulatory regime that protected them, could risk behaving antagonistically toward the Egyptian government.[61] Reporters in Port Said had the habit of showing up every day at the police offices to get their share of news. They often did not stick to the so-called facts that were reported in a special registry that the police tailor-made for them. Consular staff belittled the role of newspapers in Port Said, claiming that they did nothing but "live off gossip." They scoffed at the "absolute liberty of press in Egypt" and expressed frustration over their inability to force newspapers to retract what they had previously divulged.[62]

Old and New Days of Remembrance

The profane as well as the sacred in Port Said offered inhabitants venues and moments to congregate while also coalescing in distinct groups. The Company played a role in organizing some of these festivities, during which it made sure to highlight the allure that France exercised especially on the town's notables, school-age children, and middle-class ladies.[63] Part of the populace celebrated Port Said's foundation every year around April 25. In 1865, for example, six years into its existence, a crowd of more than three thousand people gathered around an altar erected on the beach. They attended the performance of a sacred musical composition, a mass penned by a female Port Said composer (Miss Clémentine R.), executed by a visiting artist (Miss Beneditta Pratolini) and a high-ranking Company employee (Mr. F.), and accompanied by a choir of French ladies and workshop laborers. A lunch; regattas; and tournaments of, among other things, climbing up a "greasy pole" (*mât de cocagne*) and target shooting, followed, while windy conditions got in the way of the sailing competition. This occasion included artillery fire from French armored cruisers and a speech given by the governorate representative.[64]

In April, "natives" and others feted Shamm al-Nissim. It was a day for picnics in the green patches found along the canal banks.[65] Every year the inhabitants of Port Said's "Arab quarter" also celebrated the Ottoman sultan's accession by adorning their homes with flags and animating its streets with music and torches. For the sultan's anniversary (*mawlid*), all government offices, consulates, and anchored vessels raised their pennants and cannons were fired. Fireworks illuminated all skies from Istanbul to Alexandria. Khedival anniversaries were another festive occasion for Egyptians in Port Said.[66] Some of those identifying as "Italians" in town instead celebrated the birthday of Italy's King Umberto, and they mourned his death in 1900. The Maltese and other British subjects in Port Said remembered Queen Victoria's accession to the throne with a mass in Port Said's Catholic Church of Sainte Eugénie. Port Said's Austro-Hungarian notables did the same to celebrate Emperor Francis Joseph's birthday. Every July 14 came the turn of the French to remember Bastille Day at the Minerva sea baths on Port Said's beach. In September the Maltese prepared festivities to commemorate what they interpreted as the 1565 "liberation" of Malta from the Ottoman siege.[67] Around the twentieth of that month, Italians organized festivities to remember the 1871 capture of Rome and the definitive unification of Italy. Every June some of them also celebrated the Albertine Statute, the 1848 document that had become the constitution of the unified Kingdom of Italy and a symbol of monarchic power.[68]

Religious festivities in Port Said comprised Islamic Ramadan and Ottoman Bairam, Jewish Shavuot, and Christian Pentecost and Whitsunday. A dazzling nighttime celebration was held yearly to honor the point of the Islamic calendar falling at mid-Shaʿban. Flags usually were flown at half-mast for Good Friday, and special food was prepared for Easter, such as Leghorn-style *schiacciate* and rings of braided pastry with brightly colored eggs embedded in them. Jewish traders shut down their shops and fasted for Yom Kippur.[69] In July the Maltese threw a party to remember Our Lady of Mount Carmel (Beata Vergine del Carmine) at the Minerva sea baths, with pig races, the greasy pole again (*albero di cuccagna*), and other afternoon entertainment. A banquet, fireworks, a concert by the band Musique internationale, and a grand ball concluded the evening.[70]

The groups that sources identified as distinct still had much in common. The French, Maltese, and Italians in town, for example, all liked to revel at the Minerva seaside establishments. Incidentally, organizers may have shared the awareness that festivities represented good opportunities to get revenue. Sometimes, "all nationalities" were expressly welcomed at such events.[71] However, most of the time exhibiting national or religious symbols fueled tensions and hardened group boundaries in town. The events of 1883, 1892, and 1902 in Port Said, when anti-Semitic riots occurred, were a case in point. The tradition of the Greek Orthodox clergy of bringing out, around Jewish Pesach and Orthodox Easter, an effigy of Yezou Christos and parading it through the streets around their church may have been entertaining for some and offensive to others. Reportedly, Christians in town also had the habit of preparing each Easter a straw effigy wearing Jewish attire, which they burned.[72] Festivals provided a time and a place to choreograph national or religious differences. They may have owed more to the inspiration of romantic nationalism than to the actual traditions of immigrants. "Italians" in Egypt, for example, reportedly lacked cohesion and a sense of purpose. At the same time, they regarded other communities' national celebrations with envy and vainly strove to organize similarly successful festivities.[73] As has been theorized for other contexts, "ethnic group boundaries [. . .] must be repeatedly renegotiated, while expressive symbols of ethnicity (ethnic traditions) must be repeatedly reinterpreted." By mixing old and new, the staging of such symbols produced novel concoctions suffused with nostalgia and reified perceived group boundaries.[74]

PORT SAID'S ENTERTAINMENT LANDSCAPE

In nineteenth-century Egypt, small towns generally offered few opportunities for leisure, compared to Cairo and Alexandria.[75] But they still provided a modicum of excitement. The horseback riders, snake charmers, and bear trainers working in the streets of Cairo and Alexandria, as well as the hot-air balloons traversing their skies, probably found a warm reception in the rest of the country.[76] In the southern town of Matay, near Minya, draft beer off the ice could be consumed at a Greek café. Mansura, in the Egyptian Delta, boasted music halls; Arab mimes and acrobats; a prestidigitator called Lambos; and even an entertaining *Lucia of Lammermoor* at its Theater della Gaité, owned by a Mr. Nakle.[77] The isthmus had its share of amusement. In its streets, especially at night and with no lantern on them, musicians roamed along with accordion players and others playing their instruments. A "kind of *cabaret*," or tavern, had been operating in Port Said as early as April 1860. By 1862 the Al Guisr worksite boasted a hotel with six furnished rooms, a dining hall with twelve seats, and a coffee room complete with couches. It had been established by one Antonini, a Company employee at the central warehouse (Magasin Central), to whom the Company's president himself had apparently accorded the needed authorization.[78] By the onset of the 1870s, in Qantara there was a reportedly good café with a verandah, a pretty garden, and a wide landing quay. There, a Piedmontese couple ran a large inn (the husband came from Pinerolo, while the wife came from Giaveno in the Dora Riparia valley).[79] In Ismailia, an enterprising carpenter called Aspert had transformed his workshop into a music hall (*café-concert*). The latter kind of venue, also booming in Europe between the end of the nineteenth century and the First World War, usually had a platform, a small orchestra, and singers; entertainment usually meant an increase in the price of food and drinks. Aspert rented it out to a Greek, who had a roulette wheel installed in it.[80] Compared to Port Said, Ismailia was said to be quieter. Its population was seemingly less "mixed" and violent. Brawls and homicides were reportedly rarer. Finally, compared to Port Said and Ismailia, Suez appeared to be a more respectable place and enjoyed a much better reputation in terms of modesty and morality.[81]

It was in Port Said that entertainment appeared most lively. Only a fleeting and perhaps spurious note survives about a theater in town that the Company had allegedly built in the early 1860s.[82] The Mediterranean port city materialized on the map of itinerant performers. A great number of actors and actresses were recruited or trained in Syria and opted to move to Egypt.[83] Italian theater artists and singers, among others, generally started their tours from Alexandria, then proceeded to Cairo, and then went on to Port Said.[84] In 1892, for example, a company called Duse traveled from Cairo, where it had successfully performed at the theater of Azbakiyya, to Port Said, whence it brought its "good prose." In 1893 the artists of an Italian group, led by a pair called Mazzanti and Viscardi, staged a show at the Theater of Port Said-Les-Bains. Operas and operettas were performed at both the Eldorado and the Alhambra, theaters that also functioned as music halls (or rather as gambling halls, as in the case of the Eldorado around 1897). Another theater, The Circus of A. Mitrovich, opened for business in 1892.[85] By 1893 visitors noticed that there was at least one theater that was brilliantly lit with gas at night.[86] Small and provincial stages could not afford the costs of specialized singing, acting, or dancing ensembles. Therefore, itinerant artists had to cover a wide range of performing abilities.[87] Notwithstanding their efforts, foreign performers trying their luck on the Egyptian scene were often impoverished and frustrated. Equestrian circus companies passing through the isthmus, for example, did not fare well. Some individuals, such as an Italian theatrical prompter who had gone to Egypt as a member of a troupe, blamed their destitution on their expatriate community's weak patriotism and stinginess. Others, for instance the members of a group called Tosi, were in misery and hoped to find opportunities to entertain the clients of one of the hotels in Port Said by means of their consul's compassionate intervention.[88]

Port Said's Flourishing Drinking Establishments

While entertainers' fortunes in the rest of the isthmus fluctuated, the public flocked to Port Said's growing drinking establishments. Options were available for all pockets. Steamers' passengers of "all classes" reportedly went ashore to attend nightly concerts in Port Said.[89] The town's "European quarter" in particular was reputedly rich in inns, coffeehouses, and clubs "in the style of Paris," as well as breweries, *alcazars* (from the Arabic *al-qasr*, "palace") for musical performances, and music halls with roulette tables.[90] Some of these venues, such as the Eldorado and Alcazar, may even have been named after renowned counterparts in the contemporaneous French capital, in hopes of attracting customers based on both familiarity and allure.[91] Cafés, such as the Café de la Renaissance run by the widow Valin in Place de Lesseps, generally had pool and marble tables, as well as utensils for preparing coffee and cooking. Another café in Rue de La Scierie (Sawmill Street) was composed of seven rooms, a kitchen, and a room for coffee.[92]

Port Said's venues were often located close to each other and mostly huddled on the port quays or in the de Lesseps square nearby. These were competitive businesses, to the extent that accusations of stealing customers could quickly degenerate into rows. Moreover, such venues could be rapidly outcompeted. Of the five hotels and seven cafés advertised in town in 1868, only one hotel and its attendant café (Hôtel de France et Café) remained operative in 1884.[93] Those who owned these establishments tended to expand their interest in real estate, while hiring others to run their shops. By the mid-1880s Port Said was expanding over newly reclaimed land that was growing enormously in value. "Old Port Said" had disappeared beneath acres of new buildings.[94] One Paul Lesieur, a hotelkeeper, also owned a café in de Lesseps square. A Frenchman named Bousquet ran a music hall but also owned adjoining buildings that he leased out to storekeepers, French and non-French alike.[95]

In Port Said, drinking establishments and hotels were often one and the same. By 1865 the Hôtel de la Division (apparently no longer in business by 1878) catered to the "elegant" society in town. By 1866 another hotel, called Pagnon, had made its appearance on the Quay Eugénie. A decade later, profligate visitors passing through Port Said could spend the night at the Hôtel des Pays Bas, which opened in 1876, for 20 or 25 francs a day (see right side of figure 14, chapter 4). Those on a tighter budget could check in at the "tolerable" Hôtel du Louvre for 16 francs a day. Finally, the thrifty could lodge at the more modest Hôtel de France for 12 francs a day.[96] Even the latter may have been inaccessible to many, considering that European skilled workers in Port Said at the turn of the century were making from 7 to 15 francs a day, while "Arab" workers were earning even less.[97] By 1895 other overnight options included

FIGURE 20. Boulevard Eugénie and the Eastern Exchange Hotel, n.d. *Source:* Ref193, Hisham Khatib Collection, Akkasah Photography Archive, al Mawrid, NYUAD.

the Grand Hôtel Continental and café, owned by the Simonini family; it had a pretty verandah, poor rooms, and tolerable cuisine. Finally, there was the Eastern Exchange, built in 1884 by the British and fitted with electricity by 1901. Visitors could find newspapers and view "an endless number of manufacturers' samples and framed cards." It was an impressive seven-story building in cast iron, where the first elevator in all of Egypt was installed.[98] It was known to the Arabic speakers in town as "manzil al-hadid" or the "iron house" (see figure 20).[99] Yet the fortune of Port Said's hotels had begun waning in the final years of the nineteenth century, when they developed a reputation for mediocre meals and grimy rooms. One of the minor "Greek" hotels in town barely made up for such shortcomings by offering a small-town atmosphere.[100]

In the town's entertainment venues, patrons could drink, eat, play, and be played. In late nineteenth-century Egypt, alcohol could be consumed in bars (*bars* and *buvettes*) as well as cafés and taverns. Although the law forbade managers and employees to serve drinks to individuals in inebriated state, they often turned a blind eye and poured them another glass or two.[101] In Greek wine shops especially, customers could sip sweet wine and nibble on mezze, an assortment of savories that included salted sprouting beans.[102] Water pipes, various kinds of tobacco, and possibly hashish and opium could also be found at drinking establishments, further lubricating social contact.[103] (This was so even though the Egyptian state had outlawed possession of hashish in 1891 and its consumption or trade in public establishments in 1895.)[104] Once newspaper culture became prevalent in the 1870s, papers and magazines also made their way into these venues for the reading pleasure of literate patrons as well as the aural consumption of the illiterate ones.[105] Clients at Port Said's drinking venues could also watch performances, such as that of a one-man band playing four instruments at once around 1893.[106]

Gambling was yet another enticing feature of Port Said's drinking establishments. It was in such high demand that in 1874, for example, as soon as one of the venues where it was practiced was shut down, two new ones popped up to replace it: Alcazar Lyrique and Café d'Orient. The personnel of both the military and commercial navy as well as residents themselves kept betting large sums.[107] A visitor's perception that, by the mid-1880s, "previously flourishing gambling halls and dens of infamy were being suppressed little by little," ought to have been just a fleeting impression. The Egyptian government attempted to close all public gaming tables in 1890.[108] By that year, Greek owners were running most gambling venues (as they apparently did in Cairo and Alexandria as well) in town, where they only had one German competitor.[109] Efforts to suppress gambling continued. In June 1891 the Egyptian Ministry of the Interior attempted to ban from public establishments those "games of chance" such as "Baccarat," "Lansquenet," "Trente-et-un" and "trente-et-quarante," "Pharaon," roulette, and "Petits Chevaux."[110] In 1892 a police official (an English major) yearning to catch gamblers red-handed even tried to disguise himself as a traveler, but to no avail. The town's passion for gambling continued unabated, even though authorities tried to shut gambling halls down or to limit "chance games" to the monopoly of one supervised establishment. Patrons with a penchant for gambling had to beware of the swindlers swarming cafés and breweries alike, duping the clientele into placing bets on their cardboard roulette games and sharing their earnings with complicit establishments' owners.[111]

With its numerous entertainment venues, Port Said became a terminus for alcohol flows. Alcoholic beverages streamed conspicuously into and within the city, where a relatively high demand fueled the trade in drinks or their ingredients and connected the port city to production centers in France, Italy, and Ottoman lands. As elsewhere in the Ottoman Empire, Christians and Muslims appeared to consume all kinds of alcohol.[112] In

1867, even though wine reportedly spoiled easily because of the desert heat, it was estimated that Port Said, with a population of around eight thousand, consumed one hundred hectoliters of wine every day, not including "fine wines" and liquors.[113] Of the wine imbibed in the port city, at least some was produced in Maalaka-Zahle (Syria), whence a French merchant had been importing most of his wine production into Egypt since 1873 (and storing it in Port Said, Ismailia, and Alexandria).[114] At the time, Cyprus, Crete, Tenedos, Samos, the eastern Ottoman provinces of Asia Minor, and parts of the Balkans were famous for their vineyards.[115] By 1869 "excellent" wines from the Piedmont area of Asti were available at the Garibaldi Inn in Port Said, perhaps thanks to the connections of its Piedmontese owner, who had partnered with a Tuscan man.[116] In the mid-1860s drinkers elsewhere on the isthmus could relish, alongside chocolate and cigars, sips of Champagne (the Fleury, for example), Curaçao, absinthe, vermouth, Bordeaux, Médoc, cognac, "raqui," and beer.[117] Egyptians were familiar with fermented grain beverages, both through *buza*, a "drink closer to a solid than a liquid" that resembled a "cool, slightly fermented farina, with a sort of fizzy tingle," and the small amount of beer imported into Egypt before the end of the nineteenth century. Beer had made its first appearance in the Ottoman Empire around the time of the Crimean War (1854–1856) as an imported product.[118]

Beer in particular was said to be "the beverage of choice for the Europeans living in the desert" thanks to its "hygienic and refreshing properties." On the isthmus, it was in high demand and its consumption widespread. Egypt, though rich in barley, had to import hops from, for example, Alsace and Germany.[119] By 1888–1889, beer from Vienna and Bohemia could be purchased in Port Said for double the price of common wine (*vino da pasto*). In a few years' time Port Said–based traders would also begin importing French beer, which was manufactured either in Strasbourg or in southeastern France.[120] Beer was transported by steamer. It had to be shipped in small quantities and consumed within a few days after arrival. Only after Louis Pasteur revolutionized storage technologies could beer be stored for extended periods of time under artificially cool and sustainable conditions, thus allowing large quantities to be shipped over long distances regardless of the season.[121] An advertisement that appeared in 1893 in *Le Phare de Port-Said et du Canal de Suez*, for example, read that an "Alsatian" kind of professedly alcohol-free beer, manufactured in Marseille, could be consumed all along the Suez Canal since it was "resistant to all climates."[122]

DRINKING BUDDIES AND OCCASIONAL BEDFELLOWS

Some entertainment venues in Port Said catered to specific communities. Around the mid-1860s it was mostly Greeks, Austrians, and Italians who ran Port Said's entertainment.[123] But records from the following decade show that at least two venues in town, run by an Arab and a Berber, respectively, specialized in *buza*.[124] Among the cafés in Port Said that appealed to foreigners in 1868, some had names such as Garibaldi, d'Europe, d'Italie, de Florence, and d'Algerie. Copycats in Ismailia included de France, de l'Isthme, and d'Europe.[125] The choice of name may have reflected the makeup of either the barkeepers' population or that of perspective clients. As in coeval Istanbul, patrons originating from different Ottoman provinces or ethnic backgrounds may have liked to stick together.[126] In Egyptian centers, individuals coming from the same Italian province, for instance, ostensibly hung out together in certain establishments to the exclusion of others.[127] The French were ready to do so as well. In 1875, for example, an agent working for the Messageries Maritimes circulated a note in Port Said's cafés calling upon French citizens to gather one morning in one of the halls of the Hôtel de France. This suggests that French nationals both attended multiple establishments at once and were ready to congregate in specific venues on particular occasions. That time, the particular matter to be discussed was the judiciary reform then under way in Egypt. The signatories, many of whom had been in the country since 1868, 1863, or even 1861, feared that the reform proposed by the khedive would diminish the guarantees and interests of the French colony in Egypt and place Europeans "under the yoke" of the country's laws.[128]

By 1895 a few establishments in Port Said could be clearly identified as "French" for having kept their flags lowered in mourning on the first anniversary of the death of French president Sadi Carnot, in imitation of the French consulate (the Hôtel Continental, the Hôtel de France, the Hôtel de Paris, and the Nicaleau).[129] The end of the century also registered the presence of distinctly "Greek" bars and "Jewish" bars in town. For example, in 1891 the Grand Casino, a venue for "occult gambling, music, and coffee for the public located in the beautiful center of the European city and in the most crowded and central street," was connoted as a "Greek establishment." It had become the location of two bloody fights instigated by a gang of Greek scoundrels (*bravacci*) on Greek Orthodox Christmas day.[130] Finally, in 1901 the Eastern Exchange was considered to be the only hotel in town "under English management."[131]

Drinking or work in entertainment on the isthmus often brought people together even if they supposedly belonged to groups that contemporary sources defined as discrete. In Port Said, Arabs, so-called Nègres, Europeans, and Copts could apparently all be found within the same establishment, smoking tobacco and enjoying dance shows.[132] In 1866, for example, a gambling hall in Port Said "run by Greeks" appeared to be mainly frequented by French individuals. In 1867 a supervisor of the Company works called Guibert was found in a brothel "with Greeks and French who were all drunk" and unlawfully without the prescribed lanterns on them. In 1869 one of the gambling halls in Suez was run in partnership by "Italians and Greeks." Although it employed many Italian artists passing through the port, the bulk of its clientele "was not Italian for the most part."[133] In 1878 a Jew called Moussa Grinsley and a French woman called Fanny Barthélemy waited tables side by side in one of Port Said's beer shops. A ruckus that broke out in 1879 in a wine shop in that town shows that French and Maltese subjects, among others, frequented the same establishments. A Maltese had apparently taken offense because a Frenchman, a painter, had failed to offer him some of the shellfish he was savoring.[134] Another row that occurred that same year reveals the simultaneous presence of French and Italians in a café in Port Said's Place de Lesseps. The fight involved a French subject who had slapped a Piedmontese man, who reacted by stabbing the Frenchman twice and then, with "unmatched cynicism," went to his work at a bakery and began kneading bread, paying no attention to his bloodied shirt. Cafés in the "Arab" part of town also set the stage for the partaking of pleasures among individuals who were otherwise meant to keep their distance. In 1885, for instance, constable Jean Cucco and a "native" man he encountered in Port Said's Arab village (where the former was seeking an apartment to rent) sat together to sip a cup of coffee and smoke what appeared to be simple cigarettes but would later put Cucco in a state of head-spinning giddiness.[135]

Port Said's entertainment venues were locales where group boundaries loosened up. At the same time, they remained urban spots where different kinds of belonging often dramatically played out. They were enclosed and yet open spaces, or what has been termed "transparent spaces of performative socialization," wherein patrons may have always been aware of who might be looking in from the street.[136] After all, dozens of chairs spilled out of the cafés and restaurants onto the roadside. Similarly in Suez, cafés and restaurants tended to expand into adjacent squares.[137] An example from 1867 demonstrates the haziness of the national identities proclaimed in and around drinking venues. One September day, the Italian subject Mr. Rossi, *maître* of the Café de Charenton in Port Said's De Lesseps square, led an expedition of Greeks, Italians, and perhaps Maltese to attack one Mr. Ulacacci, French subject and Corsica native. The reason for the attack, apparently, was a slight committed by Ulacacci, who had taken away Charenton's Austrian waitress Antonia. Actually, she had left voluntarily. Rossi claimed her back by virtue of a contract that consular authorities did not recognize, since they considered his establishment to be "nothing but a place of debauchery." Ulacacci declared to a friend that he was in love with Antonia, adding that he was even amenable to repaying the remarkable debt of 250 francs that she allegedly owed her boss. The lovers fled and took refuge at the domicile of a Corfu (Kerkyra)-native "Greek" commercial agent who had been in Egypt since 1864, on whose balcony they were surprised while drinking. Antonia was taken back by force, unconscious, to the Café de Charenton, "where a firearm blow was fired in sign of triumph with cries of 'Viva l'Italia' (*Long Live Italy*)." Yet the café's owner was Italian, the kidnapped waitress was Austrian, and the alleged villain was French. Assuming that Rossi's thugs were also his customers, they also claimed various nationalities. Finally, the Austrian-French couple sought help at the domicile of a "Greek" friend. Indeed, as the French consular representative declared, "provocations of this kind, given the over-excitement of the spirits, could bring serious conflicts."[138] But what were these provocations fueled by, alcohol aside? "National" identities clearly did not, at this stage, determine either clientele or love affairs. But when real or perceived violence broke out, Port Said's residents may have felt forced to take sides and acquire a sense of belonging to different social, religious, and ethnic groups.[139] Entertainment venues could thus turn into useful information conduits or rallying grounds for groups of residents identifying themselves with a certain nationality. But it would be far-fetched to picture such places in the port city as hotbeds for identity claims, at least during its first decades of life.[140]

Flows of Alcohol and Boisterous Customers

Drinking in late nineteenth-century Port Said appealed to a wide range of people, and the effects of their alcohol consumption ran the gamut. There could simply be occasional abuse, as in the case of master of vessel Samuel Shean, drunk and incapable of performing his duties from his vessel's departure from Aden to its arrival in Port Said one day in 1877. Reportedly Shean was not a chronic drinker but had consumed a large quantity of spirits in a very short time. He was found lying face down on the floor.[141] A Paris-born medical aide called Remy not only owed money for board and lodging but was also steadily accumulating further debts with his drinking habit. For example, on one single day in February 1866, he drank beer, two "raquis," and three glasses of cognac while playing pool. The following day, he consumed two "raquis." The day after, he drank one absinthe. On March 10, he drank three "raquis," and on March 13, two more. On March 14 he drank both absinthe and "raqui." Remy soon died destitute, at age forty-one, from chronic bronchitis. Impending death struck the son of Abou Saka, a Jewish tinsmith and French subject, who was found "completely drunk" and lying on the ground in a pool of blood in a Port Said street in 1880. He had stabbed himself with a knife because of family sorrows, as two coreligionists declared. A "bad subject," he was deemed "capable to commit all kinds of ill-doings," including unsuccessfully trying to commit suicide.[142] A third example of alcohol intake with bad consequences is that of Ahmet Effendi Khalil, the engineer of the *Tanzim* in Port Said, who was accused of indolence on the job because of his excessive consumption of drinks and said to be affected by delirium tremens. The same fate befell a woman who had abandoned her children in Port Said and was implicitly held responsible for their debauchery, lack of education, and poor health. The seemingly precise medical description delirium tremens really included "every variety of alcoholic poisoning, from the excited, fidgety, and prostrate condition popularly known as 'the horrors,' up to the very worst type [...] often ushered in by severe and repeated epileptic attacks."[143] All in all, perceptions of the effects of alcohol could be highly subjective. In regard to one man, for example, some declared he appeared "tipsy," while others swore he was "very drunk."[144]

Among Port Said's inhabitants, "Greeks" were said to be particularly prone to hit the bottle, especially on Sundays and during other festivities.[145] But the Egyptian governor inveighed especially against transiting outsiders and accused them of excessive drinking, committing "brutal acts" in cafés and drinking establishments, and provoking rows the minute they stepped off their vessels.[146] In particular, the British soldiers, sailors, and ship passengers passing through its port came to be closely associated with open and uncontrollable drunkenness. The English in Egypt, as in coeval Anglo-Indian society, had a long-standing reputation as heavy drinkers.[147] The local press stereotypically pictured northern Europeans as more prone to consume alcohol than "southern peoples."[148] After the British occupation in 1882, not only were British army garrisons stationed in the canal area, but many European troops passed through on their way to their colonial stations.[149] Besides the British troops shuttling to and from India, great numbers of French soldiers also regularly touched on Port Said on their way to Tonkin or Madagascar.[150] Moreover, thousands of men and women transited through town in their emigration toward the Australian region and beyond.[151] As early as 1883 the town seemingly began to swarm with troublemaking English seamen and unruly migrants. Drunken sailors and stokers, "saying nothing of Australians," were said to be constantly on shore. British mariners deserting their ships also found refuge in town and overwhelmed local prisons.[152] The commanding officer of the Marines claimed that the principal duty they had to perform as "police" was actually to keep those English sailors and visitors in order, because almost all the drunkenness and trouble in Port Said came from them.[153] Some venues, such as one called the Union Jack and another named Australian Bar (actually a "Greek cabaret"), specifically catered to this crowd.[154]

Drinking heightened the intensity of interactions among differently mobile social groups in town and frequently triggered violent brawls. The motives behind these squabbles ranged from the global to the trivial: from disagreement on international events, such as the outcome of the Russo-Japanese War in 1905, to pangs of jealousy over a local nanny's favors.[155] Drunkenness was identified as "always a starting point and sometimes the unique cause of such fights."[156] Already in 1869, Port Said's numerous cafés had been identified as "the theater of bloody fights" that needed to be shut down.[157] A traveler's account related that by the early 1890s bar brawls had become such a recurrent feature of Port Said's drinking landscape that customers hardly turned their heads when "a visitor to one of these cafes shot a fellow visitor with a revolver in a drunken brawl." These violent outbursts "seemed to create very little excitement in the place, as if the incident was an everyday occurrence to which the people were accustomed."[158] Seamen played a leading role in bar rows.[159] In vain did the governor launch appeals to prevent antagonism among mariners and soldiers of different nationalities. Upon the stationing of several dreadnoughts in port at the same time, consuls were requested to inform the local authority of the number of men who would land. This would have enabled the preparation of differentiated schedules for their landing and of police measures to prevent squabbles.[160] In 1885, for example, two "Germans," a sailor and an upholsterer, found themselves being chased and beaten up in Port Said's harbor by an unidentified group of six to ten navy sailors (*marins de guerre*) screaming "Prussiens!" and unsheathing their knives. The intervention of two guards of the Egyptian police saved the necks of the Germans, who eventually mourned only the loss of a silk tie and a hat. On another occasion, British and Russian sailors clashed in a brawl, and many were wounded.[161]

Seamen aside, anyone could unexpectedly turn boisterous in taverns. As recorded in coeval Beirut, even amicable drinking bouts could take a turn for the worse in the wake of criticism or insinuations. Drinking companions might end up quarreling and eventually harming each other.[162] In 1875, for example, a Maltese bumboatman called Farrugia and another Maltese named Mifsud entered Joseph Portelli's drinking shop in the company of two or three Greek men. Mifsud was already a little inebriated, suggesting that he had visited one or more drinking establishments prior to landing at Portelli's. He had promised his fellow countryman he would pay for the drinks of both. But the mood changed quickly. Either Mifsud provoked Farrugia by saying that he was "a pimp" for his wife and that "she was a whore," or Mifsud took Farrugia by the neck and they both fell.[163] As highlighted by Mario Ruiz in regard to contemporaneous Alexandria and Cairo, drinking venues created an atmosphere that blended male companionship and alcohol. There, some drinkers inadvertently crossed the line between the controlled play of solidarity among friends and more serious confrontations.[164]

From drinking establishments, rowdy drunkenness easily spilled over into other parts of town. The Egyptian governor railed against drunkards and their aggressiveness and decried the fact that they troubled public tranquility in the whole port city. Their brutal conduct, he claimed, harmed not just the owners or the clients of these establishments, but private citizens as well. The drunks' rowdiness, in fact, took place in the streets or in public squares and even at the doors of residents' domiciles.[165] First, alcoholic beverages were also available in places other than drinking shops proper. Enterprising sellers, for example, approached steamboats anchored in the harbor of Port Said to peddle liquor before passengers had even set foot on shore (see chapter three). European wines and liquors were also available in Port Said's grocery stores.[166] Second, individuals imbibed alcohol in one drinking establishment after the other. One morning in 1879, for instance, two Maltese who were eventually arrested for "making a disturbance" arrived at the shop of Felix Agius and demanded a glass of brandy and a glass of water. They were already drunk. Hence, they had clearly indulged in some barhopping earlier on.[167] Finally, drunkards moved about the urban space at large. On a day in

1885, for example, a French man named Victor Noyer who was in a state of drunkenness at first caused trouble in a restaurant on Rue Nouvelle Nord at 7:00 p.m. Detained and immediately set free by his consular authority, he was arrested again a few hours later at a brothel where he was stirring up a scandal by refusing to pay for his "consumption." Another example in this regard is provided by the abovementioned constable Cucco, caught one afternoon threatening the prostitutes of a brothel in the Arab village with his unsheathed bayonet, in a complete state of drunkenness.[168] The effects of drunkenness could reverberate in both place and time; the inebriation of a painter named Vaticiotis, for example, would haunt him after he had sobered up. One night Vaticiotis was "drinking with some friends" in the "drinking shop" of Vacilis Scopilitis when he suddenly threw a glass on the ground and broke it. A man passing by claimed that glass shards had struck him and threatened Vaticiotis with a knife. However, he waited until the next morning to take his revenge. The offended man then ran up behind Vaticiotis while the latter was sitting in a coffeehouse.[169]

The Plagues of Port Said

Quality and variety in Port Said's drinking scene notwithstanding, authorities in town did not appreciate these establishments' business. In the eyes of many contemporary and later observers, the town's inns and wine shops had become gathering spots for malefactors and especially foreigners. Some visitors expressed the conviction that local authorities had the power to close a bar down if a bloody altercation erupted there, thus interrupting sometimes successful businesses and causing problems for their managers.[170] However, foreign owners, managers, and patrons technically remained shielded behind the country's extant capitulatory regime, which some consequently saw as an obstacle to its "moral and physical purification."[171] Consular representatives loathed Port Said's drinking places as "the plagues of Port Said." Similarly, Catholic clergymen saw the cafés, music halls, and bars in town as fundamentally immoral.[172] The Egyptian government decreed in 1891 that such venues ought to be located only in commercial neighborhoods, away from families' residences, religious or educational institutions, and cemeteries or venerated tombs.[173] As in the 1860s, authorities persisted in their attempts to regulate the opening and closing times of places of diversion in town (see chapter two). When the Egyptian government had decreed in 1867 that these establishments in Ismailia would lock their doors at 10:00 or 11:00 p.m., it had raised Company officials' hopes that a similar measure would be taken in Port Said.[174] But come 1879, public establishments and brothels in the port city were routinely found open and operating past midnight. The Egyptian subgovernor general blamed them for compromising the safety and the repose of the city's inhabitants. He argued that such "ignoble establishments and houses" attracted bad subjects who fomented disorder and created the most regrettable scenes, in the wee hours of the night. Ostensibly, after a period of respite due to injunctions from their own consular authorities, the foreign owners of such venues had started contravening police regulations again. Each consulate, the subgovernor general pleaded, ought to place one janissary at the disposal of the police to help them close down these places at midnight sharp, have all kinds of music stopped, and make all customers leave.[175]

In 1883, attempting a crackdown, the governor decreed that cafés would close down at 10:00 p.m. Only a few could stay open until 12:30 a.m., and then exclusively under the stipulation that social peace be assured (the Eldorado, Casino, Friedmann, Palatine, Café de France [Hôtel], Bletty, Glacier, Neumann, and Fink). If foreign owners wanted to keep venues open part of the night or all night long for public festivities or other reasons, they needed to request from the governorate a special authorization via their consular authority. Consular representatives all agreed to this measure. Nonetheless, in 1884 the closing time was once again modified and was pushed to 11:00 pm rather than 10:00 p.m. For those owners who wanted to keep their establishments open past the fixed hours, the possibility to obtain special authorizations was still contemplated.[176] Accordingly, all the consuls in the city were notified, signs were hung on the city's walls, and the relevant decree was to be published in Arabic in the governmental official publication *Al-Waqa'i' Al-misriyya*. Such measures, however, did not quell the governor's anxieties. He questioned the Ministry of the Interior about the possibility of refusing permission to open a coffeeshop or a brewery to individuals of ill-repute. He also asked whether he could shut down those establishments about which neighbors had repeatedly and justifiably complained. The ministry, however, denied the governor's request: no law regulated the moral character of owners or the ability of authorities to close down drinking establishments. The only type of repression that could be exercised was, in fact, the hitherto unsuccessful surveillance of opening and closing times.[177] In 1891 the closing time was further extended: it would be midnight from mid-October to mid-April and 1:00 a.m. the rest of the year. Establishments could open their doors no earlier than 6:00 a.m. from mid-October to mid-April and 5:00 a.m. from mid-April to mid-October. Theoretically, the police were to enforce these schedules.[178]

Night owls thus won the battle for Port Said. As had repeatedly happened before, public officials could not agree on how to curtail the extent and the range of leisure opportunities in the port town or to stem the inebriation that suffused it. According to the British consul, the sailors and passengers among his subjects could not be held accountable for the bar squabbles they initiated. He claimed they were first lured into drinking venues, then forcibly inebriated, and finally paid exorbitant prices for their drinks. Moreover, he declared that whenever the Egyptian police intervened to stop brawls, the arrested individuals (especially if drunken British sailors) were ill-treated, beaten up while in prison, and stripped of their belongings. The Egyptian governor, for his part, energetically rejected these accusations and claimed that local police agents had always exercised great moderation toward British seamen and passengers. He stressed that police agents, despite being molested and having their uniforms torn, took pains to conduct the captives to their consular authorities or back on board their boats, while also preventing them from mistreating the peaceful inhabitants of the city. The governor also argued that it was not a police responsibility to restrain seamen and passengers from visiting drinking venues. Rather, it was the duty of consuls and boats' commandants.[179] Alas, no one apparently succeeded in keeping the peace in a town that both suffered from and thrived on its exclusive entertainment offer. Drunkenness and catastrophic brawls in Port Said's drinking venues continued to shape its reputation for rowdy revelry. They had aftereffects that spread elsewhere and lasted longer than the drinking frolics, damaging the city's name and creating a shared memory of disorderliness.

WHAT ARE YOU DOING HERE? WOMEN AND ENTERTAINMENT

By gambling, drinking, using coarse language, and getting into fights, men in Port Said set drinking entertainment venues apart from other hangouts. From there, they moved on to "exhibitions of masculine prowess."[180] The Ottoman coffeehouse has been defined as "a world strictly of men," a venue from which women were excluded, and "an exclusive "bastion(s) of homosociality for men."[181] As noted by Ruiz, the coffeehouse in Egyptian urban milieus may indeed have facilitated a range of masculine performances, as it was a "sanctioned non-familial site" that drew in a regular stream of male customers, including the neighborhood *shaykh*, the night watchman, or the local policeman.[182] Several authors espouse the idea that the coffeehouse came to embody a substitute for the *selamlık*, the section of the house reserved for men to greet and socialize with guests and that only the wealthiest could afford. Everybody else's gathering place had to be somewhere else.[183]

The women who worked in entertainment or enjoyed the same kind of leisure were treated as incidental or shunned altogether. It was not by chance that the abovementioned waitress Antonia (the one who run away with her paramour, was forcibly brought back to her workplace, and was reminded of the money and labor she presumably still owed to her boss) had donned men's clothes when she fled. However, neither that disguise nor her fleeing companion's declarations of love enabled her to change her path.[184] When an "Arab" woman peddling grapes came into the café of Hajj Anaissa and asked to drink some water, some soldiers had apparently asked her, "What are you doing here?" They commanded her to vacate the premises. One of the male patrons, however, intervened to defend her and got himself into a brawl with those soldiers.[185]

A case heard by the French consular court in 1886 shows how Port Said's cafés were sites where boundaries of nationally defined groups and gender relations were actively patrolled through exclusion, the use or misuse of symbols such as national flags, and shared notions of male and female respectability. The Algerian Hajj Hamza Ben Saleh, residing in Port Said since 1875, had always been in the coffeemaking business. He lamented that a less-experienced man named Hajj Ali, who had arrived in town just two years earlier, had established a café in front of his own. Both were located in the area known as the "Arab village." The longtime established coffeemaker, presumably perceiving a threat to his business, accused the latter of a number of misdeeds. First, he insinuated that Ali professed to be alternatively Algerian and Tunisian while really being Moroccan. Second, he reported that Ali's wife used to patronize her husband's establishment, hang out among the male customers, and laugh with them in front of her husband. This had ostensibly turned his competitor's café into a brothel, which Algerians as a whole, Hamza reasoned, did not find acceptable. Third, he accused Ali of hanging two French flags over his coffeehouse that "Arab" soldiers and all other patrons regularly mocked. Eventually, Hamza threatened to kill Ali and sealed his threat by proclaiming: "We Algerians are wholly pure men."

According to Hamza, Ali's willingness to become the *macro* (shortened form of *maquereau*, French for "procurer") of local soldiers would cause everybody to say that all "Algerians and French protégés are *macros* and this is a dishonor for France."[186] On the whole, Hamza's deposition presupposed the existence of clear group boundaries and the ability to determine who belonged and who did not. Moreover, it also expected members of certain groups to patronize certain establishments and not others. Further, by equating the accused's wife to a prostitute, even if it was an empty slander, the accuser articulated the view that respectable women either could not legitimately spend time in a café or had to adopt a certain mien when doing so. Finally, he sanctioned the role of "wholly pure men" to act as the café's gatekeepers and hold the monopoly on violence. Apparently, it was up to them to determine who could access those spaces and what behaviors were acceptable inside of them.

Emphasizing the role of coffeehouses and other public establishments in male sociability should not obscure the presence of women under manifold guises. While establishments where concerts were held, such as casinos, may have been unappealing to some foreign female passengers transiting via Port Said, some coffeehouses doubled as venues for the entertainment of women and children on special occasions.[187] As for their managers and owners, several bars in contemporary Cairo were run by women. Austrian subjects featured most prominently, but Egyptian women, both Christian and Muslim, also served as owners.[188] In Port Said, as already noted, female peddlers stopped by drinking establishments. Women also worked as waitresses. Others played music for the public, such as the girls hailing from a cluster of villages between Prague and Teplice, near the northwest border of Bohemia, who had traveled under the charge of a matron and performed in Port Said's cafés in 1881. Female "Bohemian" artists frequently performed in casinos at night as members of bands. Women from Austria and Romanian Wallachia played waltzes in music halls (*musicos*), advertised from afar by their vividly illuminated windowpanes. Women from Abyssinia and Upper Egypt, together with "little Arab girls," also performed as dancers in Port Said's establishments. Western men, by their own account, dallied with their "frenzied and lascivious moves."[189]

India-born British writer Rudyard Kipling, in his romanticized description of late nineteenth-century Port Said, portrayed a Madame Binat inside her drinking shop, the source of her income. While she played the piano in her "faded mauve silk always about to slide from her yellow shoulders," he wrote, "naked Zanzibari girls danced furiously by the light of kerosene lamps."[190] Young and attractive girls were thus set in contrast with the older women running businesses, portrayed as scrawny, ugly, and stingy, despairingly staring at pans and burying their noses in tobacco, or otherwise as coarse, bold, and embellished harpies with mercenary souls and plump red hands.[191] Still, especially at night, calculating older females and young seducers seemingly joined forces. They all bore responsibility for shaping Port Said's shady physiognomy as a cutthroat area along the Suez waterway, wherein women and gambling-table attendants alike lay in ambush for those who had made a fortune elsewhere.[192]

Some working women made wise use of the disreputable fame they endured in Port Said and repeatedly challenged the status quo. In the face of police intimidation, fines, imprisonment, and neighbors' censorship, they defiantly wielded their reputation against their opponents. Conversely, the inflammatory words exchanged in these establishments and brought to consular courts illustrate how domestic conceptions of respect, sexual virtue, and reputation were reinforced by legal institutions that validated their communal importance.[193] Madame Maillard, for example, was a French woman licensed to sell liquors in Port Said. She managed a refreshment stall in Rue de la Division. She stood accused of "many reprehensible acts." Maillard and her daughter had ostensibly always been "reluctant and rebellious" in the face of police orders. In July 1882 their bar was found open past midnight. When two patrolling guards ordered Maillard to shut it down, she reportedly replied in an arrogant way that she absolutely refused to comply and added more of what the authorities termed as "nonsense." Her daughter then showed up, claiming that the guards were acting that way simply because the two women were handing them no *bachish* (*baqshish,* Arabic for "tip"). Finally, when the guards told the Maillard ladies to throw the three clients out of their *buvette,* the mother "straight out said that these three men will not get out as they were sleeping with her." This was the last straw, nothing less than a "public affront" that threatened both the prestige of the police institution and the public order. Eventually, the consular tribunal charged Madame Maillard a fine of 16 francs and costs for insults against an officer exercising his functions. In 1885 her establishment was once again found open and operating beyond the time prescribed by regulations. This cost her another fine and further costs, as well as a weeklong imprisonment.[194] The testimony of Ahmed Meaouad, a thirteen-year-old "Barbarin" in their hire, suggests just how damaged the reputation of the Maillard ladies and their venue had become. Not only did he accuse mother and daughter of having beaten him up, but he also related that one of the patrons urinated on him while threatening him with a dagger. Finally, he testified that after closing the establishment shutters, the women received regulars by letting them in by a secondary door. They then ostensibly kept on with the debauchery all night long and engaged in "acts against morality and propriety." It is difficult to fathom whether Meaouad was telling the truth or slandering his employers for some real or perceived slight or hoping for gain as a police-paid snitch. Perhaps in a coordinated action of collective castigation, the police agents, the owner of the building, and the neighbors validated his grievances. They all demanded that whatever situation the Maillard ladies had created be stopped, in the interest of their collectively perceived "public morals and order."[195]

Contemporaries often saw female employment in the entertainment industry as coterminous with loose morals and even prostitution. Accounts of the time salaciously noted that in the taverns all along the canal in the making, "pretty Wallachians, Italians, Albanians waiting tables in their national costumes" charmed patrons' eyes and stirred their appetite. Their attributes figured on a par with the wine, the steaks, and the half-price roast beef displayed in the refreshment stalls.[196] Women themselves sometimes contributed to creating these impressions. Maria Ursini, for example, managing the establishment Bella Rosa in Suez in the late 1890s, conflated breweries and brothels as well as service workers and prostitutes when she advocated with the Italian consul in town in favor of an employee of hers who needed hospitalization. The latter was an Austrian woman working in Ursini's "big brewery" who had contracted syphilis through the "contacts with men" that Ursini herself admittedly organized.[197]

Egyptian as well as foreign consular authorities outright accused all females who were employed in bars and breweries of moonlighting as "clandestine" prostitutes. The Italian vice-consul in Port Said explicitly voiced his suspicion that breweries in town generally functioned as fronts for brothels. Accordingly, the girls who worked as servants or baristas (*filles de comptoir*) in taverns endured being perceived as practitioners of prostitution, especially if they were denounced as sick with venereal diseases. So did the women who were younger than thirty years of age and toiled as servants in brothels.[198] The press added fuel to the fire. In 1891, for example, a correspondent from Tanta for the newspaper *Al-Mu'ayyad* claimed that the category of public women included not just those who had been registered as such, but also those who practiced commercial sex temporarily or secretly as well as those toiling as dancers, singers in café concerts, and brewers. All were lumped into one undifferentiated group of deplorables. No alleged "loose woman," the paper claimed, was to be spared.[199] In 1904 the Cairo-based Italian newspaper *L'Imparziale* insinuated that operetta singers and performers, given how little they earned for their ten- to twelve-hour-long days of work, made the extra they needed by prostituting themselves.[200]

Women on and off Prostitution: Poverty or Bad Morals?

By the mid-1880s prostitution was reportedly a growing industry in Port Said, spreading into its "bourgeois" neighborhoods and scandalizing the city's proper "ladies."[201] Western visitors observed that the city, alongside its cafés, music offering, and shopping options "in the imitation of Paris," vaunted "special houses" where visitors could satisfy their most varied tastes. But satisfaction came at a price.[202] Most contemporaries identified prostitutes as the sole source for the diffusion of venereal diseases, even though statistics offered little support for singling out women as the principal agents of infection.[203] Especially in a busy port town such as Port Said, it was impossible to ascertain the origination point of the sexually transmitted illnesses spreading through town. Thirteen cases of syphilis, for example, were reported in 1883 by a Port Said–based medical officer on board a French transport passing through its harbor.[204]

By 1883 there were forty-five registered "public women" in Port Said. Only a few showed up for the hospital visit mandated by the regulation that the Egyptian government drafted in 1880.[205] French and Italian women appeared in strong numbers among prostitutes. There were also Greeks and Austro-Hungarians. Jewish women also prostituted themselves in Port Said. Greek Orthodox (Rumi), Syrian, "Turkish," Ethiopian, and Egyptian women hailing from Damietta and Cairo were also present in the business.[206] In 1883 the governor decreed that in Port Said as in other locations throughout Egypt, women could not practice in the streets, nor could they attract passersby from thresholds and windows, which suggests that this was their common practice. If they persisted, they could be arrested.[207]

Brothels were apparently scattered throughout different neighborhoods across Port Said and especially in its commercial areas. Through the 1880s and beyond, they continued to stir up complaints by those merchants and inhabitants who were "shocked by the constant immoral contact." The foreign consuls bemoaned the fact that the town's establishments for commercial sex constituted a veritable "attack against public morality."²⁰⁸ At the onset of the 1890s, Port Said's houses of prostitution were still disseminated in all quarters of the city, and "tavern-brothels" (*birrerie-postriboli*) had "invaded" the city center.²⁰⁹ "Shameless vestals" had taken up room along a street known as "Rue Babel," where they seemingly offered a constant and most indecent spectacle, pimps thrived, and scuffles were a regular occurrence. Perceptions of diffused chaos and cacophony most likely warranted this thoroughfare's biblical nickname, which had become an established Orientalist platitude across the nineteenth-century Mediterranean.²¹⁰ In a few years' time, Port Said would boast at least thirty houses for female prostitution, ten brothels for "boy dance" or male prostitution, and more venues for "living picture" shows.²¹¹

Egypt lay astride local, national, regional, and global circuits of recruitment for sex work. In the nineteenth century, the country had become well-known for offering this kind of female employment. It was one of a few Mediterranean destinations for prostitution, such as Crete and Malta, that housed military garrisons.²¹² With the Crimean War beginning in 1854, Russian Jewish former prisoners of war boosted Istanbul's prostitution market, as human traffickers and pimps were joined by more Jewish migrants from Russia, Austria, and Romania.²¹³ Around 1856 some of the mostly Wallachian women who had moved to Istanbul to serve the troops headed for the conflict in Crimea relocated to Egypt after the soldiers had moved on.²¹⁴ Egypt functioned as both a final stop and a stopover in women's relocation to ports farther away, from Istanbul to Buenos Aires, Bombay, or other points in the Far East.²¹⁵ The canal's official opening in 1869 in particular boosted the "peripatetic lifestyle" of those European pimps and prostitutes who had set their eyes on the Asian market. It also turned Port Said into a new and promising fairground for prostitution.²¹⁶

Like contemporary port cities elsewhere, the canal's northern port was traversed by an apparently endless stream of potential clients.²¹⁷ Once again, it functioned as both a transit point and an intended or accidental terminus for those willing or forced to cater to those customers. Some of the women destined for the Jaffa scene, for instance, stopped in Port Said.²¹⁸ Others, allegedly Polish Jews for the most part (even though Jews may have been overrepresented in such reports), used to travel from Naples and Trieste on express routes and recruit "poor young girls" whom they brought specifically to Port Said "to place them on the way to perdition."²¹⁹ While a few females were being outright "imported," others were arriving in Egypt to seek service on their own and were recruited on board by traffickers' agents on the same vessels.[220] Often they were promised employment as waitresses or stage performers.[221] Agents or even relatives, for instance, lured sixteen- and seventeen-year-old girls from Vienna as well as mostly minor Greek girls to Egypt with promises of jobs as servants. Others were decoyed by the promise of a marriage match, and local intermediaries disguised as the groom's relatives welcomed them upon arrival.[222] Finally, Port Said's prostitution industry benefited from times of war, when homeless refugees poured into the city, such as the Turkish and Greek girls fleeing the "Turkish war" of 1911–1912. They settled in town and provided their labor there because Cairo and Alexandria were already overflowing with "Arab, Syrian, and white" female prostitutes.[223]

Migration for entertainment work and commercial sex in particular highlights how Port Said had morphed into a crossroads of migrants' local, national, regional, and global trajectories. It discloses that Port Said had been forming connections to the rest of the isthmus, other Egyptian centers, and places on other Mediterranean shores such as Marseille and Naples. It also shows that, especially for women, the isthmus had itself become a circuit for local or temporary mobility. Between 1875 and 1878, for instance, a ten-year-old girl called Rosina had moved from Naples to Suez, from Suez to Port Said, from Port Said to Alexandria, and then back to Suez. In fact, she had been initially working in a brothel in Suez and residing there with a woman from Naples who declared she was her aunt. Questioned by the Italian consul in town, the woman promised she would place the girl with the nuns in Alexandria. Instead, they both relocated to Port Said in 1877, where the older woman opened a brewery and once again reportedly offered Rosina to clients for "dishonest ends." They then did spend some time in Alexandria, where they were both sighted in March 1878. But by August that year, it was rumored that the zealous "aunt" had returned to Naples, while Rosina had been sold to a brothel in Suez.[224] Naples appears to have been a fundamental link in the chain that bound Egypt to the prostitution circuits of the time, as illustrated by scholars Francesca Biancani and Laura Schettini.[225]

Networks of fictitious kin and acquaintances, streams of forged documents or the lack thereof, and a lackadaisical local police force converged to make Port Said an ideal destination for the enterprising procurers who were circulating between Istanbul, Odessa, Colombo, and Shanghai, and touching on several continents. In 1905, for example, a sixteen-year-old Romanian girl had first landed in Egypt from Istanbul. She had no papers on her and was in the company of two young men, a sailor who claimed he would marry her and an American known to the Alexandria police as concerned in the trafficking in women. The latter declared he was her uncle. Authorities found the odd threesome suspicious and sent the girl back to the Ottoman capital. However, only six weeks later she reappeared in Port Said as a bride to her "uncle." The marriage certificate was apparently false but still served the purpose of misleading the authorities in town. A sixteen-year-old Japanese girl reached Port Said from Shanghai. She had been brought to the Egyptian port town by a Russian woman allegedly "under the pretense of being [a] servant." The Russian reportedly ran a brothel in Shanghai and had unsuccessfully tried to place the girl at Colombo.[226]

Some of the women who landed in Port Said in hopes of making a living were able to negotiate the terms of their employment, either becoming brothel keepers themselves or disengaging entirely.[227] In 1905, for example, two women, aged twenty-eight and twenty-four, who had known each other from childhood, had landed in Port Said from Odessa based on the prospect of earning a living in the port city. An acquaintance had promised to introduce them to "A. A." without specifying in what sort of trade the latter was engaged. Upon their arrival, A. A. could not accommodate them: "He had plenty of girls who had long been with him and whom he could not dismiss to make room for them." They were then placed in a "drinking bar which was also a bad house" kept by an Italian woman, who wished them to be prostitutes because "that was what they had been brought to her for." The two girls refused her proposition and agreed instead to only serve in the bar. They especially wanted to wait on the Russians passing through Port Said. Unhappy with the improvised arrangement, they ended up staying in town for only four weeks, leaving the house and appealing to their consul, who sent them to Alexandria and later back to Odessa.[228]

Professional options for women differed from one locale to another within Egypt, which partly illuminates why some female migrants decided to relocate either within the isthmus itself or elsewhere in the country. In 1900 the French woman Marie Goujat, for example, traveled from Cairo to Port Said to check out job prospects there, realized that the hotel's keeper was attempting to lure her into prostitution, and decided to head back to Cairo to find herself a job as a brewer instead.[229] Migrant women in Port Said had few choices other than toiling as waitresses, prostitutes, laundry-women, ironing ladies, or servants in families. In contrast, foreign female migrants in Alexandria at the end of the 1880s and onward could easily find work as cooks, waitresses, or wet nurses (*nutrici*) and were well paid. By the late 1880s, "small offices" existed in Alexandria, wherein immigrants tried to find placement as domestics. No such bureaus yet existed at the time in Port Said or even Cairo itself, where migrants still preferred to rely on friends and acquaintances or ask their consuls for assistance. Compared to migrant women in Port Said, those in Cairo had relatively restricted prospects in laundry, ironing, and service but at least could count on the fact that such opportunities appeared to be more readily available outside the isthmus.[230]

By the end of the 1890s recruitment and placement agencies seem to have been in place within Egypt and to have contributed to organizing the employment of female and male foreign labor. In 1897, for instance, an agency in Cairo's Telephone Street for the employment of and information on "wet nurses, nannies, female cooks, waitresses, male cooks, coachmen, and waiters" publicized its services in newspapers.[231] That same year, another "renowned" hiring hall called Saverio Graziano, located in Cairo's Clot-Bey, was also in operation: a "young bride" of twenty-one years of age who had arrived in Cairo from Italy, for example, relied on it to seek employment as a nanny or wet nurse (*balia*) with a family in Egypt.[232] Agencies constructed appealing representations for both employer and employee. In wet nurses' specific case, they guaranteed to employing families that the perspective employee had undergone medical check-ups and was deemed healthy.[233] We do not know whether the abovementioned "young bride" had landed in Egypt by herself or with her newly acquired husband, but the ad she placed in the Cairo-produced Italian newspaper *L'Imparziale* informs us that she was independently pursuing employment through a formal channel. Preexisting informal networks also continued to function. In fact, they allowed Kariclia, an old woman originally from Crete and residing in Alexandria, to procure domestic laborers as well as wives for those who trusted her skill and discretion. Kariclia was apparently aware that female servants were vulnerable to sexual abuse but nonetheless exploited this vulnerability for her matching needs.[234] By 1901 there were in Egypt 32,663 Egyptian women working as domestic servants. Among the foreigners, Greek women largely prevailed (1,134), followed by similarly placed Austrian (604) and Italian women (508). Far less numerous were the French (198) and British women (135) officially figuring in this sector.[235]

In the last decades of the nineteenth century in Port Said, women took advantage of employment opportunities in its entertainment industry but paid an often hefty price for their participation. To begin with, prejudice against foreign women willing to migrate was widespread, as if they were all itching to prostitute themselves. Prescribed gender roles affected women's migration choices in their countries of origin and shaped immigration policies in receiving countries, barring the way to those seen as potentially "loose." Weaker forms of control on Mediterranean routes may explain why some unaccompanied female migrants chose to relocate to Egypt rather than across the Atlantic.[236] Once women began their drudgery at the destination, they bore the brunt of gendered and class-specific notions of "respectability." The wage earning of working women threatened the association of femininity with domesticity and the equation of masculinity with bread winning. With women at work, men's standing as "respectable" members of the working class was in question.[237] Men thus dominated as judges of women's and men's respectability and gatekeepers of their wholly pure manhood, to cite the abovementioned coffeehouse keeper in Port Said, Hajj Hamza. Women's occupations, as well as their family arrangements and sexual conduct, alleged or otherwise, all came under fire.[238]

It remains difficult and perhaps futile to determine which women worked in the sex trade. Some, in fact, "could have participated in prostitution and intimate sexual relations for pleasure or out of coercion at different points in their lives, indeed even at the same time."[239] Notwithstanding this inability to precisely identify prostitutes, policemen and journalists as well as consular officials saw all migrant working women in Port Said as potential prostitutes. By conflating certain types of employment, especially in entertainment, with prostitution, they connected labor migration with sexual license. What came first? "The ambivalent causality (poverty or bad morals?)," as Joan Scott writes, "was less important than the association itself because there was only one cure for sexual license and that was control."[240]

GONE WERE THE NIGHTS OF THE CAFÉ CONCERT

Nighttime navigation was introduced in the Suez Canal in the second half of the 1880s. Some vessels were already fitted up with electric lights in 1885, which would become compulsory by the turn of the century.[241] After that, passing ships no longer had to moor in Port Said for the night. It has been argued that as passengers were no longer forced to spend the night on shore, drinking or otherwise, the nocturnal transit of the canal harmed the trade of the town's shopkeepers. Because of electric light, one observer in 1898 harped on, gone were "the days—or nights—of café concert, and roulette-table, and dancing-saloon, bands crashing from dusk to dawn, and gold flowing in torrents into the lap of Port Said."[242] But despite the rearrangements introduced in the canal's transit schedule, changes in Port Said's fortunes and entertainment industry in particular did not occur overnight. Halts and interruptions continued to slow down the now supposedly unhindered flow of people and goods. In 1900, for example, the Port Said Lighthouse needed very urgent repairs and modifications.[243] Coaling, notwithstanding the introduction of nighttime navigation, still forced ships to stop in Port Said's harbor for as long as twenty-seven hours.[244] And even when the stopover only lasted three or four hours, some passengers still found the time to disembark to attend shows or enjoy a concert.[245] In 1901, even when voyagers disembarked from their vessels at 2:00 a.m., they could still find shops, taverns, music halls, and gambling places, "all organized on lines in accordance with the needs of modern traffic" and brightly lit up by their owners, who in spite of the unearthly hour were "quite willing to try and entice the unwary passengers into their clutches." Some twenty years later, after the First World War, Port Said was still said to "come alive at night" with the disembarkation of passengers.[246]

Through ups and downs, Port Said clung to its reputation as an intrinsically immoral and yet entertaining place. Like developing port cities elsewhere, its infamy as a den of iniquity where bawdy and drunken behavior thrived was difficult to dissipate.[247] Izmir, for example, boasted bars, cafés, and possibilities for drunken frolics. Farther away, in Bombay and Calcutta, European sailors developed a comparable reputation for drunkenness and rowdiness in drinking establishments and streets alike.[248] But by the mid-1880s, one no longer had to refrain from traversing Port Said's back streets and byways after dark or visibly winning money at the roulette tables for fear of being followed at night and stabbed in the back. By the onset of the new century, some declared that the town had ceased to be "the international dumping ground of refuse villainy," nor could it any longer claim to be "the wickedest town of two hemispheres." Its previously cumbersome reputation had shrunk to the size of "a fine beach, a boulevard along the water, a main street ... suggesting a toy bazaar, and a seven-story building of brick and iron" (the Eastern Exchange Hotel).[249] No more did Port Said house the "concentrated essence of all the iniquities and all the vices in all the continents." It had become like other ports of call, offering the same kind of "dark relaxations" that could be found elsewhere.[250] Yet Port Said still struck some visitors around 1915 as a "seething bed of infamy," "the New Babylon," and a "festering, stagnant pool of filth."[251] In 1923, someone persisted in describing it as an "unspeakably filthy, noisy, disease and vice-infected town."[252] Clearly, Port Said's fame and shame had lived on.

Neither boredom nor debauchery ever fatally discouraged immigration into town. This continued even though, according to *L'Imparziale*, by the early 1890s Egypt no longer deserved the title "promised land" that it had carried in the past in the eyes of foreigners, who had been enthralled by its alleged floridity and splendor (see chapter one). Gold was no longer flowing in the streets, ready to be picked up with little exertion.[253] Employment opportunities in Egypt's public administration for foreign bureaucrats other than the British had shrunk. Nonetheless, the country continued to attract those migrants and refugees who carried with them both their prior expertise and the ability to reinvent themselves. Some Italian subjects, for example, reached Egypt with the explicit intention of working in the taverns (*locande*) that catered to the ever-numerous foreigners who spent wintertime in Egypt. They found employment as waiters and musicians in music halls, often earning in seven months what they would have earned in a year in their homelands. Others worked as theater artists. Italian and German musicians competed against each other.[254] English-speaking nursery governesses continued to be in high demand, "especially among wealthy Jews," but also in Coptic families and Muslim households. Specifically Calabrian wet nurses as well as female cooks from either the northern or the southern regions of the Italian peninsula were highly sought after.[255] Cyprus-born Ottoman subject Pierre Philippides, for instance, landed in Alexandria in 1893 and was able to secure employment in the Egyptian Railways Administration thanks to a certificate attesting to his previous job on the line between Diacophto and Kalavryta (in the Peloponnese) between 1889 and 1893. With the job he got in Egypt, he could support his wife, daughter, and sister-in-law.[256] Around 1895, sixteen-year-old Enrico Pea left his village in the Tuscan province of Pisa and embarked for Egypt as a ship's boy. Other examples include an experienced Swiss mechanic and an unmarried German woman (a good seamstress who knew Italian and Arabic), who placed ads in Cairo-based newspapers in the hope of finding employment in the country's factories, families, or hotels.[257] As pointed out by Khuri-Makdisi, a tremendous expansion had occurred in the late nineteenth-century regional labor market, wherein raised expectations for employment were coupled with uncertainties about job prospects and heightened competition.[258]

In Port Said in particular, the "ever increasing progress" in the canal's transit, consumption options, and business opportunities persisted in attracting new workers and their families. Built-in inequalities carried over from the past, as "Arab" workers continued to earn much lower salaries than their peers.[259] Egyptian individuals continued to find employment as coal-heavers, dockers, boatmen, and petition writers (*'ardahaljiyya*), as well as translators, in and around Port Said's harbor. Coal-heaving, which experienced mass strikes in 1882, 1894, and 1907, employed workers especially from Upper Egypt.[260] There were up to fifteen thousand coal workers in town whose situation remained precarious, connected as it was to the fluctuations in the traffic of steamers.[261] Among other foreign groups, the town's "Italian population" reportedly increased throughout the 1890s. This was partly accounted for by natality within the community. But this increase was also explained by the fact that only a few of the many workers and mariners whom the Company had dismissed from their jobs between 1893 and 1895 moved elsewhere. Thanks to savings or the severance received upon dismissal, most of them stayed in the area, began practicing different professions or jobs, or found employment with other companies. For instance, Aniello Malatesta, the brother of the anarchist leader Errico (see chapter three), had relocated in 1884 from Naples to Alexandria, where "he mingled with all the internationalist and republican rascals infesting Alexandria at the time." In 1890 he moved to Port Said, where he relinquished his revolutionary ideals and tried his luck as a tailor and then as a journalist. Eventually he quietly settled to working as a lawyer for the Italian, French, and Austrian Lloyd in Port Said, even though he lacked a formal degree.[262]

CONCLUSION

Starting in the 1870s and into the following decades, Port Said developed a reputation for immorality and lowlife within Egypt and along the Mediterranean and Red Sea shipping routes. It continued to display bazaars of low character, "in which inferior British manufactures may be detected."[263] The beer that could be found there at the turn of the century was admittedly of poor quality and sold at many times its price. Equally mediocre was its musical scene. Most of the Egyptian cigarettes that were hawked in town were apparently filled with nothing but wood shavings and tobacco essence.[264] Nonetheless, many enjoyed in their spare time the amusement opportunities that Port Said offered. While some abhorred the kind of excitement that could be found in town, others appreciated the gamut of entertainment options it presented. The owners and managers of establishments tried to outdo one another, and the drinking business thrived. First-class travelers mostly saw Port Said as fundamentally immoral or profoundly boring and failed to note how those establishments successfully entertained a diverse crowd.

Leisure in the Ottoman context has often been discussed in terms of the threat it posed to public order, social hierarchies, and reliable workers. Places of popular socialization such as coffeehouses, taverns, brothels, and bathhouses have been described as sore spots for the authorities, crawling with spies on the lookout for tidbits of conversation in multiple languages. Entertainment venues have thus been approached as sources of political and cultural threats to public order and as challenges to Ottoman claims to urban control.[265] As in scholarship that has explored early twentieth-century entertainment venues in other parts of the world, the female performers and entertainment consumers of the Ottoman scene have only received partial attention. Finally, the history of alcohol in the British colonial context, of which Egypt was part, has mostly focused on either the metropolitan politics of control over the liquor revenue or the nationalistic efforts to resist such attempts.[266]

Looking closely at mobile individuals in and out of drinking establishments in Port Said suggests that the latter did not exclusively constitute hotbeds of political opposition or nationalist stirrings and that they were permeable to gendered mechanisms regulating access, consumption, and work. This kind of venue brought people together even if they supposedly belonged to groups that otherwise would have remained distinct if they were engaging in other forms of diversion, such as musical societies, newspapers, and religious or national festivals. Port Said's drinking establishments may have simultaneously functioned as places of leisure; commercial venues; nodal points of migration networks; locales of manifest resistance and opposition; and spots where people gathered to talk and exchange news, information, and opinions.[267] At the same time, they were places where social contact intensified and alcohol ignited tensions. As in other contemporaneous Eastern Mediterranean cities, what Malte Fuhrmann has called a "polyvalent society" coexisted with normative chains restraining its denizens through expectations and norms of group belonging and behavior.[268]

In late nineteenth-century Port Said, even though group boundaries may have hardened during bar brawls, they never quite settled. Port Said remained, in Kipling's words, a place that constantly broadcast a "perpetual cinematograph show of excited, uneasy travelers," where one could "watch huge steamers, sliding in and out all day and all night like railway trucks, unknowing and unsought by a single soul aboard," and wallow in the bittersweet possibility "to talk five or six tongues indifferently, but to have no country."[269]

Conclusion

IT WOULD BE WONDERFUL
IF IT WERE NOT UNHAPPY

> Port Said is waiting—one eye regretfully on the golden, riotous past, one eye dubiously on the golden industrious future. It would be wonderful if it were not unhappy. It is on the road to everywhere, and yet it is on the road to nowhere. Ships pass every day for every sea and port in the world except Port Said.
> —GEORGE W. STEEVENS, 1898

A JOKE THAT APPEARED in the folds of a newspaper in 1895 must have fallen flat with Port Said's inhabitants. But the irony was not off base. The jest suggested that Port Said, a toponym that could be translated into "happy port" given the Arabic meaning of *sa'id* (the name of the Egyptian ruler at the time of its foundation in 1859), ought to be renamed "unhappy" because of the sad state of its public services. Readers may have smiled mirthlessly in agreement with the author, who claimed that the Egyptian government treated the city "as if it were less than a village." Many were under the impression that Cairo wanted to separate this "unhappy happy port" from the rest of Egypt.[1] Puns of this sort must have circulated for a while. Already in 1875 a French author had ironically remarked that Port Said's auspicious name seemed to be quite unjustified.[2] In 1907 a visitor found the town unchanged since his stopover two decades before. Port Said continued to strike him in its "eternal precariousness, with its pulpboard houses decorated with balconies in cut-out paper and nestled between the beach sands, the Canal waters, and the Manzala lake waters, on a shifting sandy ground scorched by the sun and faced by a constantly boiling and roaring sea." It was a town of mainly wood and plaster, wearing hues of pink and light blue.[3]

This book has traced the successive metamorphoses, from the halcyon days in the early 1860s to the twilight of the 1890s and beyond, of the "small nook, remote and neglected" that Port Said originally was and that, at some point, it reverted to being.⁴ This maritime town lived through merry and tough times. At one point it had indeed become a magnet for those longing to enrich themselves, thanks to its numerous opportunities to make a living, several worksites, and incoming tourists.⁵ At first, in fact, a great flush of optimism had swept the brand-new city, especially but not exclusively among French observers, the most prolific ones.⁶ Port Said "projected faith in the future," some declared in 1864.⁷ It was destined, someone wrote in 1867, to become "a big city of trade with great traffic and high-rising buildings."⁸ At the onset of the 1870s, some swore it was called to a great destiny, and its prosperity was assured.⁹ Others even predicted that it would rival the main commercial cities of the Mediterranean to the point of undermining Alexandria's trade.¹⁰ The city was bound to become the center of human solidarity, it was proclaimed in 1875.¹¹ When, in the mid-1870s, the Egyptian government sought to have its authority recognized over the lands that Port Said's residents were occupying, the occupants signed rental agreements in the belief that the country's circumstances would remain prosperous.¹² In 1879 Muhammad Amin Fikri described it as a "recent city" whose situation pointed toward an extremely good future.¹³ Buoyancy persisted throughout the 1880s. To justify how promising its destiny looked, some cited all the vessels stopping by its port, all the stores opening in town, and all the "peoples from the whole southern Europe that were hastening to it."¹⁴ In the topographical encyclopedia he published in the 1880s, Egyptian engineer and bureaucrat 'Ali Mubarak praised Port Said as a "new and modern city" and extolled it as a technological marvel.¹⁵

Others nonetheless felt uncertain about Port Said's importance. They acknowledged its chances for future development but also stressed the precariousness of its life.¹⁶ Potential telltale signs may have been there all along. British observers eagerly presented what they saw as a French undertaking in particularly bleak terms. In 1860 one of them argued that the early laborers who had reached the Isthmus of Suez had had their "extravagant hope for profit" frustrated, their enthusiasm dissipated, and their weariness made acute by the imposed solitude. In 1865 another warned that Port Said would ultimately prove a failure.¹⁷ In 1872 Port Said appeared to some as paralyzed by a "hand-to-mouth" system, which prevented it from becoming "one of the largest commercial depots in the east." In 1876 the *Cook Tourist Handbook* unceremoniously declared that, at best, there was "nothing in particular to see in town."¹⁸ At worst, it was an "uncomfortable place, built on low sand and surrounded by sea, lakes, and sand."¹⁹ In the mid-1870s the French consul wrote that the town had remained until then "a simple entrepot of coals and staples to provision the boats transiting through the Suez Canal." Hopes that Port Said would prosper were over. Its role as an exporter of indigenous goods and as an importer of foreign ones, the consul continued, was abysmal. It was distant from agricultural land and "almost entirely isolated, because of the lack of quick means of communication."[20] A travelers' guide published in 1878 issued an ominous warning: "It was expected that the prosperity of the place would increase rapidly, but its progress has hitherto been very gradual."[21] By 1880 the abovementioned residents were trying to get out of their rental contracts. They could not keep up with the hefty governmental fees, the costly maintenance of buildings, and the preservation of stored goods in the face of the local sea and wind conditions. Further, the country's circumstances had reportedly become "very critical."[22] The locally dispatched nuns of the French order of the Bon Pasteur found that despite the art and ingenuity that had poured into Port Said to effect the city's metamorphosis into an ostensibly "European city," the town was still "in a desert."[23]

Recognizing that a railway would foster local commerce and regional prosperity, the Egyptian ruler promised in 1881 to connect Port Said to the networks of the Egyptian railways.[24] The following year de Lesseps obtained a concession from the Egyptian government to follow up on this plan. At the time, reaching Port Said involved traveling via railroad to Ismailia, the nearest point on the Cairo-Suez line, and then taking an inferior bateau-omnibus, a small steam launch that took six hours and carried the mail, to the destination. This form of locomotion was lengthy, disagreeable, and rife with "certain dangers on the Canal." Yet for many years after the British occupation, "no serious attempt was made to join Port Said and Cairo by railway."[25] The lack of communication with the Egyptian interior hindered the commercial development of the town.[26] No trains carried produce from the cotton and wheat fields to this isthmus port, where steamers offloaded coal and then had to go on empty to Alexandria for their homeward trips. According to a British observer, the problem lay with the Company and the fact that it was entitled to all customs dues at Port Said. For that reason, the Egyptian government presumably did not favor the construction of a line connecting the interior to this center.[27] This view proved to be partially correct. The Company did strive to maintain its long-standing privileges, but the Egyptian government still had an eye out for Port Said. It must have known that Port Said did not thrive on trade with the delta as Alexandria ostensibly did and that it lived off the steamers anchoring in its waters. An article published by *Al-Ahram* in 1900 clarified that reality. Those in governmental positions in Cairo ought to be aware that whenever traffic came to a halt in Port Said, the port city's commercial activities stopped, its wherewithal was interrupted, and its population was plunged into hunger and forced to beg from charitable associations (while banks and societies looked the other way).[28]

A narrow-gauge railway under the management of the Company had indeed come into service in 1892 between Port Said and Ismailia. But even though travel time had decreased by half an hour in 1895, people nonetheless mockingly compared the shuttling train to a snail.[29] Nonetheless, tourists, among others, relied on this connection. Some of them did not bother to look out the carriage windows, since their tourist guides disdained this leg of the journey. But breathtaking views repaid those who did: on one side, the vessels proceeding along the canal seemingly glided over the sand dunes; on the other, fishing boats, camels, and flocks of birds animated the surface and the banks of Lake Manzala.[30] In 1898 Muhammad Rashid Rida praised the virtues of a potential railway line connecting the northern Egyptian coast at nearby al-'Arish all the way to Basra.[31] But in 1900 the Company still seemingly refused to work toward better overland connections.[32] This was so even though around that time the need to improve the local infrastructure of Port Said had perhaps become more pressing, as it had for other Egyptian and Middle Eastern coastal towns and transportation hubs. Elsewhere in the Middle East, in fact, maritime entrepots had benefited from having connected their better port facilities to their countryside via the railroad.[33]

In 1902 representatives of the Egyptian government submitted a set of requests designed to alter Port Said's status quo to the Company, which subsequently accepted them. First, it demanded that the Company take on the expense of transforming the existing railroad into a standard-gauge railroad to be connected to the state rail network in Ismailia. At least two passenger trains would travel every day in each direction, one of which would stop at all canal stations. People and goods with a Company permit would travel for free. Second, the Egyptian government requested that the Company enlarge Port Said's harbor (this would be achieved between 1908 and 1928). The two parties also agreed to create around the port a "duty-free zone from standpoint of customs," which delegated the Egyptian government to only levy taxes for those goods that entered or exited the zone. For example, the coals that were to be deposited on the eastern bank, offloaded from some

FIGURE 21. Port Said railway station, n.d. *Source:* Archives CeAlex/CNRS, fonds J.-Y. Empereur.

ships and taken on by others, continued to be excluded from governmental purview. Additionally, the government forfeited its right to tax those materials destined for the Company's use. Hence, the government gave in to the Company and its claim that the "duty-free zone" would increase Port Said's maritime traffic and boost its role as a regional hub for goods.[34] The signing of these agreements was quickly echoed elsewhere. Parties of Italian laborers, for example, showed up in Port Said but were turned away. It was only in 1903 that the Egyptian government gradually undertook the building of the new railroad, for which it recruited mainly "indigenous" labor for paltry pay.[35] By 1906 a new standard-gauge railway was trudging through the desert expanse between Port Said and Ismailia. Travelers could now more easily move from Cairo to Port Said and back. The old train station remained in place (see figure 21).[36] People expected the new railroad to finally bring about Port Said's "splendid maritime and commercial future."[37]

Notwithstanding Port Said's early twentieth-century connection via iron tracks, contemporary and later accounts never reached an agreement about its status in relation to the rest of Egypt. On the one hand, some maintained that Port Said really was "not in Egypt, but ad Ægyptum," the same label often applied to Alexandria, Port Said's illustrious neighbor to the west.[38] This widely shared idea presumed that "Port Said was in, but not truly of,

Egypt" (at least until 1956 and the nationalization of the Suez Canal Company). In 1882, the fact that the British occupation army, after bombarding Alexandria in July, first landed in Port Said, did integrate the canal region into a broader Egyptian narrative (and so perhaps did the strikes by Arab coal-loaders in Port Said to protest Britain's invasion). Yet Port Said, as highlighted by Valeska Huber, continued to impress some as "much more intricately connected to the European, Asian, African and Australian destinations it was serving than to Egypt itself."[39] Western writers often approached the canal region as an isolated desert that French genius and the Company's technological innovations had turned into a hospitable and lush corridor. Some in Egypt also may have deemed this area marginal and under full control of foreigners. They looked at Port Said as "a gate, but a gate open on a passageway rather than on a country, a region, or a continent."[40] In 1965 scholar Janet Abu-Lughod wrote off Port Said and the other canal cities as places removed "from any hinterland except the outside world."[41]

Others highlighted the fact that Port Said built a relationship with the rest of Egypt as both a supply-base and a reportedly significant actor in the national economy.[42] Similarly positive views took on an ecumenical flair. Urban historian Fu'ad Faraj suggested that Port Said embodied a point of encounter for heterogeneous peoples and cultures, a spot where the so-called West and East merged. Author al-Sayyid Husayn Jalal, whose books were published in Cairo, as were Faraj's, defined the canal as "one of the waterways through which world trade flows," playing key roles in facilitating maritime transport and influencing "human life in the East and in the West." Accordingly, optimists narrate that the Suez Canal spawned a new world and bridged the Orient and the Occident, producing no less than a "modern utopia."[43]

This book argues that Port Said never took on one of these roles to the exclusion of all others. During its first fifty years, it lived several different lives. It first came into existence in 1859 as a labor camp for the migrant workers engaged in the Suez Canal undertaking. The terrain was forbidding, housing was temporary, and employment prospects were uncertain. Still, perspective laborers kept coming from the rest of Egypt and other countries willingly, forcibly, or out of an awkward mix of volition and compelling circumstances. As migrant labor settled in during the second half of the 1860s, hierarchies based on perceptions of ethnic and racial categories and gender relations solidified. In the Company-run encampments and nascent towns across the isthmus and Port Said in particular, workers both endured and themselves contributed to the reification of group boundaries and gender roles. Some responded to this predicament by relocating elsewhere on the isthmus or farther away. Their movements both created an isthmus-wide system of mobility and interwove this region with the rest of Egypt as well as sites overseas. Around 1869, as the excavation neared completion, Port Said morphed into an apparently lawless borderland. The drifting mass of fired workers congregating there elicited fears among Egyptian and foreign authorities alike. Scrambling state, Company, and consular officials were unable to stem the stream of people and commodities that continued transiting in and out of the isthmus, chiefly through its northern port at Port Said. Last, into the 1890s Port Said acquired a reputation as a particularly seedy yet entertaining spot. It had become a regionally and globally interconnected market for pleasure, where flows of imported alcohol and migratory routes for prostitution or other work converged. While most available options for leisure tended to separate inhabitants along national, political, and religious lines, drinking instead gave them the opportunity to blend. Small-scale movements in and out of Port Said's drinking establishments both hardened group boundaries and provided a time and a space to loosen them up. The migrant labor roaming Egypt and the Mediterranean in the second half of the nineteenth century had disparate and uneven modes of mobility to hand. These, and the attendant formation of an unequal migrant society in Port Said, undergirded the Suez Canal's genesis and its first half century of operation.

Its nineteenth-century history reveals that, as in coeval sites elsewhere in the world, a heterogenous and stratified community emerged early on and remained long after the initial rush of people ended.[44] Different subjects and groups had access to uneven and unequal kinds of mobility. Individuals who could claim affiliation with capitulatory powers could afford facilitated forms of movement. The Egyptian government tried to intervene, as it did in 1897 through a communication issued by its Ministry of the Interior: "Given the considerable number of Greeks and other foreigners who have recently come to Egypt with no means of subsistence, the disembarkation of anyone arriving in these conditions will be heretofore forbidden."[45] However, European consular officials in Cairo, Alexandria, and Port Said continued to claim that capitulations did not enable local authorities to regulate foreigners' entry into Egypt. Therefore, to land in an Egyptian port, foreigners only needed to hold a passport. To earn a living, they could freely practice the craft, trade, or profession they wished, with the only exceptions being doctors, pharmacists, and midwives (who needed an authorization from the Ministry of the Interior) and tribunal clerks (who had to enroll on an ad hoc list and go through training). They could act on a whim, as mockingly suggested by the Egyptian governor in Suez to a Warsaw-born Italian protégé, who reported being told: "If I am not content, I can go back to my homeland."[46]

Port Said's wide, dusty, unpaved streets were filled with a motley company. "Egyptians, men and women, draped in somber black, brown Arabs, ebony Sudanese, mingle[d] with Europeans of every color and clime, while English, French, and Arabic [we]re heard equally," a North American visitor related in 1901.[47] In 1907, around the conclusive benchmark of this study, Port Said had a greater proportion of people born elsewhere than any other Egyptian town: 29,756 out of 49,884, or 59.5 percent.[48] Between 1907 and 1917 Port Said's population increased steeply as the Suez Canal imposed itself as one of the main highways of global commerce.[49] As diversified as its dwellers were, it was no egalitarian or "cosmopolitan" niche, and its urban segregation persisted into the twentieth century. In 1911 the creation by decree of a municipal council in Port Said attested to both the multiplicity of its social and economic components and the dire necessity for an expansion of its services. In fact, this council was presided over by the Egyptian local governor and included, besides five appointed members (the public health inspector, the customs director, two nominees of the Company, and one representative of maritime agencies), five members from the Egyptian community and five chosen among the European residents, one from each "nation." The council imposed taxes and rendered services in the realms of public health, safety, and public works.[50] In 1920 Port Said still appeared severed into a European quarter, outwardly "as orderly and moral as an ordinary European provincial town," and an ostensibly squalid and sordid Arab quarter (see figure 22).[51] Unemployment was among the greatest economic and social problem affecting its population.[52] Even though, of the former, author Le Clézio had his protagonist declare, "This is the white city, there are no Egyptians," transgressions of that urban boundary happened time and again.[53]

Historian Zayn al-'Abidin Shams al-Din Najm argued that Port Said became a living example of European encroachment on Egyptian soil, serving as it did the interests of the West and of the "outer world." He also wrote that Port Said enshrined the power of the Company and the inability of the Egyptian administration, police, and military forces to cope with the acrimonious disputes as well as the unholy alliances between foreigners and

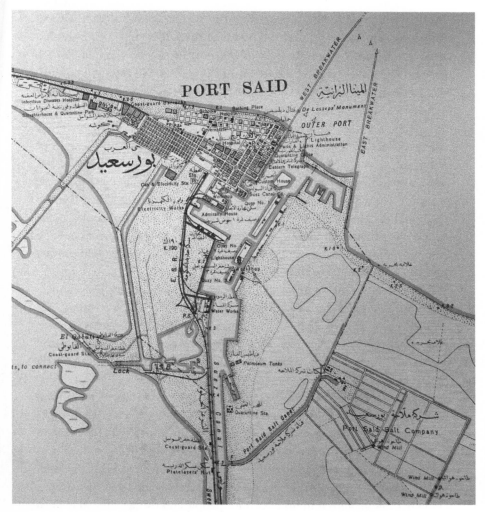

FIGURE 22. Port Said's Arab and European quarters, 1911. On the left, hai al-'Arab in the west; on the right, hai al-Afrang in the east. Scale 1:50,000. *Source: Sinai Survey Department Atlas of Lower Egypt,* 1911, IFAO.

locals.[54] This book has troubled the idea that the Company exercised absolute power in the canal region and complicated the notion that the Egyptian state failed to adequately intervene in this region. The canal excavation exemplifies how, as in comparable contexts elsewhere, state-led initiative on public works attracted both local and foreign labor and how the improved transportation infrastructure, in turn, facilitated movement.[55]

While acknowledging that massive infrastructural projects like the canal convey the economic and political power of those in charge, this book has rather privileged the prosaic lives of the migrant workers employed—or not—on and off the canal worksites. It has argued that in 1859 and onward, Port Said and the making of the whole canal project rested on two main foundations: unevenness and inequality. These translated into the various options for mobility available to distinct social groups; a social hierarchy that attempted to sort labor according to racial and gendered notions of individual and collective worth; a repressive apparatus that attempted but failed to regulate the influx and egress of clandestinely circulating people and things; and finally, a variegated entertainment scene, in which inebriation could at once bring people together and fire up their animosities. The success narratives of large infrastructural projects should not obliterate the seemingly minor accidents and prosaic acts of resistance or negotiation that were also foundational to their history.

Between its inception in 1859 and the advent of a faster railway connection in 1906, Port Said maintained a contradictory status as a terminus and a transit point, as a chokepoint and a passageway, as the isthmus gateway and as a getaway spot. It remained a point of convergence for both flows and cul-de-sacs. These contradictions came alive on multiple levels. First of all, the Suez Canal and even the 1906 Port Said-Ismailia railway line have been celebrated as tokens of the "new era of universal communication" patronized by de Lesseps and as "instruments helping in the integration of the world's economies."[56] But while praising the supposedly free-flowing circulation of capital, goods, and people, we should also take into account instances of forced mobility, uneven options for mobility, and immobility. Second, while Port Said was regionally and globally interconnected and the isthmus region became an autonomous circuit for people and things, they were both also linked to the rest of Egypt. Migratory movements tied this area to other Egyptian regions and, at the same time, authorities in Cairo constantly tried to extend their reach to the canal banks. Egyptian officials did not appear to be in full control of Port Said's population, but then neither did Company bureaucrats or foreign consular personnel. Finally, Port Said's history cannot be written by taking into account only its supposedly European foundations, which would affix its history to that of Europe and approach the latter as the sole producer of change for humanity.[57] Neither, however, should Port Said's history be discussed solely in terms of its Egyptian character. This port town was arguably a "hybrid Euro-Oriental" city, far from the "regionally authentic" locales with which older area studies have mainly been concerned.[58]

There may be serious limitations to what one can learn from a localized study: for instance, being drawn into a swamp of particulars or losing sight of larger comparative contexts.[59] If a "typical" Egyptian city ever existed, Port Said is not one. Several elements of its history make it exceptional, including its peculiar location and natural environment and the facts that its settlement began from scratch at midcentury and that all its denizens came from somewhere else. Few other places around the turn of the twentieth century seemed to embody the triumph of engineering and the conquest of distance as Port Said did.[60] Nonetheless, such anomalous characteristics make certain phenomena stand out; for example, Port Said's apparent separateness from the Egyptian interior clears the ground for examining opportunities for mobility and power struggles at the provincial level, while revealing the constraints on what Company and state administrators or workers could accomplish toward their goals. Moreover, like other provincial and nonmetropolitan Egyptian centers, Port Said provides unique angles on Egypt's urban experience.[61] Further zooming out, this port town was similar to maritime cities elsewhere on the globe, for it also embodied a point of ingress from and egress to the wider world, a site of convergence for people and things on the move, and the lightning rod for heightened contemporary anxieties relating to developing cities.[62]

The history of Port Said is, on the one hand, the story of an extraordinary place. Yet it sheds light on a tension built on the foundations of other nineteenth-century infrastructural projects, promising to enhance universal mobility while requiring a heterogenous workforce to commence and an economically exploitative and exclusionary reality to function. Through the lens of disparate forms of mobility at different scales, this book has attempted to retrace how an apparently peripheral spot changed patterns of migration, labor configurations, attempts to define lawfulness and illegality, and leisure options. In sum, the history retold here illuminates how peripheries, despite their geographic distance and social separateness from the core, still challenge and rewrite the script unfolding on the past's main stage.

Postscript

ON JULY 26, 1956, during celebrations for the fourth anniversary of the Free Officers' Revolution, Egyptian president Gamal Abdel Nasser gave a speech in Alexandria that would make waves at home and abroad. In it he roared that Egypt had contributed free forced labor (*sukhra*) to the excavation of the Suez Canal and that 120.000 workers had died in the undertaking. He also proclaimed that twentieth-century Egyptians would nationalize (*ta'mim*) the Suez Canal Company; erect the High Dam; and cement Egypt's might, freedom, and honor.[1] Drawing from both the secular and the religious domains, he hammered out a new type of human subject, a liberated one.[2] His words came on the heels of the US withdrawal, on July 18, from its participation in financing the construction of the High Dam. The revenue from the newly nationalized Suez Canal, according to Nasser, would finance that new waterwork in Aswan.[3] Thus, Nasser transformed Egyptian sacrifices from points of sorrow to sources of pride, all the while reminding his domestic and international audience of both what the country had endured and what it was now forcefully taking back. The number of the dead in the canal digs is impossible to determine; contemporaries themselves disagreed. But the content and tone of the Egyptian president's speech would still steer the course of Egypt's consciousness and memory. Street names bearing traces of the colonial, Ottoman, and monarchical eras were replaced by those of figures of the national liberation movement.[4] On December 24, 1956, a dynamite explosion brought down the statue of Ferdinand de Lesseps and ended the symbolic reign of the Frenchman over the canal, toward which he had been pointing his right index finger since the thirtieth anniversary of the waterway's inauguration on November 17, 1899 (see figure 23).[5]

FIGURE 23. Statue of Ferdinand de Lesseps at the entrance of the harbor, n.d. *Source:* Ref193, Hisham Khatib Collection, Akkasah Photography Archive, al Mawrid, NYUAD.

As Nasser had expected, the nationalization of the Suez Canal Company did indeed deliver positive short-term effects to Egypt's economy.[6] But Port Said and other canal cities paid a hefty price for this transaction: in October–November 1956, what Egyptians recall as the "tripartite aggression" by Britain, France, and Israel, organized in retaliation for the nationalization of the canal, among other reasons, devastated these centers

FIGURE 24. *Construction of the Suez Canal*, watercolor by Abdel Hadi Al-Gazzar, 1965. Source: Christie's, https://www.christies.com/en/lot/lot-5779742.

and left twenty thousand of their inhabitants dead.[7] Yet Port Said remained intact in the nation's imagination as a symbol of resistance and progress. For example, painter Abdel Hadi Al-Gazzar, in the 1962 masterpiece for which he was best known at home and abroad, *The Charter*, included ships (one of which bears the United Arab Republic acronym), warehouses, and a dredger. This tiny still life, discernible in the top left corner of his painting, resembles the canal's northern entry at Port Said and symbolizes the newly nationalized waterway (it faces, in cavalier geographic attitude, another nationalist symbol: Aswan's High Dam, begun in 1960). At the center, a female figure holds Nasser's eponymous charter (*Al-mithaq*), in which he had outlined his ideology; she is wearing a crown bearing Egypt's winged emblem and towers over a peasant and a laborer as embodiments of agriculture and industry. Only apparently celebratory, the painting offers viewers a sinister and ambivalent social commentary on the costs of nationalism and modernization.[8] In 1965, in his *Construction of the Suez Canal*, Al-Gazzar fully tackled the history of this endeavor (see figure 24). The forefront displays the faces and naked limbs of Egyptian forced laborers, compelling viewers to take in their thirst, exertion, exposure to the elements, poor tent shelters, dangerous working conditions, and inadequate tools. In the background, but well within sight, Al-Gazzar placed a couple of fez-clad foremen presiding over the worksite from the comfort of their umbrella shades. The urge to redress past injustices alongside feelings of stung national pride continued to echo in cinematic and other works about the Suez Canal through the 1960s and beyond.⁹

To this day, the Suez Canal and the canal cities occupy an ambivalent position, awkwardly straddling belonging and foreignness. When Israel occupied the Sinai in 1967, the canal became Egypt's temporary western frontier. Many departed the canal cities and surroundings to avoid the combat zone in 1967 and the war of attrition that ensued in 1969–1970. Others were evacuated. Altogether nearly one million people were displaced. When Egypt recovered the Sinai after the 1973 war, a return movement began in 1974 and continued until about 1976.¹⁰ Mostafa Mohie argued that these movements both dismantled the social fabric of Port Said and triggered in Port Saidians the need to reimagine their town.¹¹ In 1975 Port Said was declared a "free trade zone" (a move that had affected other sites before, such as Alexandria's port in 1956). On the one hand, this wove the town into Egypt's fabric, as it made it a blatant symbol of the *infitah*, the policy of economic "openness" heralded by Anwar Sadat, Nasser's successor in the Egyptian presidency. But on the other hand, its port began attracting those consumers and middlemen who wanted goods that were not available in Egypt's other markets at the time. If the "free trade zone" helped Port Saidians recover from previous displacements, it shaped for them a reputation of greed and dishonesty.¹² Even on the big screen, Port Said turned into the city of the new rich, an emblem of wealth but also corruption.¹³

Port Saidians' reputation for independence and rebelliousness carried over through the "free trade zone" establishment and beyond.¹⁴ In February 2012, roughly one year after the hope-filled uprising of January 25, 2011, against President Hosni Mubarak's rule, clashes between fans of Al-Masry, the Port Saidian sports club, and those of al-Ahly, one of Cairo's soccer teams, broke out in Port Said's stadium following one of the Egyptian League matches. Seventy-two people were killed and hundreds injured, all of whom were members of the Ultras Ahlawi, the biggest fan group of al-Ahly. Responsibility for the massacre was placed on the state's security forces and the Green Eagles, one of the fan groups of al-Masry. The reason for it was identified in the Ultras Ahlawi's participation in the 2011 revolutionary uprisings. What would be known as Port Said's "stadium massacre" created a stigma that "othered" Port Saidians as thugs. It also contributed to the demise of revolutionary hope not just among Ultras members but also for all those within the wider street-activist culture.¹⁵ It is ironic that, whenever politically convenient, Cairo has appropriated the Suez Canal and its urban centers as showcases of national pride, for example in the celebrations for the so-called New Suez Canal (a widening of the existing channel announced in 2014 by then president Abdel Fattah al-Sisi and inaugurated in August 2015). On that occasion, authorities displayed a blend of ancient and modern tropes to a carefully selected audience in an effort to project the competence of the state and the strength of Egyptians.

On the surface, Port Said may look like a sleepy provincial town or a pleasant Mediterranean destination mostly popular for its famed fish meals. For some, it may be just another small city on the margins whose destiny is regulated by the center.[16] But there are ongoing forms of ebullience. Port Saʿid ʿala Qadimo (Port Said as It Was), for example, is a group of mostly young Egyptians who, among other activities, organize walking tours designed to spotlight the town's architectural legacy. By describing their initiative as a "non-profit organization aiming to place Port Said back in its historical and cultural position," its members implicitly claim that Port Said once held a certain status in Egypt's history and culture, that this status has now been lost, and that it needs to be recovered (or reimagined).[17] Not incidentally, as noted by Mohie, most of the tours arranged by Port Saʿid ʿala Qadimo focus on the Al-ifrangi or "European" neighborhood and neglect the "Arab" parts of town, revealing the group's selectively nostalgic outlook on the past.[18] Indeed, the historic buildings in Port Said's Al-ifrangi quarter, fronted by uniquely composite three- or four-story wooden galleries or verandahs, still bear traces of European nineteenth-century architecture and are now in ruins.[19] Once the dust settles, who and how gets to be memorialized as representative of Port Said's and the canal's tumultuous and variegated history?

Seeking Bread and Fortune in Port Said has retraced the lives of migrant workers of varying provenance, language, social position, professional rank, legal status, proclivity to leisure, and overall access to mobility. Through an interplay of different and overlapping scales of analysis, it has illuminated how their mobility, ranging from the transnational to the local, both empowered and constrained them. It has accounted for the fact that Port Said both owed its genesis to the massive infrastructural undertaking of the Suez Canal and enabled its completion. Yesterday as today, the migrant workers recruited for large-scale construction projects, as well as those employed in related industries such as service or transportation, are vital to these undertakings' success and yet remain vulnerable to invisibility and systemic abuses: forced labor, passport confiscation, wage withholding, inability to change jobs or employers, prohibition against striking or joining unions, indebtedness, unsanitary living conditions, and excessive working hours, among others.[20] Lest only narratives of material triumph go down in history and hold sway over our present, migrants' past lives need to be dug up and brought to the forefront.

NOTES

INTRODUCTION

1. "On 21 Ramadan 1275 h. Corresponding to 25 April 1859," in *Taqwim al-Nil*, by Amin Sami (Cairo: Matbaʿat al-ʾAmiriyah, 1916), 3:1:326. Henri Silvestre, *L'isthme de Suez 1854–1869, avec carte et pièces justificatives* (Paris: Lacroix, Verboeckhoven, 1869), 145; Luigi Dori, "Esquisse historique de Port Said. Naissance de la ville et inauguration du Canal (1859–1869)," *Cahiers d'histoire égyptienne* 6, nos. 3/4 (October 1954): 197–98; and Paul Reymond, *Le port de Port-Saïd* (Cairo: Impr. Scribe égyptien, 1950), 33. To my knowledge, only one author argues that the inauguration took place on April 15 rather than April 25; see Leonardo Gallo da Calatafimi, *L'Egitto antico e moderno* (Catania: Tip. Eugenio Coco, 1891), 169. Leconte writes that the city was founded in April 1860. See also Casimir Leconte, *Promenade dans l'isthme de Suez* (Paris: N. Chaix, 1864), 85.

2. Nathalie Montel, *Le chantier du Canal de Suez, 1859–1869: Une histoire des pratiques techniques* (Paris: Éd. In forma/Presses de l'École Nationale des Ponts et Chaussées, 1999), 19–31; and Caroline Piquet, "The Suez Company's Concession in Egypt, 1854–1956: Modern Infrastructure and Local Economic Development," *Enterprise and Society* 5, no. 1 (2004): 107–27. Regarding the sway France held over Egypt until the 1870s, see David Todd, *A Velvet Empire: French Informal Imperialism in the Nineteenth Century* (Princeton, NJ: Princeton University Press, 2021), 227–76.

3. Edward W. Said, *Orientalism* (New York: Pantheon Books, 1978), 89.

4. J. Stephen Jeans, *Waterways and Water Transport in Different Countries: With a Description of the Panama, Suez, Manchester, Nicaraguan, and Other Canals* (London: E. & F. N. Spon, 1890), 257–58.

5. Roger Owen, *The Middle East in the World Economy, 1800–1914* (London: Methuen, 1981), 125–28; and Şevket Pamuk, "The Ottoman Empire in the 'Great Depression' of 1873–1896," *Journal of Economic History* 44, no. 1 (1984): 107–18.

6. I am borrowing this expression from Beth Baron, *The Orphan Scandal: Christian Missionaries and the Rise of the Muslim Brotherhood* (Stanford, CA: Stanford University Press, 2014), 7.

7. Faruk Bilici, *Le canal de Suez et l'Empire ottoman* (Paris: CNRS Editions, 2019), 47–48.

8. Archives du ministère de l'Europe et des Affaires étrangères, Centre des Archives Diplomatiques á la Courneuve, Paris (hereafter CADC), Affaires économiques et commerciales (hereafter AEC), Correspondance consulaire et commerciale, Port-Saïd (hereafter CCCPS), vol. 1, 1867–1877, Port Said, 19 June 1870, Pellissier, French Consul in Port Said, to French Ministry of Foreign Affairs, 82r.

9. 'Ali Mubarak, *Al-Khitat al-tawfiqiyah al-jadidah li-Misr al-Qahirah wa-muduniha wa-biladiha al-qadimah wa-al-shahirah* (Cairo: Matba'at Dar al-Kutub wa-al-Watha'iq al-Qawmiyah, 2004), 10:57. As for the first part of the city's name, it can be noted that the version indicated by Mubarak, *"purt"* (composed of what he calls "the Farsi ba' with three dots" and with a final "ta'") is seldom used. Incidentally, *sa'id* in Arabic means "happy, radiant, blissful, lucky, auspicious, felicitous."

10. Jules Charles-Roux, *L'Isthme et le canal de Suez: Historique, état actuel* (Paris: Hachette, 1901), 1:269, 442, 446.

11. Mubarak, *Al-Khitat al-tawfiqiyah al-jadidah*, 10:57; Fu'ad Faraj, *Mintaqat Qanal al-Suways* (Cairo: Matba'at al-Ma'arif wa Maktabuha, 1950), 12–17; Zayn al-'Abidin Shams al-Din Najm, *Bur Sa'id: Tarikhuha wa-tatawwuruha, mundhu nash'atiha 1859 hatta 'am 1882* (Cairo: Al-Hay'ah al-Misriyah al-'Ammah lil-Kitab, 1987), 19. Faraj notes that the Arabic for Tinnis can be rendered as either "ṭina" or "tina." The location was twenty-eight kilometers west of the ancient city of Tinnis. Najm points out that the emplacement of Port Said was far from that of the ancient city of Pelusium and therefore it ought not to be considered its extension or a substitution.

12. François-Philippe Voisin, *Le canal de Suez* (Paris: Vve C. Dunod, 1902), 4:9.

13. "Note sur la constitution géologique de l'isthme de Suez par M. Renaud," *L'Isthme de Suez*, 25 July 1856, 46; and "Note sur la constitution géologique: Suite et fin," *L'Isthme de Suez*, 10 August 1856, 62–63. Sami Amin reports that the total length of the canal was eighty-seven miles. Sami, *Taqwim al-Nil*, 3:1:326.

14. Najm, *Bur Sa'id*, 20; 'Abd al-'Aziz Muhammad Shinnawi, *Qanat al-Suways: wa-al-tayyarat al-siyasiyah allati ahatat bi-insha'iha* (Cairo: Ma'had al-Buhuth wa-al-Dirasat al-'Arabiyah), 265, 270–71.

15. On Barak, *Powering Empire: How Coal Made the Middle East and Sparked Global Carbonization* (Oakland: University of California Press, 2020), 15.

16. Ehud R. Toledano, *State and Society in Mid-Nineteenth-Century Egypt* (Cambridge: Cambridge University Press, 1990), 181–205.

17. Heinrich Schäfer, *The Songs of an Egyptian Peasant* (Leipzig: Hinrichs, 1904), 145.

18. Hilmi Ahmad Shalabi, *al-Hukm al-mahalli wa-al-majalis al-baladiyah fi Misr mundhu nash'atiha hatta 'am 1918* (Cairo: 'Alam al-Kutub, 1987), 38.

19. Ilham Khuri-Makdisi, *The Eastern Mediterranean and the Making of Global Radicalism, 1860–1914* (Berkeley: University of California Press, 2010), 5.

20. Originally, Ottoman authorities considered such privileges subservient to their own empire's interests and granted them to European "nations." They were not

initially meant to stipulate a perpetual legal agreement among equally interested parties, but rather engendered temporary privileges and immunities that the Porte liberally bestowed. See Halil İnalcık and Donald Quataert, *An Economic and Social History of the Ottoman Empire, 1300–1914* (Cambridge: Cambridge University Press, 1996), 50–51; and Domenico Gatteschi, *Manuale di diritto pubblico e privato ottomano* (Alexandria: Minasi, 1865), xx–xxi.

21. Bahi Ed Dine Barakat, *Des privilèges et immunités dont jouissent les étrangers en Egypte vis-à-vis des autorités locales* (Paris: A. Rousseau, 1912), 171; and Gabriel Baer, *Studies in the Social History of Modern Egypt*, Publications of the Center for Middle Eastern Studies, no. 4 (Chicago: University of Chicago Press, 1969), 193–98.

22. Ministero degli Affari Esteri, *Emigrazione e colonie: Raccolta di rapporti dei RR. agenti diplomatici e consolari* (Rome: Tipografia dell'Unione Cooperativa Editrice, 1906), 2:155–56, 172.

23. Julia A. Clancy-Smith, *Mediterraneans: North Africa and Europe in an Age of Migration, c. 1800–1900* (Berkeley: University of California Press, 2011), 4.

24. Ulrike Freitag et al., "Migration and the Making of Urban Modernity in the Ottoman Empire and Beyond," in *The City in the Ottoman Empire: Migration and the Making of Urban Modernity* (London: Routledge, 2011), 2.

25. Clancy-Smith, *Mediterraneans*, 9.

26. Zachary Lockman, "'Worker' and 'Working Class' in Pre-1914 Egypt," in *Workers and Working Classes in the Middle East: Struggles, Histories, Historiographies* (Albany: State University of New York Press, 1994), 84.

27. Tim Cresswell, "Towards a Politics of Mobility," *Environment and Planning D: Society and Space* 28, no. 1 (February 1, 2010): 21; and Tim Cresswell, *On the Move: Mobility in the Modern Western World* (New York: Routledge, 2006), 2–3. Creswell is here reinterpreting Henri Lefevbre's triad on lived, conceived, and perceived notions of space-place. Henri Lefebvre, *The Production of Space* (Oxford: Blackwell, 1991), 38–40.

28. Doreen B. Massey, "A Global Sense of Place," in *Reading Human Geography: The Poetics and Politics of Inquiry*, ed. Trevor J. Barnes and Derek Gregory (London: Arnold; New York: Wiley, 1997), 317.

29. Kevin Hannam, Mimi Sheller, and John Urry, "Editorial: Mobilities, Immobilities and Moorings," *Mobilities* 1, no. 1 (2006): 1–22; Cresswell, "Towards a Politics of Mobility," 29; Jaume Franquesa, "'We've Lost Our Bearings': Place, Tourism, and the Limits of the 'Mobility Turn,'" *Antipode* 43, no. 4 (2011): 1012–33; and Nina Glick Schiller and Noel B. Salazar, "Regimes of Mobility Across the Globe," *Journal of Ethnic and Migration Studies* 39, no. 2 (2013): 183–200.

30. Mimi Sheller and John Urry, "The New Mobilities Paradigm," *Environment and Planning* 38, no. 2 (2006): 214.

31. Nancy L. Green, *The Limits of Transnationalism* (Chicago: University of Chicago Press, 2020), 3; and Sallie Westwood and Annie Phizacklea, *Trans-Nationalism and the Politics of Belonging* (London: Routledge, 2000), 15. I am also drawing from scholars who complicate the idea that agency is simply a matter of recovering an individual intention or ability to act. Among others, see Walter

Johnson, "On Agency," *Journal of Social History* 37, no. 1 (2003): 113–24; C. Hughes Dayton, "Rethinking Agency, Recovering Voices," *American Historical Review* 109 (2004): 827–43; and Marisa J. Fuentes, *Dispossessed Lives: Enslaved Women, Violence, and the Archive* (Philadelphia: University of Pennsylvania Press, 2018).

32. Michel de Certeau, *The Practice of Everyday Life* (Berkeley: University of California Press, 1984). For a view of agency and structures as embedded in concrete historical practices, see Christian G. De Vito, "History without Scale: The Micro-Spatial Perspective," *Past & Present* 242 (2019): 359.

33. Gavin D. Brockett and Özgür Balkılıç, "The Ottoman Middle East and Modern Turkey," in *Handbook: The Global History of Work*, ed. Karin Hofmeester and Marcel Van der Linden (Berlin: De Gruyter Oldenbourg, 2018).

34. Pamela Sharpe, ed., *Women, Gender, and Labour Migration: Historical and Global Perspectives* (London: Routledge, 2011), 3, 12; Jan Lucassen, Leo Lucassen, and Patrick Manning, *Migration History in World History: Multidisciplinary Approaches* (Leiden: Brill, 2010), 9–10; and Leslie Page Moch, *Moving Europeans: Migration in Western Europe since 1650* (Bloomington: Indiana University Press, 1992), 14–15. On the history of migrant women and gender in the Mediterranean and Tunisian contexts, see Julia Clancy-Smith, "Women, Gender and Migration along a Mediterranean Frontier: Pre-Colonial Tunisia, c. 1815–1870," *Gender & History* 17, no. 1 (2005): 62–92; Julia A. Clancy-Smith, "Women on the Move: Gender and Social Control in Tunis, 1815–1870," in *Femmes en Villes*, ed. Dalenda Larguèche (Tunis: Centre de Publications Universitaires, 2006), 209–37; and Julia A. Clancy-Smith, "Locating Women as Migrants in Nineteenth-Century Tunis," in *Contesting Archives: Finding Women in the Sources*, ed. Nupur Chaudhuri, Sherry J. Katz, and Mary Elizabeth Perry (Urbana: University of Illinois Press, 2010).

35. I am here indebted to Jonathan Miran, "Mapping Space and Mobility in the Red Sea Region, c. 1500–1950," *History Compass* 12, no. 2 (2014): 202; and Jonathan Miran, *Red Sea Citizens: Cosmopolitan Society and Cultural Change in Massawa* (Bloomington: Indiana University Press, 2009).

36. Cresswell, *On the Move*, 26, 43; Gene Desfor and Jennefer Laidley, "Introduction: Fixity and Flow of Urban Waterfront Change," in *Transforming Urban Waterfronts: Fixity and Flow*, by Gene Desfor et al. (New York: Routledge, 2012), 5–6; and Valeska Huber, "Multiple Mobilities, Multiple Sovereignties, Multiple Speeds: Exploring Maritime Connections in the Age of Empire," *International Journal of Middle East Studies* 48, no. 4 (2016): 763–66.

37. CADC, AEC, CCCPS, vol. 1, Port Said, 20 February 1874, Saint-Chaffray, French Consul in Port Said to French Ministry of Foreign Affairs, 298r, 300r, 300v.

38. Roger Owen, "From Liberalism to Liberal Imperialism: Lord Cromer and the First Wave of Globalization in Egypt," in *Histories of the Modern Middle East: New Directions*, ed. Israel Gershoni, Y. Hakan Erdem, and Ursula Woköck (Boulder, CO: Lynne Rienner, 2002), 106; and Aaron Jakes, *Egypt's Occupation: Colonial Economism and the Crises of Capitalism* (Stanford, CA: Stanford University Press, 2020), 141–63.

39. Charles Philip Issawi, *Egypt in Revolution: An Economic Analysis* (London: Oxford University Press, 1963), 83.

40. Italian Ministry of Foreign Affairs, Commissariato dell'emigrazione, "Egitto," *Bollettino dell'emigrazione*, no. 7 (1908): 103.

41. Adam McKeown, "Global Migration 1846–1940," *Journal of World History* 15, no. 2 (2004): 155–89; Donna R. Gabaccia and Dirk Hoerder, *Connecting Seas and Connected Ocean Rims: Indian, Atlantic, and Pacific Oceans and China Seas Migrations from the 1830s to the 1930s* (Leiden: Brill, 2011); and Khuri-Makdisi, *Eastern Mediterranean*, 170.

42. Key works in this vein that influenced my thinking include Michael J. Reimer, *Colonial Bridgehead: Government and Society in Alexandria, 1807–1882* (Boulder, CO: Westview Press, 1997); Nelida Fuccaro, *Histories of City and State in the Persian Gulf: Manama since 1800* (Cambridge: Cambridge University Press, 2009); Clancy-Smith, *Mediterraneans*; Sibel Zandi-Sayek, *Ottoman Izmir: The Rise of a Cosmopolitan Port, 1840–1880* (Minneapolis: University of Minnesota Press, 2012); and Will Hanley, *Identifying with Nationality: Europeans, Ottomans, and Egyptians in Alexandria* (New York: Columbia University Press, 2017).

43. Jo Guldi, "Landscape and Place," in *Research Methods for History*, ed. Simon Gunn and Lucy Faire (Edinburgh: Edinburgh University Press, 2012), 74–75.

44. Malte Fuhrmann and Vangelis Kechriotis, "The Late Ottoman Port-Cities and Their Inhabitants: Subjectivity, Urbanity, and Conflicting Orders," *Mediterranean Historical Review* 24, no. 2 (2009): 71–78.

45. The first group includes, among others, Alexander Kitroeff, *The Greeks in Egypt 1919–1937: Ethnicity and Class* (London: Ithaca for the Middle East Centre, St. Antony's College, Oxford, 1989); Marta Petricioli, *Oltre il mito: L'Egitto degli italiani (1917–1947)* (Milan: Mondadori, 2007); and Angelos Dalachanis, *The Greek Exodus from Egypt: Diaspora Politics and Emigration, 1937–1962* (New York: Berghahn Books, 2017). For works specifically on the Lebanese diaspora in Egypt, see Thomas Philipp, *Demographic Patterns of Syrian Immigration to Egypt* (Haifa: University of Haifa, 1980); Thomas Philipp, *The Syrians in Egypt, 1725–1975* (Stuttgart: Steiner, 1985); Albert Hourani, "The Syrians in Egypt in the Eighteenth and Nineteenth Centuries," in *Colloque International Sur l'Histoire Du Caire*, vol. 229 (Cairo: Ministry of the Arab Republic of Egypt, 1972); Samir 'Abd al-Maqsud Sayyid, *al-Shawam fi Misr: Mundhu al-fath al-'Uthmani hatta awa'il al-qarn al-tasi' 'ashar* (Cairo: Al-Hay'ah al-Misriyah al-'Ammah lil-Kitab, 2003); and Hussam Eldin Raafat Ahmed, *From Nahda to Exile: A Story of the Shawam in Egypt in the Early Twentieth Century* (Montreal: McGill University Libraries, 2011). The second group includes, inter alia, Kemal H. Karpat, "The Ottoman Emigration to America, 1860–1914," *International Journal of Middle East Studies* 17, no. 2 (1985): 175–209; Albert Hourani and Nadim Shehadi, eds., *The Lebanese in the World: A Century of Emigration* (London: Centre for Lebanese Studies in association with I. B. Tauris, 1992); Akram Fouad Khater, *Inventing Home Emigration, Gender, and the Middle Class in Lebanon, 1870–1920* (Berkeley: University of California Press, 2001); Sarah Gualtieri, "Gendering the Chain Migration Thesis: Women and Syrian Transatlantic Migration, 1878–1924," *Comparative Studies of South Asia, Africa and the Middle East* 24, no. 1 (2004): 67–78; Andrew Arsan, *Interlopers of Empire: The Lebanese*

Diaspora in Colonial French West Africa (New York: Oxford University Press, 2014); Stacy Fahrenthold, "Sound Minds in Sound Bodies: Transnational Philanthropy and Patriotic Masculinity in al-Nadi al-Homsi and Syrian Brazil, 1920–32," *International Journal of Middle East Studies* 46, no. 2 (2014): 259–83; and Lily Pearl Balloffet, "From the Pampa to the Mashriq: Arab-Argentine Philanthropy Networks," *Mashriq & Mahjar* 4, no. 1 (2017): 5–30. Samples of the third group include Brian Glyn Williams, "Hijra and Forced Migration from Nineteenth-Century Russia to the Ottoman Empire: A Critical Analysis of the Great Crimean Tatar Emigration of 1860–1861," *Cahiers Du Monde Russe* 41, no. 1 (2000): 79–108; Oktay Ozel, "Migration and Power Politics: The Settlement of Georgian Immigrants in Turkey (1878–1908)," *Middle Eastern Studies* 46, no. 4 (2010): 477–96; Isa Blumi, *Ottoman Refugees, 1878–1939: Migration in a Post-Imperial World* (London: Bloomsbury, 2013); Başak Kale, "Transforming an Empire: The Ottoman Empire's Immigration and Settlement Policies in the Nineteenth and Early Twentieth Centuries," *Middle Eastern Studies* 50, no. 2 (2014): 252–71; Vladimir Hamed-Troyansky, "Circassian Refugees and the Making of Amman, 1878–1914," *International Journal of Middle East Studies* 49, no. 4 (2017): 605–23; and James H. Meyer, *Turks across Empires: Marketing Muslim Identity in the Russian-Ottoman Borderlands, 1856–1914* (Oxford: Oxford University Press, 2019).

46. Najm, *Bur Sa'id*; Valeska Huber, *Channelling Mobilities. Migration and Globalisation in the Suez Canal Region and Beyond, 1869–1914* (Cambridge: Cambridge University Press, 2013), 304–6, 319; and Huber, "Multiple Mobilities, Multiple Sovereignties, Multiple Speeds." I thank Benjamin C. Fortna for the metaphor.

47. Henri Lefebvre, *Writings on Cities* (Cambridge, UK: Blackwell Publishers, 1996), 101. On the ways in which space is both produced and productive, see also Certeau, *Practice of Everyday Life*, 117; Michel Foucault, "Space, Knowledge, and Power," in *The Foucault Reader*, ed. Paul Rabinow (New York: Pantheon Books, 1984), 245–46, 252; and Lefebvre, *Production of Space*, 11, 26. For an overview of "urban history" as a field, see Nancy Kwak, "Understanding 'Urban' from the Disciplinary Viewpoint of History," in *Defining the Urban: Perspectives across the Academic Disciplines*, ed. Deljana Iossifova, Alexandros Gasparatos, and Christopher Doll (New York: Routledge, 2018), 53–62.

48. Lucia Carminati, "Suez: A Hollow Canal in Need of Peopling; Currents and Stoppages in the Historiography, 1859–1956," *History Compass* 19, no. 5 (2021): 1–14.

49. See for example 'Abd al-'Aziz Muhammad Shinnawi, *Al-sukhrah fi hafr qanat al-suways* (Alexandria: Munsha'at al-Ma'rifa al-Hadithah, 1958); Shinnawi, *Qanat al-Suways wa-al-tayyarat al-siyasiyah*; and Mustafa Hifnawi, *Qanat al-Suways wa-mushkilatiha al-mu'asirah*, 4 vols. (Cairo: Al-Hay'ah al-Misriyah al-'Ammah lil-Kitab, 1952; repr., 2015). The inauguration of the "new" Suez Canal in August 2015 must have prompted its timely republication by the governmental agency for printing. For a critical yet restrained overview, see Emad Abou Ghazi and Xavier Daumalin, "Le Canal vu Par Les Égyptiens," in *L'épopée Du Canal de Suez* (Paris: Gallimard/Institut du monde arabe/Musée d'Histoire de Marseille, 2018), 46–51.

50. Montel, *Le chantier du Canal de Suez*; and Hubert Bonin, *History of the Suez Canal Company, 1858–1960: Between Controversy and Utility* (Geneva: Droz, 2010).

51. Céline Frémaux and Mercedes Volait, "Inventing Space in the Age of Empire: Planning Experiments and Achievements along [the] Suez Canal in Egypt (1859–1956)," *Planning Perspectives* 24, no. 2 (2009): 260; Marie-Laure Crosnier-Lecomte, Gamal Ghitani, and Naguib Amin, *Port-Saïd Architectures XIXe–XXe siècles* (Cairo: IFAO, 2006), 15; and Céline Frémaux, "Town Planning, Architecture, and Migrations in Suez Canal Port Cities," in *Port Cities: Dynamic Landscapes and Global Networks*, ed. Carola Hein (Abingdon, Oxon: Routledge, 2011), 156–73.

52. Frémaux, "Town Planning," 157; and Crosnier-Lecomte, Ghitani, and Amin, *Port-Saïd Architectures*, 322–23, 327. Among others, see Etienne Micard, *Le canal de Suez et le génie français* (Paris: Editions P. Roger, 1930), 17.

53. Edward Muir, "Introduction: Observing Trifles," in *Microhistory and the Lost Peoples of Europe*, ed. Edward Muir and Guido Ruggiero (Baltimore, MD: Johns Hopkins University Press, 1991), xxi.

54. Amy Stanley, "Maidservants' Tales: Narrating Domestic and Global History in Eurasia, 1600–1900," *American Historical Review* 121, no. 2 (2016): 438.

55. Carlo Ginzburg and Carlo Poni, "The Name and the Game," in *Microhistory and the Lost Peoples of Europe*, ed. Edward Muir and Guido Ruggiero (Baltimore, MD: Johns Hopkins University Press, 1991), 6.

56. Geoff Eley, "Labor History, Social History, 'Alltagsgeschichte': Experience, Culture, and the Politics of the Everyday; A New Direction for German Social History?," *Journal of Modern History* 61, no. 2 (1989); Alf Lüdtke, *The History of Everyday Life: Reconstructing Historical Experiences and Ways of Life* (Princeton, NJ: Princeton University Press, 1995), 8:5–6; John Brewer, "Microhistory and the Histories of Everyday Life," *Cultural and Social History* 7, no. 1 (2010): 87–109; and Francesca Trivellato, "Is There a Future for Italian Microhistory in the Age of Global History?," *California Italian Studies* 2, no. 1 (2011).

57. John-Paul A. Ghobrial, "Introduction: Seeing the World Like a Microhistorian," *Past & Present* 242 (2019): 15–16, 19; Giovanni Levi, "Frail Frontiers?," *Past & Present* 242 (2019): 37–49; and De Vito, "History without Scale."

58. Julia A. Clancy-Smith, "Twentieth-Century Historians and Historiography of the Middle East: Women, Gender, and Empire," in *Middle East Historiographies: Narrating the Twentieth Century*, ed. Israel Gershoni, Y. Hakan Erdem, and Amy Singer (Seattle: University of Washington Press, 2006), 90; Laila Parsons, "Micro-Narrative and the Historiography of the Modern Middle East," *History Compass* 9, no. 1 (2011): 84; Ulrike Freitag and Nora Lafi, "Daily Life and Family in an Ottoman Urban Context: Historiographical Stakes and New Research Perspectives," *History of the Family* 16, no. 2 (2011): 80–87; and Cyrus Schayegh, "Small Is Beautiful," *International Journal of Middle East Studies* 46, no. 2 (2014): 373–74.

59. Stanley, "Maidservants' Tales," 438. See Julia A. Clancy-Smith, "The Intimate, the Familial, and the Local in Transnational Histories of Gender," *Journal of Women's History* 18, no. 2 (2006): 174–83; and Liat Kozma, "Going Transnational:

On Mainstreaming Middle East Gender Studies," *International Journal of Middle East Studies* 48, no. 3 (2016): 574–77.

60. Khaled Fahmy, "Towards a Social History of Modern Alexandria," in *Alexandria, Real and Imagined*, ed. Anthony Hirst and Michael Silk (Aldershot, UK: Ashgate, 2004), 281.

61. Sylvia Hahn, "Nowhere at Home? Female Migrants in the Nineteenth-Century Habsburg Empire," in *Women, Gender, and Labour Migration: Historical and Global Perspectives*, ed. Pamela Sharpe (London: Routledge, 2011), 111; and Dirk Hoerder, "From Immigration to Migration Systems: New Concepts in Migration History," *OAH Magazine of History* 14, no. 1 (1999): 6–7.

62. De Vito, "History without Scale," 354–55, 370.

63. Lucassen, Lucassen, and Manning, *Migration History in World History*, 6; and Lucia Carminati, "Alexandria, 1898: Nodes, Networks, and Scales in Late Nineteenth-Century Egypt," *Comparative Studies in Society and History* 59, no. 1 (2017): 127–53. On the theoretical underpinnings of my claims, see Bernhard Struck, Kate Ferris, and Jacques Revel, "Introduction: Space and Scale in Transnational History," *International History Review* 33, no. 4 (2011): 573–84; and Sebouh Aslanian et al., "AHR Conversation How Size Matters: The Question of Scale in History," *American Historical Review* 118, no. 5 (2013): 1445.

64. For example, Najm, *Bur Sa'id*; Montel, *Le chantier du Canal de Suez*; John Chalcraft, "The Coal Heavers of Port Sa'id: State-Making and Worker Protest, 1869–1914," *International Labor and Working-Class History* 60 (October 2001): 110–24; Emily A. Haddad, "Digging to India: Modernity, Imperialism, and the Suez Canal," *Victorian Studies* 47, no. 3 (2005): 363–96; Huber, *Channelling Mobilities*; Vanja Strukelj, "Esporre l'Egitto: Viaggiatori europei all'inaugurazione del canale di Suez (1869)," *Ricerche di S/Confine* 4, no. 1 (2013): 3–23; and Lucia Carminati, "Port Said and Ismailia as Desert Marvels: Delusion and Frustration on the Isthmus of Suez, 1859–1869," *Journal of Urban History* 46, no. 3 (2019): 622–47.

65. Silvestre, *L'isthme de Suez 1854–1869*, 98; and The Franciscan Centre of Christian Oriental Studies, "Cronaca di Santa Caterina: Appendice Seconda," *Studia Orientalia Christiana, Collectanea*, nos. 26–27 (1996): 479.

66. CADC, AEC, CCCPS, vol. 1, Port Said, 19 June 1870, Pellissier, French Consul in Port Said, to French Ministry of Foreign Affairs, 80v.

67. Bonin, *History of the Suez Canal Company*, 74; and Archivio Storico del Ministero Affari Esteri e della Cooperazione Internazionale, Rome (hereafter ASMAECI), Archivi Rappresentanze Diplomatiche e Consolari all'Estero (hereafter ARDCE), Consolato Cairo (hereafter CC), Antico versamento (hereafter AV), busta (hereafter B) 12, Suez, 15 December 1869, Italian Vice-Consul, Suez, to Italian Consul in Cairo.

68. Thomas W. Gallant, *Modern Greece: From the War of Independence to the Present* (New York: Bloomsbury Academic, 2016), 118; and Mark I. Choate, *Emigrant Nation: The Making of Italy Abroad* (Cambridge, MA: Harvard University Press, 2008), 10–11.

69. Peter Gran, "Upper Egypt in Modern History: A 'Southern Question'?," in *Upper Egypt: Identity and Change*, ed. Nicholas S. Hopkins and Reem Saad (Cairo: American University in Cairo Press, 2004), 80; Zeinab Abul-Magd, *Imagined Empires: A History of Revolt in Egypt* (Berkeley: University of California Press, 2016), 7; and Nancy Y. Reynolds, "City of the High Dam: Aswan and the Promise of Postcolonialism in Egypt," *City & Society* 29, no. 1 (2017): 216.

70. Lucia Carminati and Mohamed Gamal-Eldin, "Decentering Egyptian Historiography: Provincializing Geographies, Methodologies, and Sources," *International Journal of Middle East Studies* 53, no. 1 (February 2021): 107–11.

71. Thanks to Lino Camprubí for raising this point at the workshop "Below the Surface: A New Wave of Interdisciplinary Mediterranean Studies and Environmental Changes," held in Menton in October 2018. See Fernand Braudel, *La Méditerranée et le monde méditerranéen à l'époque de Philippe II*, 1st ed. (Paris: Librairie Armand Colin, 1949); and David Abulafia, *The Great Sea: A Human History of the Mediterranean* (New York: Oxford University Press, 2011). Camprubí notes that Braudel's observation was in a chapter that would be removed in later editions.

72. Kenneth J. Perkins, *Port Sudan: The Evolution of a Colonial City* (Boulder, CO: Westview Press, 1993), 7. Alfred Thayer Mahan first postulated the idea of the Middle East as a maritime corridor and as a string of coaling depots serving British steamships; see Barak, *Powering Empire*, 2.

73. Clancy-Smith, *Mediterraneans*, 342–48.

74. Lefebvre, *Writings on Cities*, 100.

75. Zandi-Sayek, *Ottoman Izmir*, 6.

76. Julia A. Clancy-Smith, "Exoticism, Erasures, and Absence: The Peopling of Algiers, 1830–1900," in *Walls of Algiers: Narratives of the City through Text and Image*, ed. Zeynep Çelik, Julia A. Clancy-Smith, and Frances Terpak (Seattle: University of Washington Press, 2008), 22; and Clancy-Smith, *Mediterraneans*. Interestingly, British traveler Evelyn Waugh explicitly compared Port Said to Algiers, which he visited in 1929, and pointed out that "racial and colour distinctions" were much stronger in the former. Evelyn Waugh, *Labels: A Mediterranean Journal* (London: Duckworth, 1974), 186–87.

77. Charles Philip Issawi, *An Economic History of the Middle East and North Africa* (New York: Columbia University Press, 1982), 100.

78. Fuhrmann and Kechriotis, "Late Ottoman Port-Cities and Their Inhabitants." See Nezar AlSayyad, "Orchestrating Difference, Performing Identity: Urban Space and Public Rituals in Nineteenth-Century Izmir," in *Hybrid Urbanism: On the Identity Discourse and the Built Environment* (Westport, CT: Praeger, 2001); Elisabeth Kendall, "Between Politics and Literature: Journals in Alexandria and Istanbul at the End of the Nineteenth Century," in *Modernity and Culture: From the Mediterranean to the Indian Ocean*, ed. C. A. Bayly, Leila Tarazi Fawaz, and Robert Ilbert (New York: Columbia University Press, 2002), 75–95; Haris Exertzoglou, "The Cultural Uses of Consumption: Negotiating Class, Gender, and Nation in Ottoman Urban Centers during the 19th Century," *International Journal of Middle East Studies* 35, no. 1 (2003): 77–25; and Anthony Gorman, "Foreign Workers in Egypt 1882–1914:

Subaltern or Labour Elite?," in *Subalterns and Social Protest: History from below in the Middle East and North Africa*, ed. Stephanie Cronin (London: Routledge, 2008).

79. See, for example, Caroline Piquet, *Le canal de Suez, une voie maritime pour l'Egypte et le monde* (Paris: Erick Bonnier Editions, 2018). For a nuanced analysis of Port Said's position in the hydrocarbon economy, see Barak, *Powering Empire*.

80. Çağlar Keyder, Eyüp Özveren, and Donald Quataert, "Port-Cities in the Ottoman Empire: Some Theoretical and Historical Perspectives," *Review (Fernand Braudel Center)* 16, no. 4 (1993): 518.

81. Biray Kolluoğlu and Meltem Toksöz, *Cities of the Mediterranean: From the Ottomans to the Present Day* (London: I. B. Tauris, 2010), 2; and Ziad Fahmy, "Jurisdictional Borderlands: Extraterritoriality and 'Legal Chameleons' in Precolonial Alexandria, 1840–1870," *Comparative Studies in Society and History* 55, no. 2 (2013): 307.

82. Marcelo J. Borges and Susana B. Torres, *Company Towns. Labor, Space, and Power Relations across Time and Continents* (New York: Palgrave Macmillan, 2012), 23.

83. Colette Dubois, *Djibouti, 1888–1967: Héritage ou frustration* (Paris: Harmattan, 1997), 56–59; and Perkins, *Port Sudan*, 5.

84. Daniel R. Headrick, *The Tentacles of Progress: Technology Transfer in the Age of Imperialism, 1850–1940* (New York: Oxford University Press, 1988), 32, 146.

85. Jürgen Osterhammel, *The Transformation of the World: A Global History of the Nineteenth Century* (Princeton, NJ: Princeton University Press, 2014), 911; and Musab Younis, "'United by Blood': Race and Transnationalism during the Belle Époque," *Nations and Nationalism* 23, no. 3 (2017): 486.

86. Dar al-Watha'iq al-Qawmiyya, National Archives, Cairo (DWQ), 5009-000222, [Italian Archive of the Suez Canal, 1848–1931], no author, Transcription of an article that appeared in *L'Austria* in 1850.

87. Elizabeth Mitchell, *Liberty's Torch: The Great Adventure to Build the Statue of Liberty* (New York: Atlantic Monthly Press, 2018). See also "Lettres et journal du second voyage en Égypte (1869) d'Auguste Bartholdi adressés à sa mère," 13 April 1869, Colmar, Musée Bartholdi, Côte VI 3 E, cited in Hélène Braeuner, "À la frontière de l'Égypte: Les représentations du canal de Suez," *In Situ: Revue des patrimoines*, no. 38 (February 12, 2019): 1–14.

88. Waugh, *Labels*, 81.

89. Todd, *Velvet Empire*, 245.

90. Charlene L. Porsild, *Gamblers and Dreamers: Women, Men, and Community in the Klondike* (Vancouver: University of British Columbia Press, 1998), 18. Modelski instead argues that Port Said was different from presumably haphazard cities in the American West because "every yard of it was accounted for with scientific precision." Sylvia Modelski, *Port Said Revisited* (Washington, DC: FAROS, 2000), 55.

91. Reynolds, "City of the High Dam," 216.

92. Jean-Édouard Goby, *Ingénieurs et techniciens français en Egypte* (Paris: Société des ingénieurs civils de France, 1949), 702. See Miranda Frances Spieler, *Empire and Underworld: Captivity in French Guiana* (Cambridge, MA: Harvard University Press, 2012).

93. Peter Pyne, *The Panama Railroad* (Bloomington: Indiana University Press, 2021), 262.

94. Ferdinand de Lesseps, *Lettres, journal et documents pour servir à l'histoire du canal de Suez II* (Paris: Didier, 1875), 2:219.

95. Julie Greene, *The Canal Builders: Making America's Empire at the Panama Canal* (New York: Penguin Press, 2009), 21–26, 40–41, 195, 198.

96. "Recollections of Panama," *The Spectator*, March 25, 1893, 392.

97. I am here responding to one of Jonathan Miran's pleas, even if those referred to small and midsized Red Sea port-towns. Miran, "Mapping Space and Mobility in the Red Sea Region," 207–8.

98. Clare Anderson, "Subaltern Lives: History, Identity and Memory in the Indian Ocean World," *History Compass* 11, no. 7 (2013): 504; and Donna R. Gabaccia, "Is Everywhere Nowhere? Nomads, Nations, and the Immigrant Paradigm of United States History," *Journal of American History* 86 (1999): 1115.

99. Carolyn Steedman, *Dust: The Archive and Cultural History* (New Brunswick, NJ: Rutgers University Press, 2002), 9; Durba Ghosh, "National Narratives and the Politics of Miscegenation: Britain and India," in *Archive Stories: Facts, Fictions, and the Writing of History*, ed. Antoinette M. Burton (Durham, NC: Duke University Press, 2005), 27; Michelle T. King, "Working with/in the Archives," in *Research Methods for History*, ed. Simon Gunn and Lucy Faire (Edinburgh: Edinburgh University Press, 2012), 19. For a seminal reflection on multisited research, see George E. Marcus, "Ethnography in/of the World System: The Emergence of Multi-Sited Ethnography," *Annual Review of Anthropology* 24, no. 1 (1995): 95–117.

100. Lucia Carminati, "Dead Ends in and out of the Archive: An Ethnography of Dār al Wathā'iq al Qawmiyya, the Egyptian National Archive," *Rethinking History* 23, no. 1 (2019): 34–51.

101. Gabaccia, "Is Everywhere Nowhere?," 1115. The reference to fever-inducing fieldwork is inspired by Steedman, *Dust*, 17–31.

102. Craig J. Calhoun, "The Class Consciousness of Frequent Travelers: Toward a Critique of Actually Existing Cosmopolitanism," *South Atlantic Quarterly* 101, no. 4 (2002): 872; Cresswell, *On the Move*, 55; and Emily Callaci, "On Acknowledgments," *American Historical Review* 125, no. 1 (2020): 128.

103. Anna Lowenhaupt Tsing, *Friction: An Ethnography of Global Connection* (Princeton, NJ: Princeton University Press, 2005), 5–6.

104. Michel Foucault, "Lives of Infamous Men," in *The Essential Foucault: Selections from Essential Works of Foucault, 1954–1984*, ed. Paul Rabinow and Nikolas Rose (New York: New Press, 2003), 282. For an ethnographic exploration of the practices that regulate access to and research within the Egyptian National Archives, see Carminati, "Dead Ends in and out of the Archive."

105. Anderson, "Subaltern Lives," 503; and Malte Fuhrmann, "'I Would Rather Be in the Orient': European Lower-Class Immigrants into the Ottoman Lands," in *The City in the Ottoman Empire Migration and the Making of Urban Modernity*, ed. Malte Fuhrmann et al. (London: Routledge, 2011), 229.

106. David Zeitlyn, "Anthropology in and of the Archives: Possible Futures and Contingent Pasts; Archives as Anthropological Surrogates," *Annual Review of Anthropology* 41, no. 1 (2012): 464.

107. Archives du ministère de l'Europe et des Affaires étrangères, Centre des Archives Diplomatiques de Nantes (CADN), Archives Rapatriés de l'Agence Consulaire de Ismailia (ARI), carton (C) 4, Chalouf, 16 April 1866, Municipal Superintendent to Montel, Head of Section in Chalouf.

108. Beth Baron, *The Orphan Scandal: Christian Missionaries and the Rise of the Muslim Brotherhood* (Stanford, CA: Stanford University Press, 2014), 8.

109. Ussama Samir Makdisi, *Artillery of Heaven: American Missionaries and the Failed Conversion of the Middle East* (Ithaca, NY: Cornell University Press, 2008), 6.

110. Archives de la Maison-Mère du Bon Pasteur, Angers (AMMBPA), HC 43 Port Said, 26 December 1876, "De notre Monastère de Port-Saïd," 7–8.

111. Liat Kozma, *Policing Egyptian Women: Sex, Law, and Medicine in Khedival Egypt* (Syracuse, NY: Syracuse University Press, 2011), xx, xvi; Thomas W. Gallant, "Tales from the Dark Side: Transnational Migration, the Underworld and the 'Other' Greeks of the Diaspora," in *Greek Diaspora and Migration since 1700: Society, History, and Politics*, ed. Dimitris Tziovas (Farnham, UK: Ashgate, 2009), 18–19; and Ehud R. Toledano, "Law, Practice and Social Reality: A Theft Case in Cairo, 1854," *Asian and African Studies* 17, nos. 1–3 (1983): 155. See Carlo Ginzburg, "The Inquisitor as Anthropologist," in *Clues, Myths, and the Historical Method* (Baltimore, MD: Johns Hopkins University Press, 1989), 156–64.

112. Barbara Wright, "Narcisse Berchère: Peintre-écrivain dans l'Isthme de Suez," in *Écrits-voyageurs: Les artistes et l'ailleurs*, ed. Laurence Brogniez (Bruxelles: Peter Lang, 2012), 53.

113. Mohamed Gamal-Eldin, "Photography and the Politics of Erasure: A Case Study of the Suez and Isma'ilia Canals Construction and the Silencing of the Egyptian Laborer" (paper presented at the Annual History Seminar, American University in Cairo, March 30–31, 2018); and Huber, *Channelling Mobilities*, 69.

114. Huber, *Channelling Mobilities*, 63; Anne-Gaëlle Weber, "Les perroquets de Cook ou les enjeux de l'usage des lieux communs dans le récit de voyage scientifique au xixe siècle," in *La littérature dépliée: Reprise, répétition, réécriture*, ed. Jean-Paul Engélibert and Yen-Maï Tran-Gervat (Rennes: Presses universitaires de Rennes, 2008); and Derek Gregory, "Scripting Egypt: Orientalism and the Cultures of Travel," in *Writes of Passage: Reading Travel Writing*, ed. James S. Duncan and Derek Gregory (London: Routledge, 1999), 119.

115. John Pudney, *Suez: De Lesseps' Canal* (New York: Praeger, 1969), 11; and Ernest Desplaces, "Une brochure de M. De Oliveira," *L'Isthme de Suez*, 1 July 1863, 251. Antoinette J. F. A. Drohojowska authored a history of the Suez Canal bearing, alas, no explicitly biographical traces. See Antoinette Drohojowska, *L'Égypte et le Canal de Suez* (Paris: Laporte, 1870).

116. Huber, *Channelling Mobilities*, 60.

117. E. Sallior, *La vérité sur l'isthme de Suez: Lettre à messieurs les actionnaires de la société anonyme du percement de l'isthme de Suez* (Paris: E. Dentu, 1864), 12.

118. Kenneth J. Perkins, "'The Best Laid out Town on the Red Sea': The Creation of Port Sudan, 1904–09," *Middle Eastern Studies* 27, no. 2 (April 1, 1991): 294.
119. Clancy-Smith, "Exoticism, Erasures, and Absence," 21.
120. Lucie Ryzova, "The Good, the Bad and the Ugly: Collector, Dealer, and Academic in the Informal Old-Paper Markets of Cairo," in *Archives, Museums, and Collecting Practices in the Modern Arab World*, ed. Sonja Mejcher-Atassi and John Pedro Schwartz (Farnham, UK: Ashgate, 2012), 111.
121. CADN, ARI, C26, Rabbi [San Bernardo], 5 Marzo [1869], Madallena [sic] Zanoni to "My only Love"; C1, Ismailia, 27 August 1863, Marie Dutruy to Chief Engineer of Ismailia Division. See Arlette Farge, *The Allure of the Archives* (New Haven, CT: Yale University Press, 2013), 59–61.
122. Massey, "Global Sense of Place," 319–23.
123. Olivier Hambursin, ed., *Récits du dernier siècle des voyages: De Victor Segalen à Nicolas Bouvier* (Paris: Presses de l'Université Paris-Sorbonne, 2005), 128–29; and Panaït Istrati, *Vie d'Adrien Zograffi* (Paris: Gallimard, 1969), 411. In this largely autobiographical account completed in 1934, the author recollects his six visits to Egypt between 1908 and 1911.

CHAPTER 1. A UNIVERSAL MEETING POINT ON THE ISTHMUS OF SUEZ

1. Archives du ministère de l'Europe et des Affaires étrangères, Centre des Archives Diplomatiques de Nantes (hereafter CADN), Archives Rapatriés de l'Agence Consulaire de Ismailia (hereafter ARI), C9, Ismailia, 18 November 1869, Captain Arène to French Consular Agent in Ismailia; Ismailia, 21 November 1869, French Consul to Captain Arène; and ARI, C1, Seuil d'El Guisr, 28 January 1862, Imam Moustapha to Engineer in chief.
2. Dar al-Watha'iq al-Qawmiyya, National Archives, Cairo (hereafter DWQ), 5013-003229, Paris, 22 October 1873, Vice-Admiral de la Roncière; Archivio Storico del Ministero degli Affari Esteri e della Cooperazione Internazionale, Rome (hereafter ASMAECI), Archivi Rappresentanze Diplomatiche e Consolari all'Estero (hereafter ARDCE), Consolato Cairo (hereafter CC), Busta (hereafter B) 30, Port Said, 2 September 1878, Vice-Consul Zerbani, to Malmusi, Italian Consul in Cairo; CADN, Archives Rapatriés du Consulat de France à Port-Saïd (hereafter ARCFPS), C49, no place, n.d., Mr. Ch. P. to Consul in Port Said; and ARCFPS, C149, Port Said, 13 March 1889, French consul, Procès-verbal (Audibert).
3. CADN, ARCFPS, C21, Paris, 25 September 1866, Cintrat, Director of the chancellery archives, to French Consul General, Alexandria.
4. "Decesso," *L'Imparziale*, February 24, 1898, 3.
5. DWQ, 5009-000232, Cairo, 28 March 1872, Ferdinand De Lesseps, Memorandum for the khedive.
6. Ilham Khuri-Makdisi, *The Eastern Mediterranean and the Making of Global Radicalism, 1860–1914* (Berkeley: University of California Press, 2010), 170; Valeska

Huber, *Channelling Mobilities. Migration and Globalisation in the Suez Canal Region and Beyond, 1869–1914* (Cambridge: Cambridge University Press, 2013), 111; and Caroline Piquet, "La Compagnie Universelle du Canal Maritime de Suez en Égypte, de 1888 à 1956" (PhD diss., Université Paris-Sorbonne, 2006), 248, 258.

7. I am borrowing this concept from Laura Tabili, *Global Migrants, Local Culture: Natives and Newcomers in Provincial England, 1841–1939* (Basingstoke, UK: Palgrave Macmillan, 2011), 5. See Paola Corti, "L'emigrazione temporanea in Europa, in Africa e nel Levante," in *Storia dell'emigrazione italiana: Partenze*, ed. Piero Bevilacqua, Andreina De Clementi, and Emilio Franzina (Rome: Donzelli, 2009), 234.

8. Jan Lucassen, Leo Lucassen, and Patrick Manning, *Migration History in World History. Multidisciplinary Approaches* (Leiden: Brill, 2010), 10; and David Eltis, *Coerced and Free Migration: Global Perspectives* (Stanford, CA: Stanford University Press, 2004).

9. The National Archives, London (hereafter TNA), Foreign Office (hereafter FO) 423-1, Alexandria, 26 April 1860, Robert G. Colquhoun to Lord J. Russell, Report by J. Coulthard.

10. "The Suez Canal," *Times*, 12 May 1862, 6.

11. Archives Nationales du Monde du Travail, Roubaix (hereafter ANMT), Compagnie universelle du canal maritime de Suez (hereafter CUCMS), 1995 060 3600, Port Said, 18 April 1861, Laroche, Division Chief Engineer, Map no. III, "Section Port Saïd." For a discordant interpretation, see Claudine Piaton, "L'habitat Ouvrier Dans Les Villes Du Canal de Suez," in *Villages Ouvriers et Villes-Usines à Travers Le Monde* (Chambéry: Université de Savoie Mont Blanc, 2016), 160.

12. Luigi Dori, "Esquisse historique de Port Said: Naissance de la ville et inauguration du Canal (1859–1869)," *Cahiers d'histoire egyptienne* 6, nos. 3/4 (October 1954): 202; Mohamed Anouar Moghira, *L'Isthme de Suez: Passage millénaire, 640–2000* (Paris: L'Harmattan, 2002), 191; and Zayn Al-'Abidin Shams al-Din Najm, *Bur Sa'id: Tarikhuha wa-tatawwuruha, mundhu nash'atiha 1859 hatta 'am 1882* (Cairo: Al-Hay'ah al-Misriyah al-'Ammah lil-Kitab, 1987), 43.

13. 'Abd al-'Aziz Muhammad Shinnawi, *Al-Sukhrah fi hafr qanat al-suways* (Alexandria: Munsha'at al-Ma'rifa al-Haditha, 1958), 68.

14. Ferdinand de Lesseps, *Conférence sur les travaux du canal du Suez et le sort des ouvriers en Egypte* (Paris: Association polytechnique/Bureaux de l'isthme de Suez, 1862), 13–15; François-Philippe Voisin, *Le canal de Suez* (Paris: Vve C. Dunod, 1902), 6:163, 198–202; Olivier Ritt, *Histoire de l'isthme de Suez* (Paris: L. Hachette et cie, 1869), 196–98; George Percy Badger, *A Visit to the Isthmus of Suez Canal Works* (London: Smith, Elder, 1862), 1862; and TNA, FO 423-1, Alexandria, 11 March 1864, Rev. E. J. Davis, Chaplain in Alexandria, to Mr. Layard.

15. Sylvia Modelski, *Port Said Revisited* (Washington, DC: FAROS, 2000), 28.

16. Nathalie Montel, *Le chantier du Canal de Suez, 1859–1869: Une histoire des pratiques techniques* (Paris: Éd. In forma/Presses de l'École Nationale des Ponts et Chaussées, 1999), 38.

17. Ferdinand de Lesseps, *Lettres, journal et documents pour servir à l'histoire du canal de Suez* (Paris: Didier, 1875), 4:355, entry dated 29 October 1863.

18. Ritt, *Histoire de l'isthme de Suez*, 284–85.

19. TNA, FO 423-1, Constantinople, 27 June 1863, Sir H. Bulwer to Earl Russell, 210.

20. "Décret et règlement pour les ouvriers fellahs," Alexandria, July 20, 1856, in Jules Charles-Roux, *L'Isthme et le Canal de Suez: Historique, état actuel* (Paris: Hachette, 1901), 1, 469–70.

21. Lucie Duff Gordon, *Letters from Egypt* (New York: McClure, Phillips, 1902), 243, 257. See Kenneth M. Cuno, *The Pasha's Peasants: Land, Society, and Economy in Lower Egypt, 1740–1858* (Cambridge: Cambridge University Press, 1992), 122.

22. Ehud R. Toledano, *State and Society in Mid-Nineteenth-Century Egypt* (Cambridge: Cambridge University Press, 1990), 189.

23. Nubar Nubarian, *Mémoires de Nubar Pacha* (Beirut: Librairie du Liban, 1983), 227, 230, 232, 236.

24. Ehud R. Toledano, *State and Society in Mid-Nineteenth-Century Egypt* (Cambridge: Cambridge University Press, 1990), 188.

25. Roger Owen, *The Middle East in the World Economy, 1800–1914* (London: I. B. Tauris, 2009), 143.

26. On Barak, *Powering Empire: How Coal Made the Middle East and Sparked Global Carbonization* (Oakland: University of California Press, 2020), 38; and Alan Mikhail, *Nature and Empire in Ottoman Egypt: An Environmental History* (Cambridge: Cambridge University Press, 2011), 281–82.

27. P. Dubois, "Le *Constitutionnel* et l'Egypte," *L'Isthme de Suez*, 10 May 1858, 228–29.

28. TNA, FO 423-1, Alexandria, 24 January 1863, Colquhoun to Earl Russell; and "The Death of Said Pasha," *Times*, 20 January 1863, 8. Both the British Peninsular and Oriental Company and those in charge of the Alexandria-Cairo railroad had reportedly relied on the corvée system. See, among others, David S. Landes, *Bankers and Pashas: International Finance and Economic Imperialism in Egypt* (London: Heinemann, 1958), 180; and Helen Anne B. Rivlin, "The Railway Question in the Ottoman-Egyptian Crisis of 1850–1852," *Middle East Journal* 15, no. 4 (1961): 375–76, 384.

29. Cheykh Réfaah, "Chant égyptien sur le percement de l'isthme de Suez," *L'Isthme de Suez*, 25 June 1856, 16. See Albert Hourani, *Arabic Thought in the Liberal Age, 1798–1939* (London: Oxford University Press, 1962), 69–83; and Rifa'ah Rafi' Tahtawi, *Kitab manahij al-albab al-misriyah fi mabahij al-adab al-'asriyah* (Bulaq: al-Matba'ah al-Misriyah, 1869).

30. Khaled Fahmy, *All the Pasha's Men: Mehmed Ali, His Army, and the Making of Modern Egypt* (Cairo: American University in Cairo Press, 2002), 10.

31. Mikhail, *Nature and Empire in Ottoman Egypt*, 289. On forced labor on the Mahmudiyya Canal, see Alan Mikhail, "Unleashing the Beast: Animals, Energy, and the Economy of Labor in Ottoman Egypt," *American Historical Review* 118, no. 2 (2013): 346; Fahmy, *All the Pasha's Men*, 10; Kenneth M. Cuno, *The Pasha's Peasants: Land, Society, and Economy in Lower Egypt, 1740–1858* (Cambridge: Cambridge University Press, 1992), 121–22; Robert L. Tignor, *Modernization and British*

Colonial Rule in Egypt, 1882–1914 (Princeton, NJ: Princeton University Press, 1966), 121–22; and Helen Anne B. Rivlin, *The Agricultural Policy of Muhammad 'Ali in Egypt* (Cambridge, MA: Harvard University Press, 1961), 216–21.

32. Nubarian, *Mémoires de Nubar Pacha*, 226. See Faruk Bilici, *Le canal de Suez et l'Empire ottoman* (Paris: CNRS Editions, 2019), 98.

33. Lesseps, *Lettres, journal et documents*, 4:355, entry dated 29 October 1863. See Landes, *Bankers and Pashas*, 180; and Montel, *Le chantier du Canal de Suez*, 38.

34. Istituto per l'Oriente Carlo Nallino, Rome (hereafter IPOCAN), busta (hereafter B) 1956–1961, Copy of the minutes of the meeting at the Ministry of Foreign Affairs in Alexandria, 4 October 1859; and TNA, FO 423-1, Alexandria, 6 October 1859, Colquhoun to Sir H. Bulwer.

35. DWQ, 5009-000208, 6 April 1863, Ali Pacha to representatives of the sultan in Paris and London (copy of annex to no. 534-1863 of the Ministry of Commerce from the "Alte Akten des Handelsministeriums Suezkanal 1860–1875").

36. TNA, FO 423-1, London, 10 February 1863, Earl Russell to Colquhoun; and DWQ, 5009-000208, Shawwāl 1279 hijri (March–April 1863), Grand Vizier to Viceroy of Egypt.

37. Ordinance to the head of the Khedivial Da'ira, 11 sha'ban 1275 h. (16 March 1859), and Ordinance to Ja'far Pasha, governor of Damietta, 21 Dhul-Hijjah 1275 h. (22 July 1859), in Amin Sami, *Taqwim al-Nil* (Cairo: Matba'at al-Amiriyah, 1916), 3:1:317, 329.

38. Najm, *Bur Sa'id*, 22; and IPOCAN, B 1956–1961, Vienna, 21 July 1859, Ministry of Foreign Affairs to Ministry of Commerce.

39. Henri Silvestre, *L'isthme de Suez 1854–1869, avec carte et pièces justificatives* (Paris: Lacroix, Verboeckhoven, 1869); and 81 IPOCAN, B 1956–1961, Alexandria, 9 July 1859, Report by Consul General Schreiner; and B 1956–1961, Copy of the minutes of the meeting at the Ministry of Foreign Affairs in Alexandria, 4 October 1859.

40. TNA, FO 423-1, Alexandria, 6 October 1859, Colquhoun to Sir H. Bulwer; and IPOCAN, B 1956–1961, Alexandria, 9 July 1859, Report by Consul General Schreiner.

41. IPOCAN, B 1956–1961, Paris, 23 October 1859, Secretary General P. Merruare, Note presented to Emperor on behalf of the Board of Directors of the Company; and B 1956–1961, Constantinople, 19 August 1859, Prokesch to Count Rechberg Rothenlowen.

42. H. Abel, "Dernières nouvelles d'Egypte," *L'Isthme de Suez*, 15 November 1859, 338; F. Lapierre, "La France et le Canal de Suez," *L'Isthme de Suez*, 15 November 1859, 343; "Revue des journaux," *L'Isthme de Suez*, 15 November 1859, 348; Henri Silvestre, *L'isthme de Suez 1854–1869, avec carte et pièces justificatives...* (Paris: Librairie internationale/A. Lacroix, Verboeckhoven, 1869), 81; and Charles-Roux, *L'Isthme et le Canal de Suez*, 1:300–304.

43. ANMT, CUCMS, 1995 060 4136, Montbéliard (Doubs), 8 December 1858, Falker, Agent of "American liners," to Canal Company Director; 1995 060 4136, Civitavecchia, 28 August 1859, Joseph Casanova to Bourdon, Engineer in chief of

the division of the maritime works of Suez in Paris; and 1995 060 4136, Diakovar (Slavonia), 12 March 1860, Johann Steiner to De Lesseps. In the early modern era and the nineteenth century, the kingdom of Croatia-Slavonia was under the authority of the Hungarian portion of the Habsburg Empire. Dominique Kirchner Reill, *Nationalists Who Feared the Nation: Adriatic Multi-Nationalism in Habsburg Dalmatia, Trieste and Venice* (Stanford, CA: Stanford University Press, 2012), 249.

44. Nubarian, *Mémoires de Nubar Pacha*, 226.

45. TNA, FO 423-1, Alexandria, 26 April 1860, Colquhoun to Lord J. Russell, Report by J. Coulthard.

46. Najm, *Būr Saʿid*, 57.

47. ANMT, CUCMS, 1995 060 1323, n.d., Excerpts of the Reports of the Director General.

48. TNA, FO 423-1, Cairo, 23 April 1861, Colquhoun to Lord J. Russell and enclosure dated 9 March 1861, 100–101.

49. Lesseps, *Lettres, journal et documents*, 4:18, entry dated 8 February 1861; and Montel, *Le chantier du Canal de Suez*, 40.

50. Lesseps, *Lettres, journal et documents*, 4:14, 24, entries dated 23 January 1861 and 4 March 1861.

51. TNA, FO 423-1, Cairo, 23 April 1861, Colquhoun to Lord J. Russell and enclosure dated 9 March 1861, 101.

52. De Lesseps, *Lettres, journal et documents*, 4:13–15, entry dated 23 January 1861; 4:24, entry dated 4 March 1861.

53. Arthur Mangin, "Percement de l'isthme de Suez," *Journal des économistes: Révue scientifique et industrielle* 3, no. 1 (December 15, 1866): 449; and Lesseps, *Lettres, journal et documents*, 4:24.

54. Masʿud Dahir, *Hijrat al-shawam: Al-hijrah al-lubnaniyah ila Misr* (Cairo: Dar al-Shuruq, 2009), 104–5.

55. Huber, *Channelling Mobilities*, 114.

56. Ernest Desplaces, "Chronique de l'Isthme," *L'Isthme de Suez*, 1 March 1861, 66; Alfred Guillemin, *L'Égypte actuelle son agriculture et le percement de l'isthme de Suez* (Paris: Challamel, 1867), 248; Ritt, *Histoire de l'isthme de Suez*, 285; and Ferdinand de Lesseps, *Lettres, journal et documents pour servir à l'histoire du canal de Suez III* (Paris: Didier, 1875), 3:431, entry dated 7 December 1860.

57. Lesseps, *Lettres, journal et documents*, 4:49, entry dated 15 May 1861.

58. Lesseps, *Lettres, journal et documents*, 3:398; and Lesseps, *Lettres, journal et documents*, 4:12, entry dated 23 January 1861.

59. ANMT, CUCMS, 1995 060 1323, excerpts of the Reports of the Director General, n.d. See Montel, *Le chantier du Canal de Suez*, 42.

60. "The Canal Across Egypt," *Times*, 6 June 1861, 10.

61. ANMT, CUCMS, 1995 060 1323, n.d., Excerpts of the Reports of the Director General.

62. Mangin, "Percement de l'isthme de Suez," 449; and Montel, *Le chantier du Canal de Suez*, 40–41.

63. Lesseps, *Lettres, journal et documents*, 3:398, entry dated 11 November 1860.

64. ANMT, CUCMS, 1995 060 1323, n.d., Excerpts of the Reports of the Director General.

65. Montel, *Le chantier du Canal de Suez*, 42; and Hubert Bonin, *History of the Suez Canal Company, 1858–1960: Between Controversy and Utility* (Geneva: Droz, 2010), 68.

66. ANMT, CUCMS, 1995 060 1323, 8 February 1862, Division de Port-Saïd, Division de Timsah, Report of the Engineer in chief.

67. ANMT, CUCMS, 1995 060 1323, n.d., Excerpts of the Reports of the Director General; 1995 060 4136, Toussoum, Ismail Hamdy, 12 May 1863, to Sciama, Engineer in chief; IPOCAN, B1956-1961, Alexandria, 10 April 1862, Consul General Schreiner to Count Rechberg; TNA, FO 423-1, Alexandria, 1 April 1862, Colquhoun to Lord J. Russell; and Nubarian, *Mémoires de Nubar Pacha*, 227.

68. Nubarian, *Mémoires de Nubar Pacha*, 227.

69. Mikhail, "Unleashing the Beast," 345.

70. IPOCAN, B1956-1961, Alexandria, 10 April 1862, Consul General Schreiner to Count Rechberg; and Nubarian, *Mémoires de Nubar Pacha*, 227.

71. Ritt, *Histoire de l'isthme de Suez*, 286–87.

72. E. Sallior, *La vérité sur l'isthme de Suez: Lettre à messieurs les actionnaires de la société anonyme du percement de l'isthme de Suez* (Paris: E. Dentu, 1864), 57–58.

73. Sallior, *La vérité sur l'isthme de Suez*, 57; Émile Erckmann and Alexandre Chatrian, *Souvenirs d'un ancien chef de chantier à l'isthme de Suez* (Paris: Hetzel, 1877), 5, 7, 22; Robert L. Tignor, *Modernization*, 123; Evelyn Baring Cromer, *Modern Egypt* (New York: Macmillan, 1908), 1:406–7; and Lesseps, *Lettres, journal et documents*, 4:49, entry dated 15 May 1861.

74. Toledano, *State and Society in Mid-Nineteenth-Century Egypt*, 221; and Joel Beinin and Zachary Lockman, *Workers on the Nile: Nationalism, Communism, Islam, and the Egyptian Working Class, 1882–1954* (Princeton, NJ: Princeton University Press, 1987), 25. On "shaykhs" in general, see Gabriel Baer, *Studies in the Social History of Modern Egypt*, Publications of the Center for Middle Eastern Studies, no. 4 (Chicago: University of Chicago Press, 1969), 30–61. Marius Etienne Fontane and Edouard Riou, *Le canal maritime de Suez illustré: Histoire du canal et des travaux* (Paris: Aux bureaux de l'Illustration A. Marc et Cie., 1869), 130.

75. Ernest Desplaces, "Chronique de l'Isthme," *L'Isthme de Suez*, 1 January 1861, 5; and Narcisse Berchère, *Le désert de Suez: Cinq mois dans l'isthme* (Paris: Hetzel, 1863), 203.

76. CADN, ARI, C1, El Ferdane, 14 December 1861, Riche, Sanson, Guiter.

77. Lionel Wiener, *L'Egypte et ses chemins de fer* (Brussels: Weissenbruch, 1932), 70; and Rivlin, "Railway Question in the Ottoman-Egyptian Crisis of 1850–1852," 375–76.

78. L. A. Balboni, *Gl'Italiani nella civiltà egiziana del secolo XIX: Storia-biografie-monografie* (Alexandria: Tipo-litografico v. Penasson, 1906), 2:138–39; and Richard Allen, *Letters from Egypt, Syria, and Greece* (Dublin: Gunn & Cameron, 1869), 20.

79. Lesseps, *Lettres, journal et documents*, 4:346, entry dated 1 September 1863.

80. Ritt, *Histoire de l'isthme de Suez*, 429, 458, entries dated 17 January 1869 (Suez) and 23 February 1863 (El Guisr).

81. Montel, *Le chantier du Canal de Suez*, 175.

82. Wiener, *L'Egypte et ses chemins de fer*, 73, 75; Fontane and Riou, *Le canal maritime de Suez illustré*, 39; and William F. V. Fitzgerald, *The Suez Canal, the Eastern Question, and Abyssinia* (London: Spottiswoode, 1867), 54. On the antagonism between the railway and canal projects, see Andrea Leonardi and Alice Riegler, "Luigi Negrelli (1799–1858): A Tyrolean Engineer at the Heart of the Suez Canal Project," in *Italy and the Suez Canal, from the Mid-nineteenth Century to the Cold War: A Mediterranean History*, ed. Barbara Curli (Cham, Switzerland: Palgrave Macmillan, 2022), 113–28.

83. On Barak, *Powering Empire: How Coal Made the Middle East and Sparked Global Carbonization* (Oakland: University of California Press, 2020), 37–38.

84. J. Stephen Jeans, *Waterways and Water Transport in Different Countries: With a Description of the Panama, Suez, Manchester, Nicaraguan, and Other Canals* (London: E. & F. N. Spon, 1890), vii.

85. Erckmann and Chatrian, *Souvenirs d'un ancien chef de chantier*, 2. It was Erckmann who wrote this work out of the "fresh" impressions gathered in the Orient, where he had recently sojourned. The authors could not find a journal willing to publish the *Souvenirs* as a feuilleton and had to postpone its publication. This publishing feat of theirs has been considered a "mistake" by critics. Georges Benoit-Guyod and Jean-Jacques Pauvert, *La vie et l'œuvre d'Erckmann-Chatrian* (Paris: J.-J. Pauvert, 1963), 173.

86. CADN, ARI, C1, El Guisr, 10 June 1862, Viller to Chief of Recruitment.

87. Eugène de Régny, *Statistique de l'Égypte, année 1* (Alexandria: Imprimerie française Mourès, 1870), 84.

88. ANMT, CUCMS, 1995 060 4136, Toussoum, Ismail Hamdy, 12 May 1863, to Sciama, Engineer in chief.

89. Berchère, *Le désert de Suez*, 205.

90. CADN, ARI, C1, Seuil d'El Guisr, 28 January 1862, Imam Moustapha to Engineer in chief.

91. Berchère, *Le désert de Suez*, 203.

92. Najm, *Bur Saïd*, 23; Shinnawi, *Al-Sukhrah fi hafr qanat al-Suways*, 227, 236, 249; and Charles-Roux, *L'Isthme et le Canal de Suez*, 1:268, 442, 447.

93. Barthélémy-François Arlès-Dufour et al., *Le percement de l'Isthme de Suez: Enfantin—M. de Lesseps; résumé historique* (Paris: E. Dentu, 1869), 67–214; François-Philippe Voisin, *Le canal de Suez* (Paris: Vve C. Dunod, 1902), 4:17–21, 30, 78–80, 339–41, and , 7:2; and Paul Reymond, *Le port de Port-Saïd* (Cairo: Impr. Scribe égyptien, 1950), 99.

94. Nubarian, *Mémoires de Nubar Pacha*, 225.

95. John Ninet and Anwar Luqa, *Lettres d'Égypte: 1879–1882* (Paris: Éditions du Centre national de la recherche scientifique, 1979), 128, entry dated 11 April 1881 (from Zagazig).

96. J. Millie, *Isthme et Canal de Suez: Son passé, son présent et son avenir* (Milan: Civelli, 1869), 68; and Archives de la Maison-Mère du Bon Pasteur, Angers (AMMBPA), 29 December 1862, Père Germaine Récollet, in Violette Cassis, *Pages d'Histoire* (Cairo: Taba'a al-Sharika al-Misriya li-l-Nashr wa-l-I'lam, 1998), 112.

97. Nicolas Michel, "La Compagnie du canal de Suez et l'eau du Nil (1854–1896)," in *L'isthme et l'Egypte au temps de la compagnie universelle du canal maritime de Suez (1858–1956)*, ed. Claudine Piaton (Paris: IFAO du Caire, 2016), 278, 280; Montel, *Le chantier du Canal de Suez*, 175; Shinnawi, *Al-Sukhrah fi hafr qanat al-suways*, 249; and Ernest Desplaces, "Chronique de l'Isthme," *L'Isthme de Suez*, 1 May 1864, 220.

98. DWQ, 0069-004751, Alexandria, 13 June 1866, Gaudard, Secretary at the Ministry of Foreign Affairs. Following the convention of March 18, 1863, the Company returned the portion of land between Cairo (specifically, Bulaq) and the *wadi* that had not been yet excavated to the Egyptian government. The second portion of the canal-to-be, the one between the *wadi* and Suez via Ismailia, was also returned to the Egyptian government, in exchange for the hefty fee of 10 million francs, per the stipulations finalized in 1866. See Michel, "La Compagnie du canal de Suez et l'eau du Nil (1854–1896)," 280.

99. Archives du ministère de l'Europe et des Affaires étrangères, Centre des Archives Diplomatiques á la Courneuve, Paris (CADC), Affaires économiques et commerciales (AEC), Correspondance Consulaire et Commerciale Port-Saïd (CCCPS), vol. 1, Port Said, 20 March 1877, Saint-Chaffray, French Consul in Port Said, to French Ministry of Foreign Affairs, 452r.

100. 'Ali Mubarak, *Al-Khitat al-tawfiqiyah al-jadidah li-Misr al-Qahirah wamuduniha wa-biladiha al-qadimah wa-al-shahirah* (Cairo: Matba'at Dar al-Kutub wa-al-Watha'iq al-Qawmiyah, 2004), 10:57; and Barak, *Powering Empire*, 38.

101. ANMT, CUCMS, 1995 060 3600, 29 December 1860, Director General, Third Weekly Report.

102. Barak, *Powering Empire*, 41.

103. AMMBPA, Sister Marie de Sainte Elisabeth Ledoulx, to Mother General, "De notre petit Monastère de Port-Saïd," 16 May 1863, in Cassis, *Pages d'Histoire*, 120.

104. Michel, "La Compagnie du canal de Suez et l'eau du Nil (1854–1896)," 280; and Reymond, *Le port de Port-Saïd*, 99. Michel indicates April 1864 and July 1866 as times in which such pipes were finalized.

105. Doctor Aubert-Roche, "Rapport," *L'Isthme de Suez*, 15–18 July 1865, 208, 215.

106. Antoinette Drohojowska, *L'Égypte et le Canal de Suez* (Paris: Laporte, 1870), 110.

107. Najm, *Bur Sa'id*, 42–43; and Barak, *Powering Empire*, 39.

108. Aubert-Roche, "Rapport," *L'Isthme de Suez*, 15–18 July 1865, 215; Doctor Zarb, "Statistique medicale," *L'Isthme de Suez*, 15–18 July 1865, 219; and Doctor Hérouard, "Note sur l'état medical de Port-Saïd," *L'Isthme de Suez*, 15–18 July 1865, 220.

109. ANMT, CUCMS, 1995 060 4136, Ismailia, 16 January 1864, Voisin, Director General, Report.

110. TNA, FO 423-1, Alexandria, 11 March 1864, Rev. E. J. Davis, Chaplain in Alexandria, to Mr. Layard.

111. TNA, FO 423-1, Alexandria, 26 April 1860, Colquhoun to Lord J. Russell, Report by J. Coulthard.

112. François Levernay, *Guide général d'Égypte* (Alexandria: Imprimerie nouvelle, 1868), 277; Doctor Aubert-Roche, "Rapport," *L'Isthme de Suez*, 15–18 July 1865, 210–11; Doctor Aubert-Roche, "Rapport," *L'Isthme de Suez*, 15–19 June 1867, 226.

113. AMMBPA, 16 May 1863, Sister Marie de Sainte Elisabeth Ledoulx to Mother General, "De notre petit Monastère de Port-Saïd," in Cassis, *Pages d'Histoire*, 119.

114. Erckmann and Chatrian, *Souvenirs d'un ancien chef de chantier*, 6.

115. Doctor Aubert-Roche, "Rapport," *L'Isthme de Suez*, 15–18 July 1865, 207; and "Egypt," *Times*, 29 March 1864, 4.

116. Doctor Aubert-Roche, "Rapport," *L'Isthme de Suez*, 15–19 June 1867, 226; and François Levernay, *Guide général d'Égypte* (Alexandria: Imprimerie nouvelle, 1868), 277.

117. TNA, FO 423-1, Alexandria, 11 March 1864, Rev. E. J. Davis, Chaplain in Alexandria, to Mr. Layard.

118. CADN, ARI, C1, Ismailia, 9 July 1863, Valentini, Company Doctor, to Division Chief Engineer; ARI, C1, Ismailia, 9 August 1863, Voisin, Director General, to Canterini, tailor in Ismailia; and ARI, C1, Timsah, 11 February 1863, Engineer to Engineer.

119. ANMT, CUCMS, 1995 060 3600, Port Said, 18 April 1861, Division Chief Engineer Laroche, Map no. III, "Section Port Saïd."

120. TNA, FO 423-1, Alexandria, 11 March 1864, Rev. E. J. Davis, Chaplain in Alexandria, to Mr. Layard.

121. Aubert-Roche, "Rapport," *L'Isthme de Suez*, 15–18 July 1865, 215; Doctor Zarb, "Statistique medicale," *L'Isthme de Suez*, 15–18 July 1865, 219; and Doctor Hérouard, "Note sur l'état medical de Port-Saïd," *L'Isthme de Suez*, 15–18 July 1865, 220.

122. IPOCAN, B 1861, Alexandria, 10 April 1862, Consul General Schreiner to Count Rechberg.

123. Berchère, *Le désert de Suez*, 205.

124. TNA, FO 423-1, Alexandria, 11 March 1864, Rev. E. J. Davis, Chaplain in Alexandria, to Mr. Layard; and Jabez Burns and Thomas Cook (Firm), *Help-Book for Travellers to the East Including Egypt, Palestine, Turkey, Greece and Italy* (London: Cook's Tourist Office, 1872), 185.

125. Ernest Desplaces, "Chronique de l'Isthme," *L'Isthme de Suez*, 1 March 1861, 66.

126. IPOCAN, B 1861, Alexandria, 10 April 1862, Consul General Schreiner to Count Rechberg.

127. Ritt, *Histoire de l'isthme de Suez*, 286–87; and TNA, FO 423-1, Alexandria, 1 April 1862, Colquhoun to Lord J. Russell.

128. ANMT, CUCMS, 1995 060 4136, n.d., General Board of the Company Works, "Evaluation du prejudice."

129. Yusuf Nahhas, *Al-Fallah: Halatuhu al-iqtisadiyah wa-al-ijtimá'iyah* (Cairo: 'Uniya bi-Nashrihi wa-Tab'ihi Khalil Mutran, 1926), 126, cited in Nathan J. Brown, "Who Abolished Corvee Labour in Egypt and Why?," *Past & Present*, no. 144 (1994): 122.

130. Percy Fitzgerald, *The Great Canal at Suez, Its Political, Engineering, and Financial History: With an Account of the Struggles of Its Projector, Ferdinand de Lesseps* (New York: AMS Press, 1978), 2:2.

131. TNA, FO 423-1, Alexandria, 24 January 1863, Colquhoun to Earl Russell.

132. Sven Beckert, *Empire of Cotton: A Global History* (New York: Knopf, 2015), 256; Robert L. Tignor, *Egypt: A Short History* (Princeton, NJ: Princeton University Press, 2010), 229; Stanford J. Shaw and Ezel Kural Shaw, *History of the Ottoman Empire and Modern Turkey*, vol 2, *Reform, Revolution, and Republic: The Rise of Modern Turkey, 1808–1975* (Cambridge: Cambridge University Press, 1976), 145; and Roger Owen, *Cotton and the Egyptian Economy, 1820–1914: A Study in Trade and Development* (Oxford: Clarendon Press, 1969), 89–121.

133. TNA, FO 423-1, Constantinople, 27 June 1863, Sir H. Bulwer to Earl Russell; and F. Robert Hunter, "Egypt under the Successors of Muhammad Ali," in *The Cambridge History of Egypt*, vol. 2, *Modern Egypt, from 1517 to the End of the Twentieth Century*, ed. M. W. Daly (Cambridge: Cambridge University Press, 1998), 193.

134. ANMT, CUCMS, 1995 060 1323, n.d., Excerpts of the Reports of the Director General.

135. DWQ, 5009-000208, Shawwal 1279 hijri (March-April 1863), Grand Vizier to Viceroy of Egypt; QS, 5009-000208, 6 April 1863, Ali Pacha, to representatives of the Sultan in Paris and London (copy of annex to no. 534-1863 of the Ministry of Commerce from "Alte Akten des Handelsministeriums Suezkanal 1860–1875"). Another source reports that indigenous workers averaged twenty thousand. Leonardo Gallo Da Calatafimi, *L'Egitto antico e moderno* (Catania: Tip. Eugenio Coco, 1891), 170.

136. Landes, *Bankers and Pashas*, 184. DWQ, 5009-000207, no place, n.d., no author.

137. Charles W. Hallberg, *The Suez Canal, Its History and Diplomatic Importance* (New York: Columbia University Press; London: King & Son, 1931), 201; and Nubarian, *Mémoires de Nubar Pacha*, 226.

138. ANMT, CUCMS, 1995 060 1323, n.d., Excerpts of the Reports of the Director General. The institution of the corvée in Egypt would only be formally abolished throughout Egypt in January 1892; DWQ, 4003-030758, "Decret du 28 Janvier 1892." However, some forms of forced labor survived and could be revived in cases of emergency; see Brown, "Who Abolished Corvee Labour in Egypt and Why?," 135.

139. Elizabeth Mitchell, *Liberty's Torch: The Great Adventure to Build the Statue of Liberty* (New York: Atlantic Monthly Press, 2018), 55; Landes, *Bankers and Pashas*, 187; and David Todd, *A Velvet Empire: French Informal Imperialism in the Nineteenth Century* (Princeton, NJ: Princeton University Press, 2021), 257.

140. Hallberg, *Suez Canal, Its History and Diplomatic Importance*, 203–7; and François-Philippe Voisin, *Le canal de Suez* (Paris: Vve C. Dunod, 1902), 1:232.

141. IPOCAN, B 1866–1868, Cairo, 26 January 1866, Macciò, Italian Consul to Italian Ministry of Foreign Affairs.

142. Nubarian, *Mémoires de Nubar Pacha*, 238.

143. Landes, *Bankers and Pashas*, 224. In 1865 Company sources still declared that "the Ouady," a tract of land that was not situated on the isthmus itself but necessary to the establishment of the sweetwater canal, was "a vast property *belonging to the Company*." Doctor Aubert-Roche, "Rapport," *L'Isthme de Suez*, 15–18 July 1865, 215, emphasis added.

144. Ritt, *Histoire de l'isthme de Suez*, 288; and Mangin, "Percement de l'isthme de Suez," 449.

145. "Conférence de M. Borel," *L'Isthme de Suez*, 15–18 February 1867, 63; Gaston Jondet, Institut français d'archéologie orientale du Caire, and Société sultanieh de géographie, *Le port de Suez* (Cairo: Impr. de l'Institut français d'archéologie orientale, 1919), 88; and Najm, *Bur Saʿid*, 58.

146. Joel Beinin and Zachary Lockman, *Workers on the Nile*, 26.

147. Lucia Carminati, "'Improvising and Very Humble': Those 'Italians' throughout Egypt That Statisticians and Historians Have Neglected," in *On the Margins of History: Italian Subalterns between Emigration and Colonialism in the Italian Colony in Egypt (1861–1937)*, ed. Costantino Paonessa (Louvain: Université catholique de Louvain presse, 2021), 31–32.

148. Cited in Costantino Paonessa, "Introduction: Some Reflections about the History of the Forgotten from the Myth of the Italian Community in Egypt," in *On the Margins of History*, ed. Paonessa, 9.

149. "Egypt," *Times*, 29 March 1864, 4.

150. Jules Munier, *La presse en Égypte (1799–1900): Notes et souvenirs* (Cairo: Impr. de l'Institut français d'archéologie orientale, 1930), 9.

151. "Risparmii e prestiti," *L'Imparziale*, 21–22 April 1895, 3. and "Gli europei in Egitto," *Le Bosphore égyptien*, 16 August 1885. About the competition between the Italian and French press, see Angelo Sammarco, *Gli italiani in Egitto: il contributo italiano nella formazione dell'Egitto moderno* (Alexandria: Edizioni del Fascio, 1937), 152; and Munier, *La presse en Égypte*, 21.

152. Landes, *Bankers and Pashas*, 88.

153. Régny, *Statistique de l'Égypte, année 1*, 13–14. While the Egyptian population in 1846 was estimated at 4,463,244, it totaled 5,215,065 inhabitants in 1869 (78).

154. Moghira, *L'Isthme de Suez*, 201; Dori, "Esquisse historique de Port Said," 217; and Najm, *Bur Saʿid*, 58–59.

155. Erckmann and Chatrian, *Souvenirs d'un ancien chef de chantier*, 7; "Conférence de M. Borel," *L'Isthme de Suez*, 15–18 February 1867, 63; and Jondet, Institut français d'archéologie orientale du Caire, and Société sultanieh de géographie, *Le port de Suez*, 87–88.

156. Archivio della Sacra Congregazione per l'Evangelizzazione dei Popoli o de "Propaganda Fide," Rome (hereafter APF), Scritture riferite nei Congressi (hereafter Fondo S. C.), Egitto, Copti, vol. 23, Alexandria, 21 February 1874, Simeoni to Ciurcia, 778r.

157. Landes, *Bankers and Pashas*, 87–88. Landes is quoting Mohamed Sabry, *L'Empire égyptien sous Ismaïl et l'ingérence anglo-française 1863–1879* (Paris: Geuthner, 1933), 37. Sabry is quoting a letter that the French consul in Alexandria wrote in 1854 to the Ministry of Foreign Affairs in Paris.

158. Charlene L. Porsild, *Gamblers and Dreamers: Women, Men, and Community in the Klondike* (Vancouver: University of British Columbia Press, 1998).

159. Gabriel Baer, *Studies in the Social History of Modern Egypt* (Chicago: University of Chicago Press, 1969), 140; and CADN, ARI, C9, Alexandria, 27 October 1869, French Consular Chancellor to Geyler, French Consular Agent in Ismailia.

160. Baer, *Studies in the Social History of Modern Egypt*, 148; and 'Abd al-Wahhab Bakr, "Mina' Dimyat wa-dawruhu fi al-'alaqat al-tijariyah bayna Misr wa-bilad al-Lifant khilala al-qarn al-thamin 'ashar," in *Misr wa-'alam al-bahr al-mutawassit fi al-'asr al-hadith*, ed. Ra'uf 'Abbas Hamid (Cairo: Maktabat Nahdat al-Sharq, 1996), 70–71. See Ruth Kark, "The Rise and Decline of Coastal Towns in Palestine," in *Ottoman Palestine, 1800–1914*, ed. Gad G. Gilbar (Leiden: E. J. Brill, 1990), 88.

161. Régny, *Statistique de l'Égypte, année 1*, 13; Najm, *Bur Sa'id*, 58, 78–79.

162. Gaston Jondet, *Le port de Suez* (Cairo: Imprimerie de l'Institut français d'archéologie orientale, 1919), 87–88.

163. CADN, ARI, C9, Geyler, French Consular Agent in Ismailia, 3 November 1869, to French Consul General in Alexandria; ARI, C9, Sainte-Foy-l'Argentière (Rhône), 11 December 1869, no author to French Consul in Alexandria; ARI, C9, Alexandria, 7 July 1869, Contract between Clemente Buratti and Rosa Calamita.

164. Emad Abou Ghazi and Xavier Daumalin, "Le Canal vu Par Les Égyptiens," in *L'épopée du Canal de Suez*, ed. Gilles Gauthier (Paris: Gallimard/Institut du monde arabe/Musée d'Histoire de Marseille, 2018), 99; and Liat Kozma, *Global Women, Colonial Ports: Prostitution in the Interwar Middle East* (Albany: State University of New York Press, 2017), 88–89.

165. Fabien Bartolotti, "Le chantier du siècle: Les entreprises marseillaises et le creusement du canal de Suez (1859–1869)," *Revue Marseille*, no. 260 (2018): 47.

166. ANMT, CUCMS, 1995 060 4136, Ismailia, 7 November 1865, Voisin to Cadiat, Chief Engineer of Service in Paris.

167. ANMT, CUCMS, 1995 060 4136, Hong Kong, 27 May 1865, Bosman and Co. to Bertrand; and 1995 060 4136, Hong Kong, 28 June 1865, Bertrand, Maritime services of the "Messageries Impériales," to De Lesseps, Paris. "Coolie" is a pejorative term derived from Tamil and used by Europeans to describe the non-European workers they transported across the globe. It can be further defined as "unskilled laborer employed cheaply, especially one brought from Asia." Gaiutra Bahadur, *Coolie Woman: The Odyssey of Indenture* (Chicago: University of Chicago Press, 2014), xix–xx. About "coolie" migrations in Asia and to Africa, see Dirk Hoerder, *Cultures in Contact: World Migrations in the Second Millennium* (Durham, NC: Duke University Press, 2002), 384–92. British Hong Kong, in spite of Britain's humanitarian claims in matters of slavery, needed the "coolie trade" to develop. While negotiations about this form of indentured migration dragged on throughout the

1860s among Peking and foreign powers, Chinese workers continued to be shipped off by the thousands. O. J. Hui, "Chinese Indentured Labour: Coolies and Colonies," in *The Cambridge Survey of World Migration*, ed. Robin Cohen (Cambridge: Cambridge University Press, 1995), 54.

168. John Steele, *The Suez Canal: Its Present and Future* (London: Simpkin, Marshall, 1872), 15; and ANMT, CUCMS, 1995 060 4136, Cairo, 10 June 1855, Count d'Escayrac Lauteur, "Note sur l'emploi des Chinois."

169. I am here indebted to Edward W. Said, *Orientalism* (New York: Pantheon Books, 1978); and Manu Karuka, *Empire's Tracks: Indigenous Nations, Chinese Workers, and the Transcontinental Railroad* (Oakland: University of California Press, 2019), 87.

170. TNA, FO 423-1, Tangier, 26 August 1864, Sir J. D. Hay to Earl Russell.

171. "Conférence de M. Borel," *L'Isthme de Suez*, 15–18 February 1867, 63.

172. ANMT, CUCMS, 1995 060 4136, Port Said, 28 September 1865, Voisin, Director General, to De Lesseps.

173. Ernest Desplaces, "Chronique de l'Isthme," *L'Isthme de Suez*, 1 May 1864, 221; and ANMT, CUCMS, 1995 060 4136, El Guisr, 3 August 1865, Couvreux to Gioia, Engineer of El Guisr Division.

174. ANMT, CUCMS, 1995 060 4136, Civitavecchia, 28 August 1859, Joseph Casanova to Bourdon, Engineer in chief of the division of the maritime works of Suez in Paris.

175. ANMT, CUCMS, 1995 060 4136, Ismailia, 2 June 1866, Voisin, Workers' engagement contract; 1995 060 4136, Paris, 2 July 1866, Executive team to Voisin, Approval for workers sent from France; and John Pudney, *Suez. De Lesseps' Canal* (New York: Praeger, 1969), 109.

176. IPOCAN, B 1866–1868, Alexandria, 5 October 1866, Giuseppe De Martino, Italian Consul General, to Italian Ministry of Foreign Affairs.

177. Guido Cavalieri, "La emigrazione dal Polesine (1887–1901)," *La Riforma Sociale* 12, nos. 10–11 (1902): 1048.

178. Leone Carpi, *Delle colonie e dell'emigrazione d'italiani all'estero sotto l'aspetto dell'industria commercio, agricoltura, e con trattazione d'importanti questioni sociali*, vol. 3 (Milan: Tipografia editrice lombarda, 1874). On migration "agents" working for private foreign companies and being the most dangerous kind of operators in the field, see Amoreno Martellini, "Il commercio dell'emigrazione: intermediari e agenti," in *Storia dell'emigrazione italiana: Partenze*, ed. Piero Bevilacqua, Andreina De Clementi, and Emilio Franzina (Rome: Donzelli, 2009), 296.

179. ANMT, CUCMS, 1995 060 4136, El Guisr, 3 August 1865, Couvreux to Gioia, Engineer of El Guisr Division; 1995 060 4136, Civitavecchia, 24 May 1859, Joseph Casanova to De Lesseps, Paris; 1995 060 4136, Civitavecchia, 30 April 1859, Joseph Casanova to De Lesseps, Paris; and 1995 060 4136, Diakovar (Slavonia), 12 March 1860, Johann Steiner to De Lesseps.

180. CADN, ARI, Ismailia, 28 October 1869, Emile Girardeau to Consular Agent.

181. Manuel de Solá-Morales, "Ville-Port, Saint-Nazaire: The Historic Periphery," *AD Architectural Design* 78, no. 1 (2008): 89.

182. CADN, ARI, C4, Ismailia, 19 February 1866, Ritt, pour le Directeur Général des Travaux, to Geyler; ARI, C4, Saint Nazaire, 5 June 1866, Femme Chenet; ARI, C4, Lorient, 14 May 1866, Commissaire de Police to Geyler; and ARI, C9, Lesparre Gironne & Widow Lesneven, "France," to French Consul in Alexandria, forwarded to Consular Agent in Ismailia on 23 September 1869.

183. Richard Baxstrom, "Governmentality, Bio-Power, and the Emergence of the Malayan-Tamil Subject on the Plantations of Colonial Malaya," *Crossroads: An Interdisciplinary Journal of Southeast Asian Studies* 14, no. 2 (2000): 49–78; and Sunil S. Amrith, *Crossing the Bay of Bengal: The Furies of Nature and the Fortunes of Migrants* (Cambridge, MA: Harvard University Press, 2013).

184. Jondet, Institut français d'archéologie orientale du Caire, and Société sultanieh de géographie, *Le port de Suez*, 87–88. Jondet is quoting Auguste Stoecklin, *Notice sur la construction du bassin de radoub de Suez, 26 décembre 1866* (Bordeaux: Impr. de A. Bord, 1867).

185. Doctor Aubert-Roche, "Rapport," 227 *L'Isthme de Suez*, 15–19 July 1867,

186. CADN, ARI, C9, Mulhouse, 1 March 1870, Pasteur Nemagel to French Consul in Alexandria.

187. Erckmann and Chatrian, *Souvenirs d'un ancien chef de chantier*, 41.

188. ANMT, CUCMS, 1995 060 4136, Civitavecchia, 30 April 1859, Joseph Casanova to De Lesseps; Civitavecchia, 28 August 1859, Joseph Casanova to Bourdon, Engineer in chief of the division of the maritime works of Suez.

189. ANMT, CUCMS, 1995 060 4136, Montbéliard (Doubs), 8 December 1858, Falker, Agent of "American liners," to Canal Company Director.

190. IPOCAN, B1866-1868, Alexandria, 10 October 1866, Sala, Agente Superieur, Suez Canal Company, to Italian Consul, Alexandria.

191. Charles Archibald Price, *Malta and the Maltese: A Study in Nineteenth Century Migration* (Melbourne: Georgian House, 1954), xi–xii.

192. CADN, ARI, C26, Antonio Candida, 14 April 1865, to "Mio Carissimo Amico," no place.

193. Leslie Page Moch, *Moving Europeans: Migration in Western Europe since 1650* (Bloomington: Indiana University Press, 1992), 16.

194. Cavalieri, "La emigrazione dal Polesine," 1031; CADN, ARI, C26, Cherso, 17 August 1864, Marieta Salvagno to "Caro marito!"; and ARI, C26, Cherso, 25 September 1864, Maria Salvagno to "Caro marito!"

195. Adam McKeown, "Global Migration, 1846–1940," *Journal of World History* 15, no. 2 (2004): 178.

196. Dirk Hoerder, "From Immigration to Migration Systems: New Concepts in Migration History," *OAH Magazine of History* 14, no. 1 (1999): 6.

197. Olimpia Gobbi, "Emigrazione femminile: Balie e domestiche marchigiane in Egitto fra Otto e Novecento," *Proposte e ricerche. Economia e società nella storia dell'Italia centrale* 34, no. 66 (2011): 16.

198. CADN, ARI, C4, Ismailia, 16 April 1866, Greek Consular Agent to Head of Detached Offices, Ismailia; and ARI, C4, El Guisr, 22 June 1866, Gioia, Chief Engineer of Division, to Municipal Superintendent.

199. Porsild, *Gamblers and Dreamers*, 22. For thorough discussions of the literature on "chain" migration and migratory "networks," see Marlou Schrover, "Labour Migration," in *Handbook: The Global History of Work*, ed. Karin Hofmeester and Marcel Van der Linden (Berlin: De Gruyter Oldenbourg, 2018); and Marcelo J. Borges, *Chains of Gold: Portuguese Migration to Argentina in Transatlantic Perspective* (Leiden: Brill, 2009), 155–57. For insightful criticism of the notion of "migrant networks," see Leslie P. Moch, "Networks among Bretons? The Evidence for Paris, 1875–1925," *Continuity and Change* 18, no. 3 (2003): 431–56.

200. David Gutman, "Travel Documents, Mobility Control, and the Ottoman State in an Age of Global Migration, 1880–1915," *Journal of the Ottoman and Turkish Studies Association* 3, no. 2 (2016): 352; and İlkay Yilmaz, "Governing the Armenian Question through Passports in the Late Ottoman Empire (1876–1908)," *Journal of Historical Sociology* 32, no. 4 (2019): 392–93. Gutman notes that obtaining international passports was costly and difficult for Ottomans as well. Yilmaz notes that international passports for Ottomans came under regulation in 1884 and 1894. "I passaporti per l'estero," *L'Imparziale*, 16 January 1900, 3.

201. Marco Soresina, "Italian Emigration Policy during the Great Migration Age, 1888–1919: The Interaction of Emigration and Foreign Policy," *Journal of Modern Italian Studies* 21, no. 5 (2016): 726–27. Cavalieri writes that it was "military authorities" that released the "nulla osta." Cavalieri, "La emigrazione dal Polesine," 1048. Coletti notes, however, that up until 1901 "indigent people" could pay a reduced fee of 2.40 Lire rather than 12.40 Lire, see Francesco Coletti, *Dell'emigrazione italiana* (Milan: Ulrico Hoepli, 1911), 10.

202. Gur Alroey, "Journey to Early-Twentieth-Century Palestine as a Jewish Immigrant Experience," *Jewish Social Studies* 9, no. 2 (2003): 37–38.

203. Paulo G. Brenna, *L'emigrazione italiana nel periodo antebellico* (Firenze: R. Bemporad & Figlio, 1918), 205; and Gobbi, "Emigrazione femminile," 12, 20–23.

204. Edward J. Bristow, *Vice and Vigilance: Purity Movements in Britain since 1700* (Dublin: Gill and Macmillan, 1977), 178. In the last quarter of the nineteenth century, the United States similarly developed immigration laws that institutionalized assumptions about proper gender relations and approached independent female migration as a moral problem. Donna R. Gabaccia, *From the Other Side: Women, Gender, and Immigrant Life in the U.S., 1820–1990* (Bloomington: Indiana University Press, 1994), 37.

205. Schrover, "Labour Migration"; and Liat Kozma, "Women's Migration for Prostitution in the Interwar Middle East and North Africa," *Journal of Women's History* 28, no. 3 (2016): 95, 101, 106.

206. Commissariato generale dell'emigrazione, *Annuario statistico della emigrazione italiana dal 1876 al 1925 con notizie sull'emigrazione negli anni 1869–1875* (Rome: Commissariato generale dell'emigrazione, 1926), 1729. One was from

Lombardy, six were from Emilia, three from Campania, one from Calabria, and eleven from Sicily.

207. Ercole Sori, *L'emigrazione italiana dall'Unità alla seconda guerra mondiale* (Bologna: Il Mulino, 1979), 323. See Leopoldo Franchetti and Sidney Sonnino, *Inchiesta in Sicilia* (Firenze: Vallecchi, 1974), 207; and Brenna, *L'emigrazione italiana nel periodo antebellico*, 202, 205.

208. DWQ, 2001-024751, Alexandria, 9 September 1874, Alexandria Governor, Moustapha Fehmy, to Carlesino, Inspector General of the Alexandria Police; 2001-024751, paper scribbled in Italian, n.d., no author; and "Ieri all'Attarin," *L'Eco d'Italia*, 23–24 March 1890, 2.

209. McKeown, "Global Migration, 1846–1940," 178.

210. CADN, ARI, C4, Saint Nazaire, 5 June 1866, Femme Chenet; and ARI, C4, Lorient, 14 May 1866, Police Chief to Geyler.

211. Panaït Istrati, *Vie d'Adrien Zograffi* (Paris: Gallimard, 1969), 576.

212. Angelo Del Boca, *Italiani, brava gente? Un mito duro a morire* (Vicenza: Neri Pozza, 2011), 65; and Sori, *L'emigrazione italiana dall'Unità alla seconda guerra mondiale*, 321.

213. Ian Coller, "Barbary and Revolution: France and North Africa in the Revolutionary Era," in *French Mediterraneans: Transnational and Imperial Histories*, ed. Patricia M. E. Lorcin and Todd Shepard (Lincoln: University of Nebraska Press, 2016), 97–116; Maurizio Isabella and Konstantina Zanou, *Mediterranean Diasporas: Politics and Ideas in the Long Nineteenth Century* (London: Bloomsbury, 2016); Selim Deringil, *Conversion and Apostasy in the Late Ottoman Empire* (Cambridge: Cambridge University Press, 2012), 156–96; Andrew Urbanik and Joseph O'Baylen, "Polish Exiles and the Turkish Empire, 1830–1876," *The Polish Review* 26, no. 3 (1981): 43–53; and Alessandro Triulzi, "Italian-Speaking Communities in Early Nineteenth-Century Tunis," *Revue des Mondes Musulmans et de La Méditerranée*, no. 9 (1971): 153–84.

214. ASMAECI, ARDCE, CC, Antico versamento (AV), B12, Suez, 28 May 1869, Vice-Consulate of Italy to Domenico Brunenghi, Italian Consul, Cairo. CADN, ARI, C9, Ismailia, 18 October 1869, Geyler, French Consular Agent in Ismailia to Merlé, Campsite Agent in Chalouf; and ARI, C9, Ismailia, 28 October 1869, Geyler, French Consular Agent in Ismailia, to Merlé, Campsite Agent in Chalouf.

215. Julia A. Clancy-Smith, "Marginality and Migration: Europe's Social Outcasts in Pre-Colonial Tunisia, 1830–81," in *Outside in on the Margins of the Modern Middle East*, ed. Eugene L. Rogan (London: I. B. Tauris, 2002), 149, 152; Julia Clancy-Smith, "Women, Gender and Migration along a Mediterranean Frontier: Pre-Colonial Tunisia, c. 1815–1870," *Gender & History* 17, no. 1 (2005): 65–67; and Julia A. Clancy-Smith, "Women on the Move: Gender and Social Control in Tunis, 1815–1870," in *Femmes en Villes*, ed. Dalenda Larguèche (Tunis: Centre de Publications Universitaires, 2006), 215–16.

216. Beinin and Lockman, *Workers on the Nile*, 35.

217. Gobbi, "Emigrazione Femminile," 15.

218. Alroey, "Journey to Early-Twentieth-Century Palestine," 42; and Patrick Boulanger, "Témoignages sur le transport des immigrants en Méditerranée: Les rapports des capitaines des messageries maritimes (1871–1914)," in *Navigation et migrations en Méditerranée. De la Préhistoire à nos jours*, ed. Jean-Louis Miège (Paris: C.N.R.S. Editions, 1990), 352. See Torsten Feys, "The Battle for the Migrants: The Evolution from Port to Company Competition, 1840–1914," in *Maritime Transport and Migration: The Connections between Maritime and Migration Networks*, ed. Torsten Feys et al. (Oxford: Liverpool University Press, 2017), 27–48.

219. ANMT, CUCMS, 1995 060 4136, Civitavecchia, 24 May 1859, Joseph Casanova to De Lesseps; and 1995 060 4136, El Guisr, 3 August 1865, Couvreux to Gioia, Engineer of El Guisr Division.

220. Dori, "Esquisse Historique de Port Said," 205–6. Dori reports different dates. He also mentions a Compagnie Générale des Transports Maritimes de Marseille active in Port Said since 1876.

221. Giuseppe Capatti, *Indicatore Commerciale Alessandria* (Alessandria: Ottolenghi, 1874), 218; and Christopher Clay, "Labour Migration and Economic Conditions in Nineteenth-Century Anatolia," *Middle Eastern Studies* 34, no. 4 (1998): 7.

222. ANMT, CUCMS, 1995 060 4136, El Guisr, 3 August 1865, Couvreux to Gioia, Engineer of El Guisr Division.

223. Thomas Cook (Firm), *Cook's Tourist Handbook for Egypt, the Nile, and the Desert* (London: Thomas Cook & Son, 1876), 228. For hardships endured in ports of departure by children and women especially, see Augusta Molinari, "Porti, trasporti e compagnie," in *Storia dell'emigrazione italiana: Partenze*, ed. Piero Bevilacqua, Andreina De Clementi, and Emilio Franzina (Rome: Donzelli, 2009), 252–53.

224. Doctor Aubert-Roche, "Rapport," *L'Isthme de Suez*, 15–19 July 1867, 227.

225. Enrico Pea, *Vita in Egitto* (Milan: Mondadori, 1949), 44, 205–6. See Boulanger, "Témoignages sur le transport des immigrants en Méditerranée," 355–57.

226. Cavalieri, "La emigrazione dal Polesine," 1052.

227. Burns and Thomas Cook (Firm), *Help-Book for Travellers to the East*, 184; Will Hanley, "Papers for Going, Papers for Staying: Identification and Subject Formation in the Eastern Mediterranean," in *A Global Middle East: Mobility, Materiality and Culture in the Modern Age, 1880–1940*, ed. Liat Kozma, Cyrus Schayegh, and Avner Wishnitzer (London: I. B. Tauris, 2015), 177–200; and Kozma, *Global Women, Colonial Ports*, 96.

228. CADN, ARI, C1, Port Said, 27 June 1862, Deforge, Company agent in charge of general policing, Project of a regulation for the arrival and departure of individuals; and ASMAECI, ARDCE, CC, AV, B18, Suez, 18 July 1865, French Consul to Italian Consul in Cairo. For a study of residence permits in late nineteenth-century Egypt, see Hanley, "Papers for Going, Papers for Staying," 186–93.

229. Julia A. Clancy-Smith, *Mediterraneans: North Africa and Europe in an Age of Migration, c. 1800–1900* (Berkeley: University of California Press, 2011), 282; and Eugen Weber, *Peasants into Frenchmen: The Modernization of Rural France, 1870–1914* (Stanford, CA: Stanford University Press, 1976), 282.

230. Ibrahim Amin Ghali, *Sina' al-misriyah 'abra al-tarikh* (Cairo: Al-Hay'ah al-Misriyah al-'Ammah lil-Kitab, 1976), 274.
231. Régny, *Statistique de l'Égypte, année 1*, 15.
232. Price, *Malta and the Maltese*, xii; and Ramiro Vadala, *Les maltais hors de Malte: Etude sur l'émigration maltaise* (Paris: Rousseau, 1911), 59.
233. CADN, ARI, C9, Saint-Chaffray, 10 January 1870, Jacques Morand to French consul in Alexandria.
234. CADN, ARI, C26, Fiume, 25 February 1864, Anna Cobau to "Caro consorte"; ARI, C26, Fiume, 16 January 1864, Maria Baraga to "Carissimo consorte"; ARI, C26, Fiume, 27 August 1864, Maria Baraga to "Carissimo consorte"; ARI, C26, Fiume, 20 April 1865, Maria Baraga to "Carissimo consorte"; and ARI, C26, Fiume, 28 May 1865, Maria Baraga to "Caro Loise." For a map of the Austrian Empire in 1859–1867, see Pieter M. Judson, *The Habsburg Empire: A New History* (Cambridge, MA: Harvard University Press, 2018), 263.
235. CADN, ARI, C26, Cherso, 25 September 1864, Maria Salvagno to "Caro marito!"; ARI, C26, Fiume, 17 July 1864, Maria Baraga to "Carissimo Consorte;" ARI, C26, Fiume, 4 November 1867, Anna Cobau to "Carissimo consorte"; ARI, C26, Fiume, 28 May 1865, Maria Baraga to "Caro Loise"; and ARI, C26, Cherso, 17 August 1864, Marieta Salvagno to "Caro marito!"
236. Marcella Bruno, "Alcune note sulla 'emigrazione di ritorno' in Calabria," in *Calabresi sovversivi nel mondo: L'esodo, l'impegno politico, le lotte degli emigrati in terra straniera: 1880–1940*, ed. Amelia Paparazzo (Soveria Mannelli: Rubbettino, 2004), 150; and Mario Bolognari, *Rapsodia calabrese tra emigrazione e rientro* (Cosenza: Centro Editoriale e Librario, 1992), 11.
237. AMMBPA, HC 43 Port Said, "De notre Monastère de Port-Saïd, 1870."
238. Moghira, *L'Isthme de Suez*, 201.
239. AMMBPA, HC 34 Port Said, "De notre Monastère de Port Said, Circulaire, 1870?"; and Ritt, *Histoire de l'isthme de Suez*, 272.
240. Pamela Sharpe, ed., *Women, Gender, and Labour Migration: Historical and Global Perspectives* (London: Routledge, 2011), 6; and Leslie P. Moch, *Paths to the City: Regional Migration in France* (Beverly Hills, CA: Sage, 1983), 125.
241. Luigi Dori, "Esquisse historique de Port Said: Port Said après l'inauguration du Canal (1869–1900)," *Cahiers d'histoire egyptienne* 8, no. 1 (January 1956): 39; and Athanase G. Politis, *L'Hellénisme et l'Egypte moderne* (Evreux: impr. Ch. Hérissey; Paris: libr. Félix Alcan, 1929), 1:330. Politis reports that, of the seven thousand "European" workers who first landed in Port Said, as many as five thousand came from Kasos.
242. TNA, FO 423–1, Alexandria, 26 April 1860, Colquhoun to Lord J. Russell, Report by J. Coulthard.
243. ANMT, CUCMS, 1995 060 4136, Port Said, 23 December 1862, Laroche, Chief Engineer of Division, Report; and CADN, ARI, C1, Ismailia, 23 July 1863, Manuel Lepin to Viller, Division Chief.
244. Ernest Desplaces, "Chronique de l'Isthme," *L'Isthme de Suez*, 1 May 1864, 221–22.

245. Lesseps, *Lettres, journal et documents*, 3:431, entries dated 10 June 1860 and 7 December 1860.

246. Ernest Desplaces, "Chronique de l'Isthme," *L'Isthme de Suez*, 1 March 1861, 66.

247. Lesseps, *Lettres, journal et documents*, 4:18–19, 24, entries dated 8 February 1861 and 4 March 1861.

248. Politis, *L'Hellénisme et l'Egypte moderne*, 1:330.

249. "Une Conversation avec De Lesseps," *Le Bosphore égyptien*, 17 February 1885.

250. Pea, *Vita in Egitto*, 61; and Anouchka Lazarev, "La colonia italiana: Una identità ambigua," in *Italia e l'Egitto dalla rivolta di Arabi Pascià all' avvento del fascismo (1882–1922)*, ed. Romain H. Rainero and Luigi Serra (Settimo Milanese: Marzorati, 1991), 181.

251. CADN, ARCFPS, CC, C49, Perpignan, 9 June 1875, Eug. Dorche to French Consul in Port Said; and ASMAECI, Personale, serie II, Posizione P10, Port Said, B2, Primo Versamento, Port Said, 3 March 1907, Mancinelli Scotti, Italian Consul, to Malmusi, Italian Special Envoyee.

252. Khaled Fahmy, "Towards a Social History of Modern Alexandria," in *Alexandria, Real and Imagined*, ed. Anthony Hirst and Michael Silk (Aldershot, UK: Ashgate, 2004), 281; and Mario M. Ruiz, "Euro-Egyptian Romance in Turn of the Century Cairo," *International Journal of Middle East Studies* 40, no. 1 (2008): 8. For information on later exogamy, see Davide Amicucci, "La comunità italiana in Egitto attraverso i censimenti dal 1882 al 1947," in *Tradizione e modernizzazione in Egitto 1798–1998*, ed. Paolo Branca (Milan: Franco Angeli, 2000), 85–86.

253. ASMAECI, ARDCE, CC, B24, El Guisr (Ismailia, Suez Isthmus), 4 April 1870, Ludovico Alessandrini to Italian Consul in Cairo.

254. CADN, ARCFPS, CC, C49, 18 December 1877, Port Said, Widow Flandrin to Consul.

255. "Une Conversation avec De Lesseps," *Le Bosphore égyptien*, 17 February 1885.

256. Società geografica italiana, *Memorie della società geografica italiana: Indagini sulla emigrazione italiana all'estero fatte per cura della società (1888–1889)* (Rome: Presso la Società geografica italiana, 1890), 4:252.

257. For an analysis of such phenomena in the Mediterranean at large, see Paul Sant-Cassia, "Marriages at the Margins: Interfaith Marriages in the Mediterranean," *Journal of Mediterranean Studies* 27, no. 2 (2018): 111–32.

258. APF, Fondo S.C., Egitto, Copti, vol. 22Cairo, 5 April 1879, Luigi Ciurcia, Apostolic Vicar, to Cardinal Simeoni, 99.

259. ASMAECI, ARDCE, RC, B61, Port Said, 27 May 1896, Consular Agent to Italian Consul General in Cairo; RC, B61, Ismailia, 13 May 1896, (La négresse) Zafaran to Consular Agent; RC, B61, Suez, 7 December 1894, K.K. Cosma, Italian Vice Consul, to managing Italian consul in Port Said; RC, B61, Suez, 10 December 1894, K. K. Cosma to managing Italian consul in Port Said Suez; RC, B61, Suez, 20 December 1894, K. K. Cosma to managing Italian consul in Port Said; and RC, B64, Port Said, 19 August 1900, Consul to Consul General in Cairo.

260. "Ragazza greca," *L'Imparziale*, 6 November 1895, 3.
261. CADN, ARI, C26, no place, n.d.
262. Ernest Desplaces, "Chronique de l'Isthme," *L'Isthme de Suez*, 1 January 1861, 5.
263. AMMBPA, Sister Marie de Sainte Elisabeth Ledoulx to Mother General, "Quelques jours après," in Cassis, *Pages d'Histoire*, 118; and J. Heyworth-Dunne, *An Introduction to the History of Education in Modern Egypt* (London: Cass, 1968), 408.
264. The Franciscan Centre of Christian Oriental Studies, "Cronaca Di Santa Caterina: Appendice Seconda," *Studia Orientalia Christiana, Collectanea*, nos. 26–27 (1996): 478–79. On the general disregard for the education of Port Said's Muslim population, see Najm, *Bur Saʿid*, 363–64.
265. Politis, *L'Hellénisme et l'Egypte moderne*, 1:335.
266. Jacob M. Landau, *Jews in Nineteenth-Century Egypt* (New York: New York University Press, 1969), 34; Egypt, Ministry of the Interior, *Statistique de l'Egypte: Année 1873–1290 de l'hegire* (Cairoe: Imprimerie française Moures, 1873), 257; and Sydney Montagu Samuel, *Jewish Life in the East* (London: C. Kegan Paul, 1881), 12–17.
267. Muhammad Amin Fikri, *Jughrafiyat Misr* (Cairo: Matbaʿat Wadi al-Nil, 1879), 289.
268. ANMT, CUCMS, 1995 060 3600, Laroche, Engineer, Port Said, map titled "Situation of metallurgic workshops on 18 April 1861."
269. Lesseps, *Lettres, journal et documents*, 4:24, entry dated 4 March 1861.
270. ANMT, CUCMS, 1995 060 4136, Ismailia, 16 January 1864, Voisin, Director General, Report. Toledano claims that the Egyptian government had attempted, in the early 1850s, to circumscribe the population that would be subject to forced labor, which previously included men and women of all ages, children, and the elderly. Toledano, *State and Society in Mid-Nineteenth-Century Egypt*, 188. But this seems to have been flouted at the Suez worksites in the early 1860s.
271. Paolo Del Vesco, "Volti Senza Nome: I Lavoratori Egiziani Sullo Scavo," in *Missione Egitto 1903–1920: L'avventura Archeologica m.a.i. Raccontata*, ed. Paolo Del Vesco, Christian Greco, and Beppe Moiso (Modena: Franco Cosimo Panini, 2017), 206; and Georges Legrain, *Louqsor sans les Pharaons: Légendes et chansons populaires de la Haute Égypte* (Brussels: Vromant, 1914), 189–90.
272. Doctor Zarb, "Statistique medicale," *L'Isthme de Suez*, 15–18 July 1865, 220; Doctor Aubert-Roche, "Rapport," *L'Isthme de Suez*, 15 July 1869, 238–239; and CADN, ARI, C1, n.d. [1863], Geyler, Birth in Port Said. For a study of infant mortality among Egyptians, see Beth Baron, "Perilous Beginnings: Infant Mortality, Public Health and the State in Egypt," in *Gendering Global Humanitarianism in the Twentieth Century*, ed. Esther Möller, Johannes Paulmann, and Katharina Stornig (Basingstoke, UK: Palgrave Macmillan, 2020), 195–219.
273. Docteur Hérouard, "Note sur l'état medical de Port-Saïd," *L'Isthme de Suez*, 15–18 July 1865, 220.
274. ʿAli Mubarak, *Al-Khitat al-tawfiqiyah al-jadidah*, 10:57–58; and Casimir Leconte, *Promenade dans l'isthme de Suez* (Paris: N. Chaix, 1864), 59.

275. TNA, FO 423–1, Alexandria, 26 April 1860, Colquhoun to Lord J. Russell, Report by J. Coulthard.

276. ANMT, CUCMS, 1995 060 1002, Port Said, n.d. [1872–1956], "Cultes."

277. The Franciscan Centre of Christian Oriental Studies, "Cronaca di Santa Caterina: Appendice Seconda," 478–80. The Franciscan order would run it, specifically under the banner of the Custody of the Holy Land. Balboni, *Gl'Italiani nella civiltà egiziana del secolo XIX*, 1:448. A Catholic chapel was inaugurated on April 1, 1866, at the Chalouf worksite. By 1869 in Al Guisr, both a Catholic chapel consecrated to Sainte Marie du Désert and a mosque had been erected. CADN, ARI, C4, Chalouf, 31 March 1866, Representative of Borel Lavalley & Cie. to Lavissier, Municipal Superintendent in Chalouf; and Marius Etienne Fontane and Edouard Riou, *Le canal maritime de Suez illustré: Histoire du canal et des travaux* (Paris: Aux bureaux de l'Illustration A. Marc et Cie., 1869), 130, 132.

278. ʿAli Mubarak, *Al-Khitat al-tawfiqiyah al-jadidah*, 10:57–59.

279. ANMT, CUMCS, 1995 060 3600, no place, 1 December 1860, Director General to President; and Ernest Desplaces, "Chronique de l'Isthme," *L'Isthme de Suez*, 1 January 1861, 5.

280. ANMT, CUCMS, 1995 060 1002, Port Said, n.d. [1872–1956], "Cultes"; 1995 060 1002, 23 March 1881, Board of Directors, "Communauté Israëlite de Port-Saïd"; and Samuel, *Jewish Life in the East*, 14.

281. ASMAECI, ARCDE, RC, B50, *La phare de Port Said*, 8 August 1893; and Sylvia Modelski, *Port Said Revisited* (Washington, DC: FAROS, 2000), 73, 79.

282. Khaled Fahmy, "The Police and the People in Nineteenth-Century Egypt," *Die Welt Des Islams* 39, no. 3 (1999): 350; Giuseppe Augusto Cesana, *Da Firenze a Suez e viceversa: impressioni di viaggio* (Firenze: Tipografia Fodratti, 1870), 38; Nubarian, *Mémoires de Nubar Pacha*, 554; and Erckmann and Chatrian, *Souvenirs d'un ancien chef de chantier*, 1877, 8–9.

283. F. Robert Hunter, "Egypt under the Successors of Muhammad Ali," in *The Cambridge History of Egypt*, vol. 2, 189; and Byron Cannon, *Politics of Law and the Courts in Nineteenth-Century Egypt* (Salt Lake City: University of Utah Press, 1988), 54.

284. Todd, *Velvet Empire: French Informal Imperialism in the Nineteenth Century*, 265–69.

285. ANMT, CUCMS, 1995 060 3601, Ismailia, 23 September 1868, Ritt, Director General of the Works, to Mourad Pacha, Governor of the Isthmus in Ismailia.

286. ASMAECI, ARDCE, CC, AV, B18, Suez, 4 May 1866, Italian Vice-Consul in Suez to Italian Consul in Cairo; "Conference de M. Ferdinand De Lesseps a Lyon," *L'Isthme de Suez*, 15 November 1865, 381; and Ritt, *Histoire de l'isthme de Suez*, 318. For a discussion of the labeling and appropriation of "Italian" migrants in politics and historiography, see Carminati, "'Improvising and Very Humble.'"

287. Robert Ilbert, "Qui est Grec? La nationalité comme enjeu en Égypte (1830–1930)," *Relations internationales*, no. 54 (1988): 141; and Thomas W. Gallant, *Modern Greece* (London: Hodder Education, 2001), 13–14. For a discussion of Greeks

within Egypt as an "ethnic or national community" whose boundaries were "fluid and constantly changing," see Angelos Dalachanis, *The Greek Exodus from Egypt: Diaspora Politics and Emigration, 1937–1962* (New York: Berghahn Books, 2017), 5.

288. Thomas W. Gallant, "Tales from the Dark Side: Transnational Migration, the Underworld and the 'Other' Greeks of the Diaspora," in *Greek Diaspora and Migration since 1700. Society, History, and Politics*, ed. Dimitris Tziovas (Farnham, UK: Ashgate, 2009), 19; Gallant, *Modern Greece*, 13–14; and Frances Kraljic, "Round Trip Croatia, 1900–1914," in *Labor Migration in the Atlantic Economies: The European and North American Working Classes during the Period of Industrialization*, ed. Dirk Hoerder (Westport, CT: Greenwood Press, 1985), 406. Hungary administered Croatia by way of an agreement stipulated in 1867 with Austria, thus gaining access to the sea. Kraljic, *Labor Migration in the Atlantic Economies*, 410.

289. CADN, ARI, C26, Turinetto Giacinto, born in Giaveno, going to Marseille, passport released in Philippeville on 17 June 1867; ARI, C26, Alexandria, 16 October 1869, Passport; and ASMAECI, ARDCE, CC, AV, B12, Suez, 4 November 1869, Italian Vice-Consul to Italian Consul in Cairo.

290. Shana Elizabeth Minkin, *Imperial Bodies: Empire and Death in Alexandria, Egypt* (Stanford, CA: Stanford University Press, 2020), 94.

291. Adam McKeown, "Global Migration 1846–1940," *Journal of World History* 15, no. 2 (2004): 185; and Eric R. Wolf, *Europe and the People without History* (Berkeley: University of California Press, 1982), 23.

292. Leo Lucassen and Lex Heerma van Voss, "Introduction: Flight as Fight," in *A Global History of Runaways Workers, Mobility, and Capitalism, 1600–1850*, ed. Marcus Rediker, Titas Chakraborty, and Matthias Van Rossum (Oakland: University of California Press, 2019), 15; Rubén Hernández León, "Conceptualizing the Migration Industry," in *The Migration Industry and the Commercialization of International Migration Book*, ed. Thomas Gammeltoft-Hansen and Ninna Nyberg Sorensen (London: Routledge, 2013), 24–44; and Hoerder, *Cultures in Contact*, 15–22, 564.

293. Janet Henshall Momsen, *Gender, Migration, and Domestic Service* (London: Routledge, 1999), 9.

294. Engin Özendes, *Photography in the Ottoman Empire: 1839–1923* (Istanbul: YEM Yayin, 2013), 49, 146, 336. Özendes writes that Louis Dodero of Marseille invented this product in 1851, and that in 1854 André-Adolphe-Eugène Disdéri developed a method of taking six or eight portraits at once with a single plate, which significantly reduced time and costs.

295. CADN, ARI, C4, Serapeum, 20 June 1866, Municipal Superintendent; and ARI, C1, Anonymous photograph [1861–1870].

296. Jerry H. Bentley, "The Human Web: A Bird's-Eye View of World History by J. R. McNeill and William H. McNeill," *History and Theory* 44, no. 1 (2005): 103; and Liisa Malkki, "National Geographic: The Rooting of Peoples and the Territorialization of National Identity among Scholars and Refugees," *Cultural Anthropology* 7, no. 1 (1992): 268.

297. I am borrowing this expression from Toledano, *State and Society in Mid-Nineteenth-Century Egypt*, 196.

CHAPTER 2. LIKE A BEEHIVE

1. Doctor Aubert-Roche, "Rapport," *L'Isthme de Suez*, 15–18 July 1865, 205; and Tesner, "Une correspondance de Suez," *L'Isthme de Suez*, 1 June 1867, 167. For more on the tension between human toil and mechanical exertion on the Suez worksites, see Lucia Carminati, "Of Machines and Men: The Uneasy Synergy of Mechanization and Migrant Labor on the Suez Canal Worksites, 1859–1864," in *Oxford Handbook of Modern Egypt*, ed. Beth Baron and Jeffrey Culang (Oxford: Oxford University Press, forthcoming).

2. Zachary Karabell, *Parting the Desert: The Creation of the Suez Canal* (New York: A. A. Knopf, 2003), 211–12; Zayn al-'Abidin Shams al-Din Najm, *Bur Sa'id: Tarikhuha wa-tatawwuruha, mundhu nash'atiha 1859 hatta 'am 1882* (Cairo: Al-Hay'ah al-Misriyah al-'Ammah lil-Kitab, 1987), 80–81, 161; Jules Charles-Roux, *L'Isthme et le Canal de Suez: Historique, état actuel* (Paris: Hachette, 1901), 2:265–66; Archives du ministère de l'Europe et des Affaires étrangères, Centre des Archives Diplomatiques de Nantes (hereafter CADN), Archives Rapatriés de l'Agence Consulaire de Ismailia (hereafter ARI), C1, El Guisr, 26 September 1862, Viller to Director General; ARI, C1, Timsah ville, 5 February 1863, Philipe [illegible last name] to Viller; ARI, C1, Ismailia, 27 August 1863, Cohen to Campsite Agent and Campsite Agent to Cohen; and ARI, C1, Port Said, 15 February 1864, Zarb, Head of Health Service in Port Said.

3. R. H. Miles, "Variétés: Voyage d'une mer à l'autre à travers l'isthme de Suez," *L'Isthme de Suez*, 15–21 January 1867, 25.

4. Ferdinand de Lesseps, *Lettres, journal et documents pour servir à l'histoire du canal de Suez* (Paris: Didier, 1875), 4:13–14, 19, 49, entries dated 23 January 1861 and 8 February 1861.

5. Émile Erckmann and Alexandre Chatrian, *Souvenirs d'un ancien chef de chantier à l'isthme de Suez* (Paris: Hetzel, 1877), 7, 46; Lesseps, *Lettres, journal et documents*, 4:13, 19, entries dated 12 January 1861 and 8 February 1861; and Archivio Storico del Ministero degli Affari Esteri e della Cooperazione Internazionale, Rome (hereafter ASMAECI), Archivi Rappresentanze Diplomatiche e Consolari all'Estero (hereafter ARDCE), Consolato Cairo (hereafter CC), Antico versamento (hereafter AV), Busta (hereafter B) 18, sottofascicolo (hereafter s.f.) 6, Suez, 4 May 1866, Italian Vice-Consul in Suez to Italian Consul in Cairo.

6. Istituto per l'Oriente Carlo Nallino, Rome (hereafter IPOCAN), B 1866–1868, Alexandria, 28 January 1866, Giuseppe De Martino, Italian Consul in Cairo to Italian Ministry of Foreign Affairs; B 1866–1868, Alexandria, 5 October 1866, Italian Consul in Cairo to Italian Ministry of Foreign Affairs; B 1866–1868, 13 October 1866, Italian Consul in Cairo to Italian Ministry of Foreign Affairs;

ASMAECI, ARDCE, CC, AV, B18, s.f. 6, Suez 4 May 1866, Italian Vice-Consul in Suez to Italian Consul in Cairo; and Ministero degli Affari Esteri, *Emigrazione e colonie: Raccolta di rapporti dei RR. agenti diplomatici e consolari* (Roma: Tipografia dell'Unione Cooperativa Editrice, 1906), 2:199. For more on the recruitment of stonecutters in the northern Italian peninsula, see Lucia Carminati, "'Improvising and Very Humble': Those 'Italians' throughout Egypt That Statisticians and Historians Have Neglected," in *On the Margins of History: Italian Subalterns between Emigration and Colonialism in the Italian Colony in Egypt (1861–1937)*, ed. Costantino Paonessa (Louvain: Université catholique de Louvain presse, 2021), 48.

7. Lesseps, *Lettres, journal et documents*, 4:18–19, entry dated 8 February 1861.

8. Hubert Bonin, *History of the Suez Canal Company, 1858–1960: Between Controversy and Utility* (Geneva: Droz, 2010), 67; and Nathalie Montel, *Le chantier du Canal de Suez, 1859–1869: Une histoire des pratiques techniques* (Paris: Éd. In forma/ Presses de l'École Nationale des Ponts et Chaussées, 1999), 120.

9. François Levernay, *Guide général d'Égypte* (Alexandria: Imprimerie nouvelle, 1868), 277. Apparently Greeks still dominated retail trade, at least in Port Said, from the mid-1890s through 1914–1916; see L. A. Balboni, *Gl'Italiani nella civiltà egiziana del secolo XIX; Storia-biografie-monografie* (Alexandria: Tipo-litografico V. Penasson, 1906), 1:424; and Ronald Storrs, *The Memoirs of Sir Ronald Storrs* (New York: Putnam, 1937), 142.

10. "Conférences de M. De Lesseps à Lyon," *L'Isthme de Suez*, 1 December 1865, 391.

11. Ernest Desplaces, "Chronique de l'Isthme," *L'Isthme de Suez*, 1 May 1864, 221.

12. Doctor Aubert-Roche, "Rapport," *L'Isthme de Suez*, 15–18 July 1865, 215; Doctor Aubert-Roche, "Rapport," *L'Isthme de Suez*, 15–19 July 1867, 225.

13. CADN, ARI, C1, El Guisr, n.d., Sanson, Company Agent, Report of auction sale; ARI, C4, Sérapeum, 8 June 1866, Municipal Superintendent, Note and annexes; and Archives Nationales du Monde du Travail, Roubaix (hereafter ANMT), Compagnie universelle du canal maritime de Suez (hereafter CUCMS), 1995 060 3601, Damietta, 13 March 1862, Voisin, Director General, Report.

14. Emile de La Bédollière, *De Paris à Suez: Souvenirs d'un voyage en Égypte* (Paris: G. Barba, 1870), 32.

15. John MacGregor, *The Rob Roy on the Jordan, Nile, Red Sea, and Gennesareth, etc.: A Canoe Cruise in Palestine and Egypt, and the Waters of Damascus* (London: John Murray, 1869), 13.

16. Library of Congress Prints and Photographs Division, Washington DC, Bain Collection, *Street in Port Said*, 17 January 1916.

17. Harry Alis, *Promenade en Égypte* (Paris: Librairie Hachette, 1895), 9.

18. Narcisse Berchère, *Le désert de Suez: Cinq mois dans l'isthme* (Paris: Hetzel, 1863), 20; and Pierre-Henri Couvidou, *Itinéraire du canal de Suez* (Port Said: A. Mourès, 1875), 55.

19. Eric R. Wolf, *Europe and the People without History* (Berkeley: University of California Press, 1982), 380. While Wolf uses all three kinds of distinctions

("cultural," "racial," and "ethnic") to build his argument, I privilege the terms found in the sources, namely "nationality" and "race."

20. John T. Chalcraft, *The Invisible Cage: Syrian Migrant Workers in Lebanon* (Stanford, CA: Stanford University Press, 2009), 84.

21. J. Clerk, "Le Canal de Suez," *L'Isthme de Suez*, 15–18 April 1869, 142–43.

22. Valeska Huber, *Channelling Mobilities: Migration and Globalisation in the Suez Canal Region and Beyond, 1869–1914* (Cambridge: Cambridge University Press, 2013), 111; and Tesner, "Une correspondance de Suez," *L'Isthme de Suez*, 1 June 1867, 167.

23. Najm, *Bur Saïd*, 354; Gaston Jondet, Institut français d'archéologie orientale du Caire, and Société sultanieh de géographie, *Le port de Suez* (Cairo: Impr. de l'Institut français d'archéologie orientale, 1919), 87–88; Erckmann and Chatrian, *Souvenirs d'un ancien chef*, 10, 7, 37; Olivier Ritt, *Histoire de l'isthme de Suez* (Paris: L. Hachette, 1869), 270–71; Archives du ministère de l'Europe et des Affaires étrangères, Centre des Archives Diplomatiques á la Courneuve, Paris (hereafter CADC), Affaires économiques et commerciales (hereafter AEC), Correspondance Consulaire et Commerciale Port-Saïd (hereafter CCCPS), vol. 1, Port Said, 20 July 1867, Mr. Flesch, Vice-Consul in Port Said, to French Ministry of Foreign Affairs; and Tesner, "Une correspondance de Suez," *L'Isthme de Suez*, 1 June 1867, 167.

24. CADN, ARI, C4, Ismailia, 15 March 1866, Ismail Hamdy, Governor of the Isthmus, to Geyler, Chef des bureaux détachés (hereafter CdBD) in Ismailia; and ARI, C4, Tell el Kebir, 31 March 1866, Ibrahim Nagib, Doctor of Circonscription, to CdBD.

25. Sydney Montagu Samuel, *Jewish Life in the East* (London: C. Kegan Paul, 1881), 16.

26. Wolf, *Europe and the People without History*, 381.

27. Geoff Eley and Ronald Grigor Suny, "From the Moment of Social History to the Work of Cultural Representation," in *Becoming National: A Reader*, ed. Geoff Eley and Ronald Grigor Suny (Oxford: Oxford University Press, 1996), 9–12.

28. Darwin, who originally published his *Origin of the Species* in 1859, the same year the canal excavation began, first heard the expression "survival of the fittest" in the mid-1860s from Herbert Spencer, the father of the theory of social selection, and introduced it in the *Origin*'s fifth edition in 1869, the year the canal was inaugurated. In 1871 Darwin published *The Descent of Man*, the chief source for his ambiguous views on social evolution. See Diane B. Paul, "The Selection of the 'Survival of the Fittest,'" *Journal of the History of Biology* 21, no. 3 (1988): 412, 420–21; John C. Greene, "Darwin as a Social Evolutionist," *XIV International Congress of the History of Science; Proceedings*, no. 3 (1975): 26; and John S Haller, *Outcasts from Evolution: Scientific Attitudes of Racial Inferiority, 1859–1900* (Carbondale: Southern Illinois University Press, 1997), 86–88.

29. Edward W. Said, *Orientalism* (New York: Pantheon Books, 1978), 231–32.

30. Cemil Aydin, *The Idea of the Muslim World: A Global Intellectual History* (Cambridge, MA: Harvard University Press, 2019), 69.

31. Marwa Elshakry, *Reading Darwin in Arabic* (Chicago: University of Chicago Press, 2013); Omnia S. El Shakry, *The Great Social Laboratory: Subjects of Knowledge in Colonial and Postcolonial Egypt* (Stanford, CA: Stanford University Press, 2007), 56–57; and Eve Troutt Powell, *A Different Shade of Colonialism: Egypt, Great Britain, and the Mastery of the Sudan* (Berkeley: University of California Press, 2003), 17. For a discussion of the awareness within Africa and Asia, including Egypt, of European racial thinking and its implications for the colonized world, see Musab Younis, "'United by Blood': Race and Transnationalism during the Belle Époque," *Nations and Nationalism* 23, no. 3 (2017): 484–504.

32. James S. Duncan, *In the Shadows of the Tropics: Climate, Race and Biopower in Nineteenth-Century Ceylon* (Aldershot, UK: Ashgate, 2007); David N. Livingstone, "Race, Space and Moral Climatology: Notes toward a Genealogy," *Journal of Historical Geography* 28, no. 2 (2002): 159–80; David N. Livingstone, "The Moral Discourse of Climate: Historical Considerations on Race, Place and Virtue," *Journal of Historical Geography* 17, no. 4 (1991): 413–34; and David Arnold, *The Problem of Nature: Environment, Culture and European Expansion* (Oxford: Blackwell, 1996), 142, 189.

33. Arnold, *Problem of Nature*, 7, 189.

34. On Barak, *Powering Empire: How Coal Made the Middle East and Sparked Global Carbonization* (Oakland: University of California Press, 2020), 98.

35. Arthur Mangin, "Percement de l'isthme de Suez," *Journal des économistes: Revue scientifique et industrielle* 3, no. 1 (December 15, 1866): 449; *Conférence de M. Ferdinand de Lesseps à Lyon sur les travaux de l'isthme de Suez* (Paris: Impr. de N. Chaix, 1865), 14; The National Archives, London (hereafter TNA), Foreign Office (hereafter FO) 423-1, London, 10 February 1863, Earl Russell to Colquhoun.

36. Doctor Aubert-Roche, "Rapport," *L'Isthme de Suez*, 15–18 July 1865, 214.

37. IPOCAN, B1861, Alexandria, 10 April 1862, Austrian Consul General Schreiner to Count Rechberg, Report. This racial and demographic argument had already been present in Lane's account, first published in 1836. Edward William Lane, *The Manners & Customs of the Modern Egyptians* (J. M. Dent, 1908), 161.

38. Leone Carpi, *Delle colonie e dell'emigrazione d'italiani all'estero sotto l'aspetto dell'industria, commercio, agricoltura, e con trattazione d'importanti questioni sociali*, (Milan: Tipografia editrice lombarda, 1874), 3: 84.

39. Paul, "Selection of the 'Survival of the Fittest,'" 420–21.

40. Doctor Aubert-Roche, "Rapport," *L'Isthme de Suez*, 15–18 July 1865, 213–14, 217; ANMT, CUCMS, 1995 060 4136, El Guisr, 3 August 1865, Couvreux to Gioia, Chief Engineer of El Guisr Division. For a later example of racialized labor organization in a Middle Eastern context, wherein some workers were reputedly endowed with greater physique and others with higher intelligence, see Nimrod Ben Zeev, "Building to Survive: The Politics of Cement in Mandate Palestine," *Jerusalem Quarterly* 79 (2019): 41.

41. Barak, *Powering Empire*, 17.

42. Kathleen Neils Conzen et al., "The Invention of Ethnicity: A Perspective from the U.S.A.," *Journal of American Ethnic History* 12, no. 1 (1992): 12.

43. Antoinette Drohojowska, *L'Égypte et le Canal de Suez* (Paris: Laporte, 1870), 127; *Conférence de M. Ferdinand de Lesseps à Lyon*, 14; ANMT, CUCMS, 1995 060 1323, n.d., Excerpts of the Reports of the Director General; and 1995 060 3603, Damietta, 27 December 1861, Director General, Report.

44. TNA, Public Record Office (PRO) 30/22/93/57, 10 April 1865, Sir H. Bulwer to Lord Russell, folios 336–38; Leslie P. Moch, "Domestic Service, Migration, and Ethnic Stereotyping. Bécassine and the Bretons in Paris," *Journal of Migration History* 1, no. 1 (2015): 32–53; and Catherine Bertho Lavenir, "L'invention de la Bretagne: Genèse sociale d'un stereotype," *Actes de la recherche en sciences sociales*, no. 35 (1980): 45–62.

45. Rinaldo De Sterlich, *Sugli italiani d'Egitto: Lettera aperta di Fausto* (Cairo: Imprimerie parisienne J. Cèbe, 1888), 25; and IPOCAN, B 1866–1868, Alexandria, 10 October 1866, Sala, Company Agent, to Italian Consul in Alexandria.

46. ANMT, CUCMS, 1995 060 3599, Kantara, 22 May 1868, Head of Section to Chief Engineer of El Guisr Division; Sayyid 'Ashmawi, "Perceptions of the Greek Money-Lender in Egyptian Collective Memory at the Turn of the Twentieth Century," in *Money, Land and Trade an Economic History of the Muslim Mediterranean*, ed. Nelly Hanna (London: I. B. Tauris, 2002), 247–53; and Jabez Burns and Thomas Cook (Firm), *Help-Book for Travellers to the East Including Egypt, Palestine, Turkey, Greece and Italy* (London: Cook's Tourist Office, 1872), 184.

47. Julia A. Clancy-Smith, *Mediterraneans: North Africa and Europe in an Age of Migration, c. 1800–1900* (Berkeley: University of California Press, 2011), 181. For contemporary testimony, see Ramiro Vadala, *Les maltais hors de Malte: Etude sur l'émigration maltaise* (Paris: Rousseau, 1911), 66.

48. Donald Quataert, *Social Disintegration and Popular Resistance in the Ottoman Empire, 1881–1908: Reactions to European Economic Penetration* (New York: New York University Press, 1983), 90. For an analysis that is similar but based on later data, see Caroline Piquet, "La Compagnie Universelle du Canal Maritime de Suez en Égypte, de 1888 à 1956" (PhD diss., Université Paris-Sorbonne, 2006), 248–58.

49. ASMAECI, Personale, Serie II, Posizione P10, Port Said, B2, Primo Versamento, "Affare Giammugnai," Ancona, 15 November 1906, Giuseppe Giammugnai to Italian Ministry of Foreign Affairs.

50. ANMT, CUCMS, 1995 060 3600, Damietta, 22 December 1860, Director General, Report; and Doctor Aubert-Roche, "Rapport," *L'Isthme de Suez*, 15–19 July 1867, 227. For a similar claim within the Ottoman context of railroad employment, see Peter Mentzel, "The 'Ethnic Division of Labor' on Ottoman Railroads," *Turcica* 37 (2005): 232.

51. CADN, ARI, C4, Ismailia, 10 January 1866, Companyo, Health Service, Certificate; ARI, C4, Ismailia, 12 January 1866, Companyo, Certificate; ARI, C4, Chalouf, 24 March 1866, Company Doctor, Certificate; ARI, C4, Ismailia, 30 March 1866, Police Generale to Companyo; and "Nouvelles interpellations a la chambre des communes," *L'Isthme de Suez*, 15 November 1865, 106.

52. Doctor Aubert-Roche, "Rapport," *L'Isthme de Suez*, 15–18 May 1868, 150.

53. CADN, ARI, C9, Ismailia, 19 December 1869, Ismail Bey, Governor General of the Isthmus, to Geyler.

54. Will Hanley, *Identifying with Nationality: Europeans, Ottomans, and Egyptians in Alexandria* (New York: Columbia University Press, 2017), 5, 9; Julia A. Clancy-Smith, "Marginality and Migration: Europe's Social Outcasts in Pre-Colonial Tunisia, 1830–81," in *Outside in on the Margins of the Modern Middle East*, ed. Eugene L. Rogan (London: I. B. Tauris, 2002), 154; and Mary Dewhurst Lewis, "Europeans before Europe? The Mediterranean Prehistory of European Integration and Exclusion," in *French Mediterraneans: Transnational and Imperial Histories*, ed. Patricia M. E. Lorcin and Todd Shepard (Lincoln: University of Nebraska Press, 2016), 232–64.

55. Doctor Aubert-Roche, "Rapport," *L'Isthme de Suez*, 15–18 July 1865, 217; and Doctor Aubert-Roche, "Rapport," *L'Isthme de Suez*, 15–18 May 1868, 150.

56. ANMT, CUCMS, 1995 060 3600, Port Said, Laroche, Engineer, Map titled "Situation of metallurgic workshops on 18 April 1861." Laroche keeps "Greeks" distinct from "Europeans."

57. CADC, AEC, CCCPS, vol. 1, Port Said, 13 July 1870, Pellissier, French Consul in Port Said, to French Ministry of Foreign Affairs, 87r–87v. Spelling varies; both "barberin" and "berberin" are used. Socrates Spiro, *An Arabic-English Dictionary of the Colloquial Arabic of Egypt: Containing the Vernacular Idioms and Expressions, Slang Phrases, Vocables, etc., Used by the Native Egyptians* (Beirut: Librairie du Liban, 1999). This dictionary was first published in Cairo in 1895.

58. El-Said Badawi and Martin Hinds, *A Dictionary of Egyptian Arabic: Arabic-English* (Beirut: Librairie du Liban, 1986).

59. Baedeker, *Egypt, Handbook for Travellers: Part First, Lower Egypt* (Leipzig: K. Baedeker, 1878), 49. Doctor Aubert-Roche, "Rapport," *L'Isthme de Suez*, 15–18 July 1865, 217.

60. Mangin, "Percement de l'isthme de Suez," 449.

61. Y. Hakan Erdem, "Magic, Theft, and Arson: The Life and Death of an Enslaved African Women in Ottoman Izmit," in *Race and Slavery in the Middle East: Histories of Trans-Saharan Africans in Nineteenth-Century Egypt, Sudan, and the Ottoman Mediterranean*, ed. Kenneth M. Cuno and Terence Walz (Cairo: American University in Cairo Press, 2010), 140.

62. Julie Greene, "Spaniards on the Silver Roll: Labor Troubles and Liminality in the Panama Canal Zone, 1904–1914," *International Labor and Working-Class History* 66, no. 1 (2004): 83.

63. Raoul Lacour, *L'Égypte d'Alexandrie à la seconde cataracte* (Paris: Hachette, 1871), 459; and Charles-Roux, *L'Isthme et le Canal de Suez*, 2:237, 242.

64. John Torpey, *The Invention of the Passport: Surveillance, Citizenship, and the State* (Cambridge: Cambridge University Press, 2018), 118.

65. Joel Beinin and Zachary Lockman, *Workers on the Nile: Nationalism, Communism, Islam, and the Egyptian Working Class, 1882–1954* (Princeton, NJ: Princeton University Press, 1987), 41–42.

66. Najm, *Bur Saʿid*, 78–79; Serge Jagailloux, *La médicalisation de l'Égypte au XIXe siècle: 1798–1918* (Paris: Éd. Recherche sur les civilisations, 1986), 105; John Pudney, *Suez: De Lesseps' Canal* (New York: Praeger, 1969), 95; and Halford Lancaster Hoskins, *British Routes to India* (New York: Octagon Books, 1966), 369.

67. Beinin and Lockman, *Workers on the Nile*, 26–27; and Jennifer L. Derr, *The Lived Nile: Environment, Disease, and Material Colonial Economy in Egypt* (Stanford, CA: Stanford University Press, 2019).

68. Conzen et al., "Invention of Ethnicity," 4–5; and Arnold, *Problem of Nature*, 189.

69. See, for instance, Vittorio Briani, *Italiani in Egitto* (Rome: Istituto poligrafico e Zecca dello Stato, 1982), 119.

70. CADN, ARI, C1 El Ferdane, 14 December 1861, Riche & Sanson & Guiter, Inquiry.

71. Mangin, "Percement de l'isthme de Suez," 449; ANMT, CUCMS, 1995 060 1323, n.d., Excerpts of the Reports of the Director General; and 1995 060 4136, n.d., General Board of the Company Works, "Evaluation du prejudice."

72. Ritt, *Histoire de l'isthme de Suez*, 318; and Fabien Bartolotti, "Le chantier du siècle: Les entreprises marseillaises et le creusement du canal de Suez (1859–1869)," *Revue Marseille*, no. 260 (2018): 48.

73. Ritt, *Histoire de l'isthme de Suez*, 380–81; and Alis, *Promenade en Égypte*, 8. Arabic words made up half of the *sabir*, from the Latin verb "to know." One-quarter consisted of "more or less French" words, while the remainder derived from Italian, Spanish, or Latin. See sources quoted in Jocelyne Dakhlia, *Lingua franca* (Arles: Actes Sud, 2008), 444, 450, for the Algerian military context in which the *sabir* developed, see 443–54.

74. Alis, *Promenade en Égypte*, 9. The Volapük was invented by the Konstanz priest Johann Martin Schleyer in 1879–1880; see Jürgen Osterhammel, *The Transformation of the World: A Global History of the Nineteenth Century* (Princeton, NJ: Princeton University Press, 2014), 511.

75. Erckmann and Chatrian, *Souvenirs d'un ancien chef de chantier*, 46; and ANMT, CUCMS, 1995 060 3603, Damietta, 9 August 1862, Report de l'Ingénieur en Chef, Voisin.

76. ANMT, CUCMS, 1995 060 3603, Damietta, 9 August 1862, Director General, "Report"; and Damietta, 14 March 1862, Director General, "Report." Sources employ the term "drogman" (or "dragoman"), a distortion of the Arabic word "targuman," indicating a translator, someone explicating obscure terms, or a mediator; see Marie de Testa and Antoine Gautier, *Drogmans et diplomates européens auprès de la porte ottomane* (Istanbul: Editions Isis, 2003), 9.

77. CADN, ARI, C1, El Guisr, 26 September 1862, Viller, Chief Engineer, to Director General; ANMT, CUCMS, 1995 060 3603, Damietta, 9 August 1862, Director General, Report; Archivio della Sacra Congregazione per l'Evangelizzazione dei Popoli o de "Propaganda Fide," Rome (hereafter APF), Scritture riferite nei Congressi (hereafter Fondo S.C.), Egitto, Copti, vol. 20, 79–80, Suez, 21 August

1862, Fr. Alfonso da Cava, Apostolic Missionary and Suez Parish Priest, to Cardinal Barnabò, Prefect of Propaganda Fide in Rome, 79r; and ASMAECI, ARDCE, CC, AV, B12, Suez, 24 October 1869, English Consul, French Consul, Austrian Vice-Consul, and Italian Vice-Consul, to Governor. In twenty years' time, French seems to have overtaken the other languages in Port Said. By 1895 most signboards on shops appeared to be in French. French also dominated street talk, followed by Greek and Italian. Only a little English was in use. Nunda Lall Doss, *Reminiscences, English and Australasian: Being an Account of a Visit to England, Australia, New-Zealand, Tasmania, Ceylon, Etc.* (Calcutta: M. C. Bhowmick, 1893), 25; and Alis, *Promenade en Égypte*, 9.

78. TNA, FO 423-1, Alexandria, 26 April 1860, Robert G. Colquhoun to Lord J. Russell, Report by J. Coulthard.

79. CADN, ARI, C4, Ismailia, 11 February 1865, Companyo, Note; ARI, C4, Ismailia, 18 April 1865, CdBD to Director General; ARI, C4, Ismailia, 2 September 1865, Companyo, Note; ARI, C4, Ismailia, 17 November 1865, Companyo, Certificate; ARI, C4, Ismailia, 20 January 1866, Companyo, Certificate; Ismailia, 20 February 1866, Companyo, Note; ARI, C4, Ismailia, 15 March 1866, CdBD to Director General; ARI, C4, Ismailia, 18 March 1866, CdBD to Director General; ARI, C4, Ismailia, 21 March 1866, CdBD to Director General; and ARI, C4, Ismailia, 22 March 1866, CdBD to Director General.

80. CADN, ARI, C4, Ismailia, 12 October 1865, CdBD to Director General; ARI, C4, Circonscription du Ouady, 19 December 1865, Ibrahim Nagib, Certificate; ARI, C4, Ismailia, 26 January 1866, CdBD to Director General; ARI, C4, Ismailia, 16 March 1866, CdBD to Director General; ARI, C4, Ismailia, 25 March 1866, CdBD to Director General; ARI, C4, Tell el Kebir, 31 March 1866, Ibrahim Nagib, Doctor of Circonscription, to CdBD; and ARI, C4, Ismailia, 8 July 1866, CdBD to Director General.

81. I am drawing this expression from Julie Greene, *The Canal Builders: Making America's Empire at the Panama Canal* (New York: Penguin, 2009), 84.

82. CADN, ARI, C26, Ismailia, 19 September 1867, Nicola Castello to "Dear father"; ARI, C1, El Guisr, 31 January 1862, Jean Fraghasty to "Excellence" [Larousse, Chief Engineer of the Timsa Division at El Guisr]; and ARI, C1, El Guisr, 19 February 1862, J. B. Borneff to Larousse.

83. APF, Fondo S.C., Egitto, Copti, vol. 21, Jerusalem, 15 February 1877, Fr. Gaudenzio, Custodian of Terra Santa, to F. Bernardino da Portogruaro, Ministry General of the Order of Frati Minori in Rome, 1197v.

84. ASMAECI, ARDCE, CC, AV, B12, Suez, 3 January 1868, Enrico Ivaniski, Italian Protégé, to Italian Vice-Consul in Suez; CC, AV, B12, Suez, 13 July 1869, Eugenio Valli, Italian subject resident in Suez, to Italian Vice-Consul in Suez; and CADN, ARI, C26, Ismailia, 12 August 1869, Lorenzo Spano to [unspecified] Consul.

85. CADN, ARI, C9, Mulhouse, 1 March 1870, Pasteur Nemagel to French Consul in Alexandria; and Ana Barbič and Inga Miklavčič-Brezigar, "Domestic Work Abroad: A Necessity and an Opportunity for Rural Women from the Goriška

Borderland Region of Slovenia," in *Gender, Migration, and Domestic Service*, ed. Janet Henshall Momsen (London: Taylor & Francis, 1999), 162.

86. CADN, ARI, C26, Ismailia, 19 September 1867, Nicola Castello to "Caro padre"; and ARI, C26, Fiume, 24 May 1865, Maria Baraga to "Carissimo consorte."

87. Huber, *Channelling Mobilities*, 112.

88. Ehud R. Toledano, *State and Society in Mid-Nineteenth-Century Egypt* (Cambridge: Cambridge University Press, 1990), 192. On slowness as resistance among Egyptian workers at a later time, see On Barak, *On Time: Technology and Temporality in Modern Egypt* (Oakland: University of California Press, 2016), 13–14, 174.

89. Georges Legrain, *Louqsor sans les Pharaons: Légendes et chansons populaires de la Haute Égypte* (Brussels: Vromant, 1914), 189–90; Paolo Del Vesco, "Volti Senza Nome. I Lavoratori Egiziani Sullo Scavo," in *Missione Egitto 1903–1920: L'avventura Archeologica m.a.i. Raccontata*, ed. Paolo Del Vesco, Christian Greco, and Beppe Moiso (Modena: Franco Cosimo Panini, 2017), 210–11; and Anne Clément, "Rethinking 'Peasant Consciousness' in Colonial Egypt: An Exploration of the Performance of Folksongs by Upper Egyptian Agricultural Workers on the Archaeological Excavation Sites of Karnak and Dendera at the Turn of the Twentieth Century (1885–1914)," *History and Anthropology*, May 26, 2010, 93–94.

90. ANMT, CUCMS, 1995 060 1323, 8 February 1862, Division de Port-Saïd, Division de Timsah, Report of the Engineer in Chief.

91. CADN, ARI, C1, El Ferdane, 14 December 1861, Riche, Head of Section in El Guisr & Sanson, Company clerk & Guiter, employee of the contracting firm, Inquiry done on 14 September 1861.

92. ANMT, CUCMS, 1995 060 3600, Ismailia, 19 August 1863, Sciama, Engineer-in-Chief of the Works, to Gioia, Chief Engineer of Second Division; 1995 060 3600, Ismailia, 9 September 1863, Sciama to Gioia; Gaston Jondet, *Les Travaux d'extension du port de Suez* (Paris: Publications du journal Le Génie civil, 1919), 87; and Erckmann and Chatrian, *Souvenirs d'un ancien chef de chantier*, 7.

93. ANMT, CUCMS, 1995 060 4136, Port Said, 6 February 1865, Chief Engineer Laroche to Voisin, Director General in Ismailia; and 1995 060 4136, El Guisr, 3 August 1865, Couvreux to Gioia, Chief Engineer of El Guisr Division.

94. Ernest Desplaces, "Chronique de l'Isthme," *L'Isthme de Suez*, 1 May 1864, 221; and ANMT, CUCMS, 1995 060 4136, Port Said, 30 January 1865, Clément to Laroche, Chief Engineer of First Division, Port Said. The "poached" workers' names, as originally spelled, were Vassili Atanachi, Mikael Constantino, Giorgi Elia, Nicolas Papa Dimitrio, Antonio Vassilli, Giorgi Constantini, Nina Gerafi, Giorgi Civaselli, Paoli Manolli, Mikaël Zenzéfilé, Jouanin Mikaël, Elias Démitri, and Miquali Theophile.

95. Mangin, "Percement de l'isthme de Suez," 449; "Conférence de M. Borel," *L'Isthme de Suez*, 15–18 February 1867, 63; and CADN, ARI, C9, Port Said, 16 October 1869, Representative of Borel Lavalley and Co. to French Consular Agent.

96. Jondet, Institut français d'archéologie orientale du Caire, and Société sultanieh de géographie, *Le port de Suez*, 87–88.

97. ASMAECI, ARDCE, CC, AV, B12, Suez, 26 April 1869, Lambertenghi, Italian Vice-Consul, to Director General; CC, AV, B12, Suez, 11 October 1869, Italian Vice-Consul to Italian Consul in Cairo; CC, AV, B12, 4 November 1869, Italian Vice-Consul to Italian Consul in Cairo; and CC, AV, B18, folder 6, Suez, 3 October 1866, Italian Vice-Consul to Italian Consul in Cairo.

98. Ritt, *Histoire de l'isthme de Suez*, 319–21; and Antonio Colucci-Bey, *Le choléra en Égypte* (Paris: P. Dupont, 1866), 11.

99. ASMAECI, ARDCE, CC, AV, B18, 6 March 1866, French Consul in Suez to Italian Consul, Telegraph; and s.f. 6, Suez, 2 June 1866, Doctor Augusto Terra to Italian Vice-Consul.

100. CADN, ARI, C1, El Guisr 15 July 1861, Montant to Director General.

101. Erckmann and Chatrian, *Souvenirs d'un ancien chef de chantier*, 63; and CADN, ARI, C26, Canal de Hiènes, 10 August 1869, Poni, employee of Monsieur Junique, to [unspecified] Consul.

102. ANMT, CUCMS, 1995 060 3601, Port Said, 28 February 1866, Representative of Borel Lavalley and Co., to Director General; 1995 060 3601, Ismailia, 7 February 1866, Governor of the Isthmus of Suez to Director General; and Luigi Torelli, *Descrizione di Porto Said, del canale marittimo di Suez* (Venice: Gius Antonelli, 1869), 12.

103. Katharine Rollwagen, "'That Touch of Paternalism': Cultivating Community in the Company Town of Britannia Beach, 1920–58," *BC Studies* no. 151 (2006): 53.

104. CADN, ARI, C4, Chalouf, 3 March 1866, Larisiére, Municipal Superintendent, to Geyler; and Antonio Colucci-Bey, *Procès-verbaux des séances du Conseil de l'intendance générale sanitaire d'Egypte, présidé par M. le Docteur Ant. Colucci Bey* (Paris: Typ. De Renou, 1866), 102, 204.

105. Walter Johnson, "On Agency," *Journal of Social History* 37, no. 1 (2003): 116; and C. Hughes Dayton, "Rethinking Agency, Recovering Voices," *American Historical Review* 109 (2004): 834.

106. Leo Lucassen and Lex Heerma van Voss, "Introduction: Flight as Fight," in *A Global History of Runaways Workers, Mobility, and Capitalism, 1600–1850*, ed. Marcus Rediker, Titas Chakraborty, and Matthias Van Rossum (Oakland: University of California Press, 2019), 5; and Matthias van Rossum and Özgür Balkılıç, "Desertion," in *Handbook: The Global History of Work*, ed. Karin Hofmeester and Marcel Van der Linden (Berlin: De Gruyter Oldenbourg, 2018).

107. ANMT, CUCMS, 1995 060 4136, Paris, 24 November 1858, Captain Brout, Report.

108. ANMT, CUCMS, 1995 060 3601, El Guisr, 18 July 1861, Chief Engineer of Timsa Division, Report; CADN, ARI, C1, Port Said, 6 March 1862, Laroche to Barrellier, Director of the Works of the [Contracting] Firm in Port Said; and ARI, C1, Port Said, 7 March 1862, Ritt, Inspector of the General Administration, to Laroche.

109. ANMT, CUCMS, 1995 060 3599, El Guisr, 23 May 1867, Gioia to Director General; and 1995 060 3601, Ismailia, 10 October 1867, Director General to Governor.

110. "Egypt: From Our Own Correspondent," *Times*, 29 March 1864, 4.

111. IPOCAN, B 1866–1868, Alexandria, 28 January 1866, Italian Consul in Cairo to Italian Ministry of Foreign Affairs; B 1866–1868, Alexandria, 5 October 1866, Italian Consul in Cairo to Italian Ministry of Foreign Affairs; and B 1866–1868, 13 October 1866, Italian Consul in Cairo to Italian Ministry of Foreign Affairs.

112. IPOCAN, B 1866–1868, Alexandria, 13 October 1866, Italian Consul in Cairo to Italian Ministry of Foreign Affairs.

113. CADN, ARI, C1, Port Said, 7 March 1862, Ritt to Laroche; ARI, C1, Port Said, 11 March 1862, Schmitt to Laroche; and ARI, C1, Port Said, 27 June 1862, Deforge, Company Agent, Project of a regulation for the arrival and departure of individuals.

114. CADN, ARI, C1, Alexandria, 10 March 1862, Beraud to Laroche; and ARI, C1, Port Said, 11 March 1862, Schmitt to Laroche.

115. CADN, Archives Rapatriés du Consulat de France à Port-Saïd (hereafter ARCFPS), C149, 18 May 1868, Governor to French Vice-Consul in Port Said.

116. IPOCAN, B 1866–1868, Alexandria, 13 October 1866, Italian Consul in Cairo to Italian Ministry of Foreign Affairs.

117. CADN, ARI, C1, Port Said, 26 May 1863, Municipal Superintendent Delforge to Borney, Municipal Superintendent; ARI, C1, Ismailia, 23 November 1863, CdBD to Director General; and ARI, C1, Ismailia, 14 September 1863, A. Grouaz, Trader, to French Consul in Alexandria.

118. Francesca Biancani, "Globalisation, Migration, and Female Labour in Cosmopolitan Egypt," in *From Slovenia to Egypt: Aleksandrinke's Trans-Mediterranean Domestic Workers' Migration and National Imagination*, ed. Mirjam Milharčič Hladnik (Göttingen: VetR Unipress, 2015), 208.

119. Sylvia Hahn, "Nowhere at Home? Female Migrants in the Nineteenth-Century Habsburg Empire," in *Women, Gender, and Labour Migration: Historical and Global Perspectives*, ed. Pamela Sharpe (London: Routledge, 2011), 123.

120. ASMAECI, ARDCE, CC, AV, B18, folder 6, Suez, 9 May 1865, Dragoman and Chancellor at the French Consulate in Suez, Copy of death certificate by doctor at the French hospital in Suez; and 25 November 1866, Suez, Italian Vice-Consul, Declaration.

121. Hahn, "Nowhere at Home?," 110; Jose Moya, "Domestic Service in a Global Perspective: Gender, Migration, and Ethnic Niches," *Journal of Ethnic and Migration Studies* 33, no. 4 (2007): 562; Louise A. Tilly and Joan Scott, *Women, Work, and Family* (New York: Routledge, 1987), 69; Mirjam Milharčič Hladnik, "Trans-Mediterranean Women Domestic Workers: Historical and Contemporary Perspectives," in *From Slovenia to Egypt*, 18; and Janet Henshall Momsen, *Gender, Migration, and Domestic Service* (London: Routledge, 1999), 9.

122. Najm, *Bur Sa'id*, 161; ANMT, CUCMS, 1995 060 3603, Damietta, 27 December 1861, Director General, Report; Baedeker, *Egypt, Handbook for Travellers*, 50–51; and "Violenza al Cairo," *L'Imparziale*, 15 October 1895, 3.

123. Erckmann and Chatrian, *Souvenirs d'un ancien chef de chantier*, 6.

124. CADN, ARI, C1, no place, 22 February 1862, illegible signature, no recipient; ARI, C4, no place, n.d. [1866], Khalil to French Consul; ARI, C9, Ismailia, 6 October 1869, Chevenet to Agent Consulaire; and ARI, C9, Ismailia, 7 October 1869, Chevenet to Tétard.

125. Erckmann and Chatrian, *Souvenirs d'un ancien chef de chantier*, 6.

126. Julia A. Clancy-Smith, "'Making It' in Pre-Colonial Tunis: Migration, Work, and Poverty in a Mediterranean Port-City, c. 1815–1870," in *Subalterns and Social Protest: History from below in the Middle East and North Africa*, ed. Stephanie Cronin (London: Routledge, 2008), 217; Madeline C. Zilfi, "Servants, Slaves and the Domestic Order in the Ottoman Middle East," *HAWWA: Journal of Women of the Middle East and the Islamic World* 2, no. 1 (2004): 27–28; and Judith E. Tucker, *Women in Nineteenth-Century Egypt* (Cambridge: Cambridge University Press, 1985), 93. Narratives of the pitfalls of female domestic employment in global perspective, especially in regard to sexual predation and confinement, are analyzed in Amy Stanley, "Maidservants' Tales: Narrating Domestic and Global History in Eurasia, 1600–1900," *American Historical Review* 121, no. 2 (2016): 456.

127. CADN, ARI, C26, no place, n.d. [1857–1957], Luigia Frigieri to Italian Consul in Ismailia.

128. Rachel G. Fuchs and Leslie Page Moch, "Pregnant, Single, and Far from Home: Migrant Women in Nineteenth-Century Paris," *American Historical Review* 95, no. 4 (1990): 1009.

129. CADN, ARI, C1, Ismailia, 29 June 1863, Viller, Engineer of Division to Director General; and ARI, C1, Ismailia, 10 July 1863, Director General to Viller.

130. CADN, ARI, C9, Ismailia, 7 October 1869, Geyler, CdBD, Summary of births.

131. Ministero degli Affari Esteri, *Emigrazione e colonie* (1906), 2:225.

132. On the wide-ranging debate on whether prostitution should be discussed in terms of labor migration, see Marlou Schrover, "Labour Migration," in *Handbook: The Global History of Work*, ed. Karin Hofmeester and Marcel Van der Linden (Berlin: De Gruyter Oldenbourg, 2018). On women's entrepreneurship and other tactics in this field, see Lucia Carminati, "'She Will Eat Your Shirt': Foreign Migrant Women as Brothel Keepers in Port Said and Along the Suez Canal: Prostitution as Business and Survival, 1880–1914," *Journal of the History of Sexuality* 30, no. 2 (2021): 161–94.

133. Lesseps, *Lettres, journal et documents*, 4:24, entry dated 4 March 1861.

134. Tucker, *Women in Nineteenth-Century Egypt*, 91.

135. Liat Kozma, *Policing Egyptian Women: Sex, Law, and Medicine in Khedival Egypt* (Syracuse, NY: Syracuse University Press, 2011), 81; John Rodenbeck, "'Awalim; or, the Persistence of Error," in *Historians in Cairo: Essays in Honor of George Scanlon*, ed. George T. Scanlon and Jill Edwards (Cairo: American University in Cairo Press, 2002), 107–21; Khaled Fahmy, "Prostitution in Egypt in the Nineteenth Century," in *Outside In: On the Margins of the Modern Middle East*, ed. Eugene L. Rogan (London: I. B. Tauris, 2002), 82; Karin van Nieuwkerk, *A Trade Like Any Other: Female Singers and Dancers in Egypt* (Austin: University of Texas

Press, 1995), 32, 35–36; Toledano, *State and Society in Mid-Nineteenth-Century Egypt*, 237; and Francis Steegmuller, *Flaubert in Egypt: A Sensibility on Tour; A Narrative Drawn from Gustave Flaubert's Travel Notes and Letters* (London: Bodley Head, 1972), 83. On the obsession of Flaubert and others with the Orient as a place for sexual experience unobtainable in Europe, see Said, *Orientalism*, 190.

136. Charles-Roux, *L'Isthme et le Canal de Suez*, 1:398.

137. M. L. M. Carey, "The Almé Dancers, 1861," in *Women Travelers on the Nile: An Anthology of Travel Writing through the Centuries*, ed. Deborah Manley (Cairo: American University in Cairo Press, 2012), 104–6.

138. Ferdinand de Lesseps, *Souvenirs de quarante ans dédiés à mes enfants* (Paris: Nouvelle Revue, 1887), 2:44, entries dated 18 November 1854 and 4 December 1855.

139. Khaled Fahmy, *All the Pasha's Men: Mehmed Ali, His Army, and the Making of Modern Egypt* (Cairo: American University in Cairo Press, 2002), 230; Nieuwkerk, *Trade Like Any Other: Female Singers and Dancers in Egypt*, 36; and Tucker, *Women in Nineteenth-Century Egypt*, 153.

140. Hanan Hammad, "Regulating Sexuality: The Colonial-National Struggle over Prostitution after the British Invasion of Egypt," in *The Long 1890s in Egypt: Colonial Quiescence, Subterranean Resistance*, ed. Anthony Gorman and Marilyn Booth (Edinburgh: Edinburgh University Press, 2014), 197; and ʿImād Aḥmad Hilal, *Al-Al-Baghaya fi Misr: Dirasah tarikhiyah ijtimāʿīyah, 1834–1949* (Cairo: Al-ʿArabi lil-Nashr wa-al-Tawziʿ, 2001), 160–64.

141. Bruce W. Dunne, "Sexuality and the 'Civilizing Process' in Modern Egypt" (PhD diss., University of Michigan, 1996), 104–5; and Liat Kozma, *Global Women, Colonial Ports: Prostitution in the Interwar Middle East* (Albany: State University of New York Press, 2017), 77.

142. Christelle Taraud, *La prostitution coloniale: Algérie, Tunisie, Maroc (1830–1962)* (Paris: Payot, 2003), 19–22; and Clancy-Smith, *Mediterraneans*, 188–89.

143. ANMT, CUCMS, 1995 060 3600, Ismailia, 12 March 1865, Voisin, Instructions concernant la police des filles publiques.

144. CADN, ARI, C9, Ismailia, [illegible] 1869, Geyler, French Consular Agent in Ismailia, to D'Angelis, Campsite Agent at Sérapeum; and ARI, C9, Ismailia, 5 October 1869, Joseph César to French Consular Agent in Ismailia.

145. CADN, ARCFPS, C42, Port Said, 11 December 1879, Sub-Governor General, Aly Sabit, to Saint-Chaffray, Consul of France.

146. ANMT, CUCMS, 1995 060 3601, Damietta, 13 March 1862, Engineer-in-Chief and Director General, Report; 1995 060 3601, Kantara, 6 June 1867, M. Aladenize, Engineer, to Gioia, Chief Engineer of El Guisr Division; 1995 060 3601, Ismailia, 27 October 1867, Governor Ismail Hamdy to Director General; 1995 060 3601, Chalouf, 18 March 1868, Head of Section at Chalouf to Larousse, Chief Engineer of the Fourth Division in Suez; CADN, ARI, C4, Kantara, 28 March 1866, Municipal Superintendent in Kantarah to Geyler, CdBD, Ismailia; and ARI, C9, Ismailia, [illegible] 1869, French Consular Agent in Ismailia to Campsite Agent at Sérapeum.

147. APF, Fondo S.C., Egitto, Copti, vol. 21, 6 January 1869, Suez, Pancrazio da Luicciana, President, to Eminence, 2–3; and Dar al-Watha'iq al-Qawmiyya, National Archives, Cairo (hereafter DWQ), 0069–027010, 25–29 April 1889, Extract of the report of el-Kaimakam Bewley Bey, Suez.

148. CADN, ARI, C1, Ismailia, 27 August 1863, Cohen to Campsite Agent and Campsite Agent to Cohen; and ARI, C1, Ismailia, 28 August 1863, Y. Cohen.

149. DWQ, 2001–022821, Port Said, 20 August 1870, Boucherie Française, Receipt; CADN, ARI, C9, Ismailia, 15 February 1870, Rental Contract; and "Fatto di sangue," *L'Imparziale,* 21 December 1898, 3.

150. ASMAECI, ARDCE, CC, B24, Livorno, 11 aprile 1870, Claudio Soldaini to "Dearest Sister."

151. CADN, ARCFPS, C38, 29 December 1886, Port Said, French Consul to Cossery, consular agent in Damietta; and ARCFPS, C44, Port Said, 1 September 1889, Mikeli Clementi, Captain and Mahoun of the police (European section).

152. CADN, ARI, C26, Ismailia, 17 December 1869, Antonia Campi to Italian Consular Agent in Ismailia; Port Said, Vice-Consulate of H. H. of Italy, 22 December 1869, to Italian Consular Agent in Ismailia; and C4, El Guisr, 18 March 1866, Claudine Lapeyrouse to Brunetière, Municipal Superintendent in El Guisr.

153. CADN, ARI, C4, Port Said, 23 March 1866, Fiévre, Vinot, and Cie. to Municipal Superintendent in El Guisr.

154. For a similar claim in a dissimilar context, see Hahn, "Nowhere at Home?," 109–10.

155. Christiane Harzig, "Domestics of the World (Unite?): Labor Migration Systems and Personal Trajectories of Household Workers in Historical and Global Perspective," *Journal of American Ethnic History* 25, no. 2 (2006): 68.

156. CADN, ARI, C1, Ismailia, 27 August 1863, Marie Dutruy to Chief Engineer of Ismailia Division; ARI, C4, El Guisr, 18 March 1866, Claudine Lapeyrouse to Brunetiere, Municipal Superintendent; El Guisr, 13 March 1866, Lapeyrouse to Blanchet; and El Guisr, 18 March 1866, Blanchet to Lapeyrouse.

157. ANMT, CUCMS, 1995 060 3601, Ismailia, 25 November 1868, Ritt, on behalf of Director General, to President, Paris; and Ismailia, 24 November 1868, Chief of Detached Offices Division to Voisin.

158. CADN, ARCFPS, C21, Km. 95 of the Bitter Lakes, 14 September 1870, Rainouard, Station Chief, to French Consular Agent in Ismailia.

159. I am borrowing this expression from Pamela Sharpe, ed., *Women, Gender, and Labour Migration,* 8.

160. On the later history of so-called corporate familism within the Canal Company, see Angelos Dalachanis, "Ordinary People in Extraordinary Contexts: Italian and Greek Employees of the Suez Canal Company in Moments of Crisis" (paper presented at the conference "Italy and the Suez Canal: A Global History," Turin, 23–24 May 2019); and Barbara Curli, "Dames Employés at the Suez Canal Company: The 'Egyptianization' of Female Office Workers, 1941–56," *International Journal of Middle East Studies* 46, no. 3 (2014): 553–76.

161. Henri Silvestre, *L'isthme de Suez 1854–1869, avec carte et pièces justificatives* (Paris: Lacroix, Verboeckhoven, 1869), 82; and Charles-Roux, *L'Isthme et le Canal de Suez*, 2:247–49.

162. Jacquelyn Dowd Hall, *Like a Family: The Making of a Southern Cotton Mill World* (Chapel Hill: University of North Carolina Press, 1987), xxiii; and Marcelo J. Borges and Susana B. Torres, *Company Towns: Labor, Space, and Power Relations across Time and Continents* (New York: Palgrave Macmillan, 2012), 10, 17.

163. CADN, ARI, C1, Ismailia, 23 July 1863, no author, to "Dear aunt."

164. Erckmann and Chatrian, *Souvenirs d'un ancien chef de chantier*, 51.

165. CADN, ARI, C4, Serapeum, 26 September 1865, Dechon, Note.".

166. Charles-Roux, *L'Isthme et le Canal de Suez*, 2:247, 249, 251.

167. Ferdinand de Lesseps, *Lettres, journal et documents pour servir à l'histoire du canal de Suez* (Paris: Didier, 1875), 5:380; and Charles-Roux, *L'Isthme et le Canal de Suez*, 1:256. For an analysis of this rhetoric, see Lucia Carminati, "Port Said and Ismailia as Desert Marvels: Delusion and Frustration on the Isthmus of Suez, 1859–1869," *Journal of Urban History* 46, no. 3 (2019): 5.

168. Doctor Hérouard, "Note sur l'état medical de Port-Saïd," *L'Isthme de Suez*, 15–18 July 1865, 220.

169. Hélène Braeuner, "À la frontière de l'Égypte: Les représentations du canal de Suez," *In Situ: Revue des patrimoines*, no. 38 (February 12, 2019): 5.

170. Sarga Moussa, "Un Canal Pour Rire: Le Chantier de Suez vu Par Le Charivari," *Sociétés & Représentations* 48, no. 2 (2019): 55, 66. For a history of gendered images of the Egyptian nation, see Beth Baron, *Egypt as a Woman: Nationalism, Gender, and Politics* (Berkeley: University of California Press, 2005).

171. Henri de Bornier, *L'Isthme de Suez: Poème qui a remporté le prix proposé par l'Académie française* (Paris: E. Dentu, 1861), 10; and Louis de Trogoff de Kerbiguet, *À travers mon époque: Satires, poésies diverses, la question musicale* (Paris: Challamel aîné, 1874), 163–74, cited in Braeuner, "À la frontière de l'Égypte."

172. Erckmann and Chatrian, *Souvenirs d'un ancien chef de chantier*, 37.

173. CADN, ARCFPS, C149, 20 September 1866, Vice-Consul of France in Port Said to Moustapha Bey, Governor of Port Said; no place, n. d. [September–October 1866], no author to Consul General; and ANMT, CUCMS, 1995 060 3601, Ismailia, 21 May 1869, Representative of Borel Lavalle and Co. to Director General.

174. Different approaches were attempted to control gambling specifically in Port Said; it had been at first permitted but then was circumscribed to "bigger" establishments. Later, at the onset of the 1890s, there would be talk of turning Port Said into a "second Monte Carlo" by allowing one single venue to hold the monopoly on hazard. In 1904 a khedivial decree established harsher punishments for the foreign managers of those establishments where gambling was exposed. In 1906 Greek consular authorities in Egypt agreed to no longer protect their subjects who became proprietors of gaming houses. These entrepreneurs would henceforth be subject to Egyptian law. "L'Elettrico di Port Said," 16 January 1895, 2; "Le bische," *L'Imparziale*, 30 June 1904, 3; and "Gambling in Egypt," *The Egyptian Gazette*, 11 July 1906, 2.

175. CADN, ARCFPS, C149, Port Said, 29 September [1866], Laroche, French Consular Agent and Agent Secretary, Testimony; ARCFPS, C149, Port Said, 16 September 1866, Regnier, Campsite Agent, to French Vice-Consul in Port Said; ARCFPS, C149, Port Said, 20 September 1866, French Vice-Consul in Port Said, to Moustapha Bey, Governor; and ANMT, CUCMS, 1995 060 3601, Damietta, 13 March 1862, Director General, Report.

176. ASMAECI, ARDCE, CC, AV, B12, Suez, 24 October 1869, English Consul, French Consul, Austrian Vice-Consul, Italian Vice-Consul, to Governor; CC, AV, B12, Suez, 30 December 1869, Italian Vice-Consul, to Italian Consul in Cairo; and CC, AV, B12, Suez, Italian Vice-Consul, 30 December 1869, to Italian Consul in Cairo.

177. CADN, ARCFPS, C149, no place, n. d. [September–October 1866], no author to French Consul General.

178. CADN, ARCFPS, C149, no place, n. d. [September–October 1866], no author to French Consul General.

179. ANMT, CUCMS, 1995 060 3601, El Guisr, 18 July 1861, Chief Engineer of Timsa Division, Report; 1995 060 3601, Damietta, 13 March 1862, Director General, Report. CADN, ARI, C1, El Guisr, 23 August 1862, Petros Ribao, Complaint; and ARI, C1, El Guisr, 9 November 1862, Richard Wofs, Pharmacist, Response to Complaint.

180. ANMT, CUCMS, 1995 060 3599, El Guisr, 23 May 1867, Gioia to Director General; ASMAECI, ARDCE, CC, AV, B12, Suez, 24 October 1869, English Consul, French Consul, Austrian Vice-Consul, Italian Vice-Consul, to Governor.

181. CADN, ARCFPS, C149, no place, n.d. [September–October 1866], no author to French Consul General.

182. CADN, ARI, C1, El Guisr, 22 February 1862, Director General to Larousse.

183. ANMT, CUCMS, 1995 060 3601, Damietta, 13 March 1862, Director General, Report; and CADN, ARCFPS, C149, Alexandria, 22 October 1863, Memorandum to Consuls General issued by the Egyptian Ministry of Foreign Affairs.

184. ANMT, CUCMS, 1995 060 3601, Chalouf, November 1866, Henry, Head of Section in Chalouf, to Geyler, CdBD & response by Geyler; 1995 060 3601, Ismailia, 27 October 1867, Governor Ismail Hamdy to Director General; and 1995 060 3601, Ismailia, 10 March 1868, De Lesseps to Governor.

185. ASMAECI, ARDCE, CC, AV, B12, Suez, 30 December 1869, Italian Vice-Consul to Italian Consul in Cairo.

186. ANMT, CUCMS, 1995 060 3601, Ismailia, 21 May 1869, Representative of Borel Lavalley and Co. to Director General.

187. Rebecca Rogers, "Telling Stories about the Colonies: British and French Women in Algeria in the Nineteenth Century," *Gender & History* 21, no. 1 (2009): 54; and Julia A. Clancy-Smith, "Twentieth-Century Historians and Historiography of the Middle East: Women, Gender, and Empire," in *Middle East Historiographies: Narrating the Twentieth Century*, ed. Israel Gershoni, Y. Hakan Erdem, and Amy Singer (Seattle: University of Washington Press, 2006).

188. Ernest Desplaces, "Chronique de l'Isthme," *L'Isthme de Suez*, 1 February 1862, 37; and ANMT, CUCMS, 1995 060 4126, Board of Directors, Meeting of 1 September 1874.

189. Rollwagen, "'That Touch of Paternalism,'" 52, 56; and Ritt, *Histoire de l'isthme de Suez*, 272.

190. CADN, ARI, C1, El Guisr, 11 November 1862, Hassan Ismail, Chief Engineer of Timsa Division; ARI, C1, Ismailia, 9 July 1863, Valentini, Company Doctor, to Chief Engineer of El Guisr Division; and ARI, C1, Kantara, 14 November 1863, Head of Section & Company Doctor & and Municipal Superintendent, Report.

191. Marcelo J. Borges and Susana B. Torres, *Company Towns*, 22–23.

192. CADN, ARI, C1, Seuil 28 June 1862, Ahmad Effendia to Engineer; and Erckmann and Chatrian, *Souvenirs d'un ancien chef de chantier*, 3, 6, 61.

193. Gaston Maspero, *Chansons populaires recueillies dans la Haute-Egypte de 1900 à 1914 pendant les inspections du Service des antiquités* (Cairo: Service des antiquités de l'Egypte: Impr. de l'Institut français d'archéologie orientale, 1914), 232.

194. CADN, ARI, C9, Alexandria, 7 July 1869, Contract between Mr. Buratti Clemente and M.me Rosa Calamita; and Ismailia, 20 October 1869, Rosa Calamita, Receipt. Every fifteen days, the sisters would receive the so-called interest of 800 francs paid in advance.

195. CADN, ARI, C9, Carcassonne 23 October 1869, Rogilion to French Consul in Alexandria. On the side is scribbled "look for her." But the consular decision may have come too late, on April 9, 1870.

196. CADN, ARI, C4, Ismailia, 30 April 1866, CdBD to Governor; ARI, C1, Timsahville, 8 July 1862, six signatures including that of the Chief of Section; and ARI, C9, Alexandria, 4 April 1870, Chancellor at French Consulate in Alexandria to Geyler, French Consular Agent in Ismailia.

197. CADN, ARI, C4, Ismailia, 23 October 1866, Director General to Geyler.

198. ASMAECI, ARDCE, CC, AV, B12, Suez, 6 July 1869, Cross of Domenico Giorgini to Italian Vice-Consul in Suez; and Suez, 7 July 1869, Italian Vice-Consul in Suez to Italian Consul in Cairo.

199. Mary Lether Wingerd, "Rethinking Paternalism: Power and Parochialism in a Southern Mill Village," *Journal of American History* 83, no. 3 (1996): 901.

200. CADN, ARI, C26, Ismailia, 10 September 1869, Gaspare Lucia to Italian Consul; ANMT, CUCMS, 1995 060 3601, Port Said, 11 May 1868, Jeanne Felicite to Director General; and CADN, ARCFPS, C21, Paris, 25 September 1866, Director of the Chancellery Archives to French Consul General, Contrat.

201. Joan Wallach Scott, *Gender and the Politics of History* (New York: Columbia University Press, 1999).

202. CADN, ARI, C26, Ismailia, 25 September 1869, Benedetto Abiuri to Italian Consular Agent in Ismailia.

203. "Stamattina a Ismailia," *L'Imparziale*, 11–12 December 1892, 2.

204. CADN, ARCFPS, C21, Km. 95 of the Bitter Lakes, 14 September 1870, Rainouard, Station Chief, to French Consular Agent in Ismailia; and ARI, C1, Seuil, 28 June 1862, Ahmad Effendia to Engineer.

205. CADN, ARCFPS, C334, Cairo, 3 October 1884, Clemence Vial, Midwife in Cairo, to French Consul; and ARCFPS, C334, 15 December 1884, Clemence Vial to French Consul.

206. Moch, "Domestic Service, Migration, and Ethnic Stereotyping," 33; Sheila Rowbotham, foreword to *Women, Gender, and Labour Migration*, xvi; Harzig, "Domestics of the World (Unite?)," 48; and Momsen, *Gender, Migration, and Domestic Service*, 1.

207. Bruna Bianchi, "Lavoro ed emigrazione femminile (1880–1915)," in *Storia dell'emigrazione italiana: Partenze*, ed. Piero Bevilacqua, Andreina De Clementi, and Emilio Franzina (Rome: Donzelli, 2009), 257.

208. Milharčič Hladnik, "Trans-Mediterranean Women," 17–18; Sylvia Hahn, "Migration and Career Patterns of Female Domestic Servants," in *From Slovenia to Egypt*, 202; and Marta Verginella, "Le Aleksandrinke Tra Mito e Realtà," in *Le Rotte di Alexandria: Po Aleksandrijskih Poteh*, ed. Franco Però and Patrizia Vascotto (Trieste: EUT, 2011), 172.

209. Sharpe, *Women, Gender, and Labour Migration*, 9; and Olimpia Gobbi, "Emigrazione femminile: Balie e domestiche marchigiane in Egitto fra Otto e Novecento," *Proposte e ricerche: Economia e società nella storia dell'Italia centrale* 34, no. 66 (2011): 13.

210. Mohamed Gamal-Eldin, "Doing Environmental, Infrastructural, and Urban Histories along the Suez Canal," *Jadaliyya*, October 22, 2020, www.jadaliyya.com/Details/41886.

211. ANMT, CUCMS, 1995 060 3601, Port Said, 26 July 1869, Thouzet to the Ingenieur Chef de la Division, Laroche.

212. Mostafa Mohie, *Biographies of Port Said*, vol. 36, no. 1, Cairo Papers in Social Science (Cairo: American University in Cairo Press, 2021), 10; Lucia Carminati and Mohamed Gamal-Eldin, "Decentering Egyptian Historiography: Provincializing Geographies, Methodologies, and Sources," *International Journal of Middle East Studies* 53, no. 1 (February 2021): 107; see Dia'al-Din Hassan Qadi, *Mawsu'at tarikh Bur Sa'id* (Cairo: Al-Hay'ah al-Misriyah al-'Ammah lil-Kitab, 2015).

213. DWQ, 2001-018593, Doctor F. Flood, "Gouvernatorat d'Ismailiah," in *Rapports des Inspecteurs Sanitaires des Gouvernatorats et Provinces* (1890), 19. See Mohamed Gamal-Eldin, "Cesspools, Mosquitoes and Fever: An Environmental History of Malaria Prevention in Ismailia and Port Said, 1869–1910," in *Seeds of Power: Explorations in Ottoman Environmental History*, ed. Onur İnal and Yavuz Köse (Winwick UK: White Horse Press, 2019), 184–207.

214. Najm, *Bur Sa'id*, 31; Céline Frémaux, "Town Planning, Architecture, and Migrations in Suez Canal Port Cities," in *Port Cities: Dynamic Landscapes and Global Networks*, ed. Carola Hein (Abingdon, UK: Routledge, 2011), 167; and

Lucia Carminati, "Dividing and Ruling a Mediterranean Port-City: The Many Boundaries within Late Nineteenth Century Port Said," in *Multi-Ethnic Cities in the Mediterranean World: Controversial Heritage and Divided Memories, from the Nineteenth through the Twentieth Centuries*, ed. Marco Folin and Heleni Porfyriou (London: Routledge, 2021), 2:30–44.

215. Rajiyah Isma'il Abu Zayd, *Tarikh madinat al-Isma'iliyah: Min al-nash'ah ila muntasaf al-qarn al-'ishrin* (Cairo: Maktabat al-Adab, 2012), 401–2.

216. Gyan Prakash, "The Urban Turn," in *Sarai Reader 02: The Cities of Everyday Life*, ed. Ravi S. Vasudevan (Delhi: Center for the Study of Developing Societies, 2002), 6.

217. ANMT, CUCMS, 1995 060 3601, El Guisr, 18 July 1861, Chief Engineer of Timsa Division, Report; and CADN, ARCFPS, C149, no place, n.d. [September–October 1866], no author to French Consul General.

218. ANMT, CUCMS, 1995 060 4136, Ismailia, 2 June 1866, Voisin, Workers' engagement contract; Charles-Roux, *L' Isthme et le Canal de Suez*, 2:237–38; and J. Clerk, "Le Canal de Suez," *L'Isthme de Suez*, 15–18 April 1869, 142–43.

219. CADN, ARI, C1, Ismailia, 14 September 1863, A. Grouaz, Trader, to French Consul in Alexandria.

CHAPTER 3. A SEMILAWLESS BORDERLAND

1. L. A. Balboni, *Gl'Italiani nella civiltà egiziana del secolo XIX: Storia-biografie-monografie* (Alexandria: Tipo-litografico v. Penasson, 1906), 2:145.

2. Archivio Storico del Ministero degli Affari Esteri e della Cooperazione Internazionale, Rome (hereafter ASMAECI), Archivi Rappresentanze Diplomatiche e Consolari all'Estero (hereafter ARDCE), Consolato Cairo (hereafter CC), Antico versamento (hereafter AV), Busta (hereafter B) 12, Suez, 7 July 1869, Italian Vice-Consul to Italian Consul in Cairo.

3. Archives du ministère de l'Europe et des Affaires étrangères, Centre des Archives Diplomatiques de Nantes (hereafter CADN), Archives Rapatriés du Consulat de France à Port-Saïd (ARCFPS), C149, Port Said, 16 October 1869, Sous-Governor of Port Said Circulaire to French Vice-consul in Port Said.

4. ASMAECI, ARDCE, CC, AV, B12, Suez, 7 November 1869, Italian Vice-Consul to Italian Consul in Cairo; and "Nel passato di de Lesseps," *L'Imparziale*, 26 December 1892, 2.

5. Kenneth J. Perkins, *Port Sudan: The Evolution of a Colonial City* (Boulder, CO: Westview Press, 1993), 13–14.

6. Archives du ministère de l'Europe et des Affaires étrangères, Centre des Archives Diplomatiques á la Courneuve, Paris (hereafter CADC), Affaires économiques et commerciales (hereafter AEC), Correspondance Consulaire et Commerciale Port-Saïd (hereafter CCCPS), vol. 1, Port Said, 20 February 1874, Saint-Chaffray, French Consul in Port Said, to French Ministry of Foreign Affairs of Foreign Affairs, 299v;

Archivio della Sacra Congregazione per l'Evangelizzazione dei Popoli o de "Propaganda Fide," Rome (hereafter APF), Scritture riferite nei Congressi (hereafter Fondo S.C.), Egitto, Copti, vol. 21, Constantinople, 20 January 1877, F. Lodovico Arc. Di Fiunia, Apostolic Vicar of Syria, Cardinal Le Franch, Prefect, 1155v.

7. The National Archives, London (hereafter TNA), Foreign Office (hereafter FO) 846-3, Port Said, 1 January 1875, Royle Margossian v. James Testaferrata; and Mabel Caillard, *A Lifetime in Egypt, 1876–1935* (London: G. Richards, 1935), 16.

8. Olivier Ritt, *Histoire de l'isthme de Suez* (Paris: L. Hachette, 1869), 288.

9. Dar al-Watha'iq al-Qawmiyya, National Archives, Cairo (hereafter DWQ), 5009-000232, Cairo, 28 March 1872, Ferdinand De Lesseps, Memorandum for the Khedive.

10. Nubar Nubarian, *Mémoires de Nubar Pacha*, ed. M. B. Ghali (Beirut: Librairie du Liban, 1983), 362–63; The Franciscan Centre of Christian Oriental Studies, "Cronaca di Santa Caterina: Appendice Seconda," *Studia Orientalia Christiana, Collectanea*, nos. 26–27 (1996): 479–80; and Balboni, *Gl'Italiani nella civiltà egiziana del secolo XIX*, 2:147.

11. William George Hamley, *A New Sea and an Old Land: Being Papers Suggested by a Visit to Egypt at the End of 1869* (Edinburgh: W. Blackwood, 1871), 113–14; and Balboni, *Gl'Italiani nella civiltà egiziana del secolo XIX*, 2:156.

12. Sylvia Modelski, *Port Said Revisited* (Washington, DC: FAROS, 2000), 39; Nubarian, *Mémoires de Nubar Pacha*, 364; and Jabez Burns and Thomas Cook (Firm), *Help-Book for Travellers to the East Including Egypt, Palestine, Turkey, Greece and Italy* (London: Cook's Tourist Office, 1872), 11.

13. Roberto Morra di Lavriano, Alberto Siliotti, and Alain Vidal-Naquet, *Giornale di viaggio in Egitto: Inaugurazione del Canale di Suez* (Verona: Archeologia Dossier, 1995), 82.

14. CADN, Archives Rapatriés de l'Agence Consulaire de Ismailia (hereafter ARI), C9, Ismailia, 5 December 1869, Elisa Boschi to Geyler, agent consulaire de France (see list of witnesses); and DWQ, 5009-000219, Alexandria, 6 September 1869, Antoine Rannucci to Riaz Pacha. Had they implemented the hotel registration regulations, they would have been precursors of later developments in Istanbul driven by anti-anarchist sentiments. İlkay Yılmaz, "Propaganda by the Deed and Hotel Registration Regulations in the Late Ottoman Empire," *Journal of the Ottoman and Turkish Studies Association* 4, no. 1 (May 2017): 137–56.

15. DWQ, 5009-000219, Ghizeh, 30 April 1869, Khedive, no addressee; 5009-000219, July 1869, Khedive to Nubar; Modelski, *Port Said Revisited*, 35; and Emily A. Haddad, "Digging to India: Modernity, Imperialism, and the Suez Canal," *Victorian Studies* 47, no. 3 (2005): 364.

16. Samuel Selig de Kusel, *An Englishman's Recollections of Egypt* (London: John Lane; John Lane, 1915), 75; and Qallini Fahmi Basha, *Mudhakkirat Qallini Fahmi Basha* (Al-Minya: Matba'at Sadiq, 1934), 11.

17. Burns and Thomas Cook (Firm), *Help-Book for Travellers to the East*, 9–10; Kusel, *Englishman's Recollections of Egypt*, 75–76; Nubarian, *Mémoires de Nubar*

Pacha, 363; and The Franciscan Centre of Christian Oriental Studies, "Cronaca di Santa Caterina," 479. The reference to the Shaykh of Al-Azhar is by Nubarian. It was rumored that the empress received a special kind of treatment by the Khedive in person. Fahmi, *Mudhakkirat*, 11. Huber notes that the Greek Orthodox archbishop of Jerusalem was also in attendance; Valeska Huber, *Channelling Mobilities. Migration and Globalisation in the Suez Canal Region and Beyond, 1869–1914* (Cambridge: Cambridge University Press, 2013), 41.

18. Burns and Thomas Cook (Firm), *Help-Book for Travellers to the East*, 9–10; Kusel, *Englishman's Recollections of Egypt*, 76–77; and Huber, *Channelling Mobilities*, 40.

19. Ahmad Shafiq, *Mudhakkirati fi nisf qarn* (Cairo: Matba'at Misr, 1934), 25.

20. Omnia S. El Shakry, *The Great Social Laboratory. Subjects of Knowledge in Colonial and Postcolonial Egypt* (Stanford, CA: Stanford University Press, 2007), 24; and Carmen M. K. Gitre, *Acting Egyptian Theater: Identity, and Political Culture in Cairo* (Austin: Texas University Press, 2019), 23.

21. Burns and Thomas Cook (Firm), *Help-Book for Travellers to the East*, 11.

22. Fahmi, *Mudhakkirat*, 12.

23. Robert L. Tignor, *Egypt: A Short History* (Princeton, NJ: Princeton University Press, 2010), 224; Mia Carter and Barbara Harlow, eds., *Archives of Empire*, vol. 1, *From the East India Company to the Suez Canal* (Durham, NC: Duke University Press, 2003), 557; and Adam Mestyan, *Arab Patriotism: The Ideology and Culture of Power in Late Ottoman Egypt* (Princeton, NJ: Princeton University Press, 2017), 84. For a discussion of *Aida*, opera production in Egypt, and new frameworks for the country's elite to imagine themselves as a coherent whole, see Gitre, *Acting Egyptian Theater*, 16–41.

24. Burns and Thomas Cook (Firm), *Help-Book for Travellers to the East*, 5; and Fahmi, *Mudhakkirat*, 12. For currency references, see "Money Table" in Karl Baedeker, *Egypt: Handbook for Travellers* (Leipzig: Karl Baedeker, 1898).

25. David S. Landes, *Bankers and Pashas: International Finance and Economic Imperialism in Egypt* (London: Heinemann, 1958), 324; and Tignor, *Egypt*, 224.

26. Carter and Harlow, *Archives of Empire, Volume 1*, 556.

27. Hilmi Namnam and Muhammad Musa Nifin, *Qanat al-Suways, watha'iq al-hulm al-misri* (Cairo: Dar al-Kutub wa-al-Watha'iq al-Qawmiyah, 2015), 63.

28. Burns and Thomas Cook (Firm), *Help-Book for Travellers to the East*, 5.

29. CADC, AEC, CCCPS, vol. 1, Port Said, received 9 October 1867, Flesch, Vice-Consul in Port Said, to French Ministry of Foreign Affairs, 23r.

30. CADN, ARI, C9, 31 December 1869, Inventory; and ARI, C9, Serapeum, 24 October 1869, D. Angelys to Consul Geyler, Ismailia.

31. Balboni, *Gl'Italiani nella civiltà egiziana del secolo XIX*, 2:141; and ASMAECI, ARDCE, CC, B12, Suez, 13 December 1869, Italian Vice-Consul to Italian Consul in Cairo.

32. CADC, AEC, CCCPS, vol. 1, Port Said, 9 June 1870, Pellissier, French Consul in Port Said, to French Ministry of Foreign Affairs, 78v, 79r; and vol. 1, Port Said, 6 December 1872, Consul to Ministry Foreign Affairs, Paris, 221v.

33. ASMAECI, ARDCE, CC, AV, B12, Suez, 26 August 1869, 13 December 1869, Italian Vice-Consul to Italian Consul in Cairo; CC, AV, B12, Suez, 13 December 1869, Italian Vice-Consul to Italian Consul in Cairo; and F. Barham Zincke, *Egypt of the Pharaohs and of the Kedivé* (London: Smith, Elder, 1871), 484.

34. APF, Fondo S.C., Egitto, Copti, vol. 23, Alexandria, 1 August 1891, F. Guido Corbelli, Apostolic Vicar, to Simeoni, 925r; and vol. 23, Suez, 1 October 1885, Cardinal Moran, Archbishop of Sydney, to Cardinal Prefect, 185v.

35. CADC, AEC, CCCPS, vol. 1, Port Said, 20 March 1877, Saint-Chaffray, French Consul in Port Said, to French Ministry of Foreign Affairs, 452r, 455v, 456r; and vol. 1, Port Said, 9 June 1870, Pellissier, French Consul in Port Said, to French Ministry of Foreign Affairs, 78v, 79r.

36. CADC, AEC, CCCPS, vol. 1, Port Said, 20 March 1877, Saint-Chaffray, French Consul in Port Said, to Ministry of Foreign Affairs, 456r; and Richard Tangye, *Reminiscences of Travel in Australia, America, and Egypt* (London: Sampson Low, Marston, Searle, & Rivington, 1883), 232. On Port Said's position in the geography of coal, see On Barak, *Powering Empire: How Coal Made the Middle East and Sparked Global Carbonization* (Oakland: University of California Press, 2020).

37. François Levernay, *Guide général d'Égypte* (Alexandria: Imprimerie nouvelle, 1868), 310; and CADC, AEC, CCCPS, vol. 1, Port Said, 10 January 1871, Pellissier, French Consul in Port Said, to Ministry of Foreign Affairs, 123–26.

38. 'Ali Mubarak, *Al-Khitat al-tawfiqiyah al-jadidah li-Misr al-Qahirah wa-muduniha wa-biladiha al-qadimah wa-al-shahirah* (Cairo: Matba'at Dar al-Kutub wa-al-Watha'iq al-Qawmiyah, 2004), 10:58.

39. Doctor Aubert-Roche, "Rapport," *L'Isthme de Suez*, 15–19 June 1867, 225–26; and "Une correspondence du *Dauphiné*, Cairo, 2 January 1869," *L'Isthme de Suez*, 15 January 1869, 24.

40. 'Ali Mubarak, *Al-Khitat al-tawfiqiyah*, 10:58; and Luigi Dori, "Esquisse historique de Port Said: Naissance de la ville et inauguration du Canal (1859–1869)," *Cahiers d'histoire egyptienne* 6, nos. 3/4 (October 1954): 216.

41. CADC, AEC, CCCPS, vol. 1, Port Said, 20 July 1867, Flesch, Vice-Consul in Port Said, to Ministry of Foreign Affairs in Paris, 6r.

42. Marius Etienne Fontane and Edouard Riou, *Le canal maritime de Suez illustré: Histoire du canal et des travaux* (Paris: Aux bureaux de l'Illustration A. Marc et Cie., 1869), 111; "Une correspondence du Dauphiné, Cairo, 2 January 1869," *L'Isthme de Suez*, 15 January 1869, 24; and Pietro Perolari-Malmignati, *Su e giù per la Siria; note e schizzi* (Milan: Fratelli Treves, 1878), 21.

43. CADC, AEC, CCCPS, vol. 1, Port Said, 9 June 1870, Pellissier, French Consul in Port Said, to French Ministry of Foreign Affairs, 78v, 79r; Egypt, Ministry of the Interior, *Statistique de l'Egypte: Année 1873–1290 de l'hegire* (Cairo: Imprimerie française Moures, 1873), 20; and John Steele, *The Suez Canal: Its Present and Future* (London: Simpkin, Marshall, 1872), 16.

44. CADC, AEC, CCCPS, vol. 1, Port Said, 6 December 1872, Consul to Ministry of Foreign Affairs, Paris, 221v; and Port Said, 20 February 1874, Consul to French Ministry of Foreign Affairs, 298r, 298v, 299r.

45. Archives de la Maison-Mère du Bon Pasteur, Angers (hereafter AMMBPA), HC 43 Port Said, De notre Monastère de Port-Saïd, 6 January 1881, 16; TNA, FO 78/2409, Bath, 27 May 1875, Consul Charles Perceval to Earl of Derby, Foreign Office; and CADN, ARCFPS, C49, Porte des Galets (Réunion), 22 June 1879, no author (owner of property), to Consul.

46. CADC, AEC, CCCPS, vol. 1, Port Said, 20 February 1874, Saint-Chaffray, French Consul in Port Said to French Ministry of Foreign Affairs, 299v.

47. CADN, ARI, C9, 11 December 1869, Ismail Bey, Governor General of the Isthmus, to French Consular Agent in Ismailia; and Marius Fontane, *Voyage pittoresque à travers l'isthme de Suez* (Paris: Dupont-Lachaud, 1870), 30.

48. CADN, ARCFPS, C149, no place, n. d. [September–October 1866], no author to Consul General. See Mahmud Salih Ramadan, *Al-Haya al-ijtima'iya fi Misr fi 'asr Isma'il* (Alexandria: Munsha'at al-Ma'rifa al-Haditha, 1977), 307; and Huber, *Channelling Mobilities*, 136, 295.

49. CADC, AEC, CCCPS, vol. 1, Port Said, 20 July 1867, Flesch, Vice-Consul in Port Said, to Ministry of Foreign Affairs, 8v; and vol. 1, Port Said, received 9 October 1867, Flesch to French Ministry of Foreign Affairs, 22r.

50. CADN, ARCFPS, C149, Port Said, January 1867, Laroche, French Vice-Consul, to Outrey, Agent et Consul General de France in Alexandria; and ARCFPS, C46, Port Said, 19 March 1876, Saint-Chaffray, French Consul, to Governor of Port Said.

51. CADN, ARCFPS, C49, Port Said, 7 January 1878, G. Colomb to French Consul; ARCFPS, C42, Port Said, 8 August 1878, Vignocchi to Ismail Hamdy Pacha, Governor; and ARCFPS, C49, Port Said, 30 May 1878, Baun to Laporte, Vice-Consul, Port Said.

52. Arjun Appadurai, ed., *The Social Life of Things: Commodities in Cultural Perspective* (Cambridge: Cambridge University Press, 1986), 3–5.

53. ASMAECI, ARDCE, Rappresentanza Cairo (hereafter RC), B39, 1891, folder 34, Port Said, 30 August 1889, Rouchdy, Governor General, to Leoni, Italian Consul in Port Said.

54. CADC, AEC, CCCPS, vol. 1, Port Said, 20 December 1871, Pellissier, French Consul in Port Said, to French Ministry of Foreign Affairs.

55. CADN, ARCFPS, C42, Port Said, 11 April 1880, Aly Sabit, Sub-Governor, to Saint-Chaffray, French Consul.

56. ASMAECI, ARDCE, RC, B86, Port Said, 16 January 1901, Italian Consul to Consul General; and RC, B86, Port Said, 3 January 1902, Italian Consul to Consul General.

57. CADN, ARI, C9, Elisa Boschi, no place, n.d. [1869].

58. "Sotto il titolo—Molestie," *L'Imparziale*, 29 July 1895, 2.

59. CADN, ARCFPS, C21, Port Said, 14 June 1898, French Consul to Ries, French Managing Consul in Aden.

60. Zincke, *Egypt of the Pharaohs and of the Kedivé*, 424.

61. DWQ, 2001-025100, Port Said, 13 February 188[?], Ibrahim Tewfik, Governor General of the Canal and Port Said, to Sabet Pacha, Ministry of the Interior,

Cairo; and CADN, ARCFPS, C42, Port Said, 14 June 1884, Ibrahim Tewfik, Governor, to Dobignie, French Consul. For a discussion of West African pilgrims traveling clandestinely to Mecca, see Jonathan Miran, "'Stealing the Way' to Mecca: West African Pilgrims and Illicit Red Sea Passages, 1920s–50s," *Journal of African History* 56, no. 3 (2015): 389–408.

62. Archives Nationales du Monde du Travail, Roubaix (hereafter ANMT), Compagnie universelle du canal maritime de Suez (hereafter CUCMS), 1995 060 3603, Ismailia, 13 August 1866, Voisin to De Lesseps, President.

63. Perkins, *Port Sudan*, 13.

64. CADC, AEC, CCCPS, vol. 1, Port Said, 20 February 1874, Saint-Chaffray, French Consul in Port Said, to French Ministry of Foreign Affairs of Foreign Affairs, Paris, 300; and Zayn al-'Abidin Shams al-Din Najm, *Bur Sa'id: Tarikhuha watatawwuruha, mundhu nash'atiha 1859 hatta 'am 1882* (Cairo: Al-Hay'ah al-Misriyah al-'Ammah lil-Kitab, 1987), 217. Controversy raged in town for decades, especially around the issue of a railing that Customs wished to erect all along the quay to prevent contraband. Opposers lamented that strollers in Port Said would no longer be able to enjoy the unobstructed view of the sea. DWQ, 2001-010773, Alexandria, 20 June 1885, Caillard, Director General of Egyptian Customs, to Moustafa Pacha Fehmy, Ministry of Finance; "La Verità di Port Said," *L'Imparziale*, 16–17 June 1895, 3; and Luigi Dori, "Esquisse historique de Port Said: 1900–1914," *Cahiers d'histoire egyptienne* 8 (July 1956): 335.

65. CADN, ARI, C1, Ismailia, 31 December 1866, Mustapha Nithary, Governor of Port Said, to Chief Engineer of Port Said Division.

66. Eric Tagliacozzo, *Secret Trades, Porous Borders: Smuggling and States along a Southeast Asian Frontier, 1865–1915* (New Haven, CT: Yale University Press, 2005), 16–17.

67. DWQ, 2001-014127, Port Said, 12 February 1883, Synge, Colonel, Inspector of Gendarmerie, to General Baker Pacha.

68. Ramadan, *Al-Haya al-ijtima'iya fi Misr*, 307; Sayyid 'Ashmawi, "Perceptions of the Greek Money-Lender in Egyptian Collective Memory at the Turn of the Twentieth Century," in *Money, Land and Trade: An Economic History of the Muslim Mediterranean*, ed. Nelly Hanna (London: I. B. Tauris, 2002), 246; and Najm, *Bur Sa'id*, 202.

69. Philippe Gélat, *Répertoire général annoté de la législation et de l'administration egyptiennes: Part I, 1840–1908* (Alexandria: Impr. J.C. Lagoudakis, 1906), 2:813.

70. Cyrus Schayegh, "The Many Worlds of 'Abud Yasin; or, What Narcotics Trafficking in the Interwar Middle East Can Tell Us about Territorialization," *American Historical Review* 116, no. 2 (2011): 274; and Liat Kozma, "Cannabis Prohibition in Egypt, 1880–1939: From Local Ban to League of Nations Diplomacy," *Middle Eastern Studies* 47, no. 3 (2011): 449. Opium was for the most part produced in Turkey and Lebanon. However, in the mid-nineteenth century at least some opium was grown locally within Egypt. CADC, Entrées exceptionnelles et collections, Mémoires et documents, Egypte, vol. 15, Paris, 24 February 1863, A. Le Moyne to French Ministry of Foreign Affairs, 16r.

71. TNA, FO 846-3, Port Said, 29 May 1875, Customs v. Vincenzo Salomone & Giovanni Bessina; FO 846-4, Port Said, 18 July 1879, Testaferrata Abela (customs) v. Joseph Caruana (bumboatman); DWQ, 2001-025611, 8 May 1880, Governor of Port Said to Ministry of Interior; and 2001-020157, Alexandria, 1 July 1880, Caillard, Egyptian Customs Director General, to Ministry of Finances.

72. Ronald Storrs, *The Memoirs of Sir Ronald Storrs* (New York: Putnam, 1937), 44, entry dated 1905–1907.

73. "Bur Saʿid," *Al-Ahram*, 17 March 1892, 2; "Akhbarna khadra makatibna," *Al-Ahram*, 13 April 1895, 3; and "Astuzie di contrabbandieri," *L'Imparziale*, 18 April 1895, 3.

74. "Contrabbando di gioiellerie," *L'Imparziale*, 23 November 1900, 3.

75. Kusel, *Englishman's Recollections of Egypt*, 232; and Najm, *Bur Saʿid*, 209.

76. Steele, *Suez Canal*, 19; J. Millie, *Isthme et Canal de Suez: Son passé, son présent et son avenir* (Milan: Civelli, 1869), 67; and DWQ, 2001-020157, Alexandria, 1 July 1880, Caillard, Egyptian Customs Director General, to Ministry of Finances.

77. Henry de Monfreid, *Mer rouge* (Paris: Grasset, 2002), 393; and Enrico Pea, *Vita in Egitto* (Milan: Mondadori, 1949), 203.

78. Tagliacozzo, *Secret Trades, Porous Borders*, 18.

79. CADC, AEC, CCCPS, vol. 1, Port Said, 20 July 1867, Flesch, Vice-Consul in Port Said, to French Ministry of Foreign Affairs, 8v.

80. Huber, *Channelling Mobilities*, 112.

81. Leo Lucassen, "Eternal Vagrants? State Formation, Migration, and Travelling Groups in Western-Europe, 1350–1914," in *Gypsies and Other Itinerant Groups: A Socio-Historical Approach*, ed. Annemarie Cottaar, Leo Lucassen, and Wim Willems (New York: St. Martin's Press, 1998), 240.

82. CADN, ARCFPS, C49, Port Said, 30 May 1878, Baun to Laporte, Vice-Consul in Port Said; and C46, Port Said, 19 June 1877, French Consul to Governor of Port Said.

83. "Un tiro birbone," *L'Imparziale*, 3 May 1895, 3.

84. ASMAECI, ARDCE, CC, AV, B12, Suez, 10 December 1869, Italian Vice-Consul to Italian Consul in Cairo; and ANMT, CUCMS, 1995 060 3601, Cairo, 17 March 1863, President De Lesseps to Viceroy Ismail.

85. ANMT, CUCMS, 1995 060 3603, Damietta, 13 March 1862, Voisin, Engineer-in-Chief, proposition; and 1995 060 3603, Damietta, 14 March 1862, Voisin, Engineer-in-Chief, Report.

86. Juan Ricardo Cole, *Colonialism and Revolution in the Middle East: Social and Cultural Origins of Egypt's ʿUrabi Movement* (Princeton, NJ: Princeton University Press, 1993), 216.

87. ʿAbd al-ʿAziz Muhammad Shinnawi, *Al-Sukhrah fi hafr qanat al-Suways* (Alexandria: Munshaʾat al-Maʿrifa al-Haditha, 1958), 141. By 1870 a governor general residing in Ismailia ruled over the Isthmus of Suez. He responded directly to the Ministry of the Interior and in turn supervised the two governors who were specifically responsible for Port Said and for Suez. See Eugène de Régny, *Statistique de l'Égypte, année 1* (Alexandria: Imprimerie française Mourès, 1870), 9.

88. ANMT, CUCMS, 1995 060 3601, Cairo, 17 March 1863, President De Lesseps to Viceroy Ismail.

89. Najm, *Bur Saïd*, 191; and DWQ, 2001-022821, Alexandria, 31 August 1870, Director of the Central Office of Port Said Police, Report "About the Bruzzi sentence."

90. Najm, *Bur Saïd*, 205–7. For a discussion of police responsibilities in Alexandria, see Michael J. Reimer, *Colonial Bridgehead: Government and Society in Alexandria, 1807–1882* (Boulder, CO: Westview Press, 1997), 142.

91. ANMT, CUCMS, 1995 060 3601, Ismailia, 4 August 1866, Voisin, Director General, to Ismail Bey, Governor of the Isthmus; and 1995 060 3601, 24 August 1866, Voisin to Campsite Agent in Serapeum.

92. "Fuggito di gabbia," *L'Imparziale*, 21–22 July 1895, 3; and TNA, FO 78/2409, Bath, 27 May 1875, British Consul Charles Perceval to Earl of Derby, Foreign Office.

93. B. A. Roberson, "The Emergence of the Modern Judiciary in the Middle East: Negotiating the Mixed Courts of Egypt," in *Islam and Public Law*, ed. Chibli Mallat (London: Graham & Trotman, 1993), 119.

94. ANMT, CUCMS, 1995 060 3601, Cairo, 22 November 1866, French Consul to Voisin; and Leone Carpi, *Delle colonie e dell'emigrazione d'italiani all'estero sotto l'aspetto dell'industria, commercio, agricoltura, e con trattazione d'importanti questioni sociali* (Milan: Tipografia editrice lombarda, 1874), 3:165.

95. Istituto per l'Oriente Carlo Nallino, Rome (IPOCAN), busta (hereafter B) 1861, Alexandria, 5 December 1861, Consul General Schreiner to Count Rechberg, Report.

96. CADN, ARCFPS, C149, Alexandria, 27 September 1863, French Consul to Laroche, French Vice-Consul in Port Said; and ARI, C4, Tell El Kebir, 22 March 1866, Chief of Section to Geyler, Chief of Division, Ismailia.

97. ANMT, CUCMS, 1995 060 3599, El Guisr, 23 May 1867, Gioia, Chief of Division, to Voisin, Director General.

98. CADN, ARCFPS, C149, Port Said, 29 May 1866, Prefect of Port Said to French Vice-consul in Port Said.

99. ANMT, CUCMS, 1995 060 3601, Ismailia, 18 September 1866, Ismail Hamdy, Governor of the Isthmus of Suez, to Ritt, Representative of the Director General.

100. ANMT, CUCMS, 1995 060 3601, Ismailia, 29 October 1864, Chief Engineer of Ismailia Division to Voisin, Director General; 1995 060 3601, Ismailia, 29 October 1868, Ismail Bey, Governor of the Isthmus, to Voisin; and 1995 060 3601, Ismailia, 29 October 1869, Voisin to Company Superior Agent in Alexandria.

101. ASMAECI, ARDCE, CC, AV, B18, 1864–1866, folder 6, Suez, 23 September 1866, Italian Vice-Consul to Italian Consul in Cairo.

102. CADC, AEC, CCCPS, vol. 1, Port Said, 20 July 1867, Flesch, French Vice-Consul in Port Said to French Ministry of Foreign Affairs, 8; ASMAECI, ARDCE, RC, B50, 1893, Port Said, 23 November 1893, Leoni, Italian Consul, to "Commendatore"; CC, AV, B12, Suez, 26 April 1869, Italian Vice-Consul to Italian Consul in Cairo; and DWQ, 2001-025100, Port Said 13 February 188[?], Ibrahim

Tewfik, Governor General of the Canal and of Port Said, to Sabet Pacha, Ministry of the Interior.

103. CADC, AEC, CCCPS, vol. 1, Port Said, 12 July 1870, Pellissier, Consul in Port Said, to French Consul in Alexandria, 203r.

104. CADN, ARCFPS, C149, Port Said, January 1867, Laroche to Outrey, French Consul in Alexandria; and CADC, AEC, CCCPS, vol. 1, Port Said, 11 September 1867, Champonillo, Pr. Borel, Lavalley, et Cie, to French Vice-Consul in Port Said, 25r.

105. John Torpey, *The Invention of the Passport: Surveillance, Citizenship, and the State* (Cambridge: Cambridge University Press, 2018), 159. On documentary controls as a central component of the coeval Ottoman modernization project, see İlkay Yilmaz, "Governing the Armenian Question through Passports in the Late Ottoman Empire (1876–1908)," *Journal of Historical Sociology* 32, no. 4 (2019): 388–403; David Gutman, "Travel Documents, Mobility Control, and the Ottoman State in an Age of Global Migration, 1880–1915," *Journal of the Ottoman and Turkish Studies Association* 3, no. 2 (2016): 351; Cristoph Herzog, "Migration and the State: On Ottoman Regulations Concerning Migration since the Age of Mahmud II," in *The City in the Ottoman Empire: Migration and the Making of Urban Modernity*, ed. Ulrike Freitag et al. (London: Routledge, 2011), 130; and Reşat Kasaba, *A Moveable Empire: Ottoman Nomads, Migrants, and Refugees* (Seattle: University of Washington Press, 2009), 1–12.

106. CADN, ARCFPS, C149, Port Said, 19 March 1869, Governor of Port Said to French Vice-Consul in Port Said; and ARCFPS, C149, Port Said, 9 August 1869, Governor of Port Said to French Managing Consul.

107. Amy Fullerton, *A Lady's Ride through Palestine and Syria: With Notices of Egypt and the Canal of Suez* (London, 1872), 9–10.

108. CADN, ARCFPS, C149, Alexandria, 24 July 1869, Zoulfikar, Minister of Foreign Affairs, to Tricou, French Managing Consul; and ARCFPS, C21, Alexandria, 14 March 1871, Agent et Consul General to Pelissier de Renaud, Consul in Port Said.

109. Yilmaz, "Governing the Armenian Question through Passports in the Late Ottoman Empire," 393.

110. "Per gli egiziani senza passaporto," *L'Imparziale*, 27 June 1895, 3.

111. CADN, ARCFPS, C149, Alexandria, 24 July 1869, Zoulfikar, Minister of Foreign Affairs, to Tricou, French Managing Consul; and ARCFPS, C149, Port Said, 9 August 1869, Governor of Port Said to French Managing Consul.

112. CADN, ARCFPS, C46, Port Said, 17 January 1874, Saint-Chaffray, French Consul, to Governor of Port Said; ARCFPS, C46, Port Said, 23 December 1876, Saint-Chaffray, French Consul, to Governor of Port Said; and ANMT, CUCMS, 1995 060 3138, Ismailia, 10 February & 20 March 1888, meetings minutes, annexed list.

113. CADN, ARCFPS, C42, Port Said, 5 December 1880, Aly Sabit, Sub-Governor, to Gilbert, French Consul.

114. TNA, FO 846-3, Port Said, 11 February 1875, El Hodry v. Wirlte.

115. H. Couvidou, *Itinéraire du canal de Suez* (Port Said: A. Mourès, 1875), 45.

116. APF, Fondo S.C., Egitto, Copti, vol. 23, Suez, 1 October 1885, Cardinal Moran, Archbishop of Sydney, to Cardinal Prefect, 185v. Couvidou, *Itinéraire du canal de Suez*, 45–46.

117. Fullerton, *Lady's Ride through Palestine and Syria*, 9–10; and TNA, FO 846-3, Port Said, 11 February 1875, El Hodry V Wirlte.

118. DWQ, 2001-011478, Port Said, 6 August 1888, Ibrahim Tewfik, Governor General, to Riaz Pacha, Ministry of the Interior; and 2001-025132, Port Said, 24 October 1883, Governor General (attached to Port Said, 30 October 1883, Ibrahim Tewfik, Governor General, Notification to Consuls).

119. DWQ, 2001-019760, Port Said, 5 April 1882, Ibrahim Tewfik to Mustapha Pacha Fehmy, Minister of the Interior. On *drogmans* as both assets and liabilities to their employers, see Maya Jasanoff, "Cosmopolitan: A Tale of Identity from Ottoman Alexandria," *Common Knowledge* 11, no. 3 (2005): 398.

120. DWQ, 2001-025132, Port Said, 24 October 1883, Governor General to Consuls; and 2001-010755, Cairo, 27 December 1884, Pietri, Director of Legal Department, to Nubar Pacha, Minister of Justice.

121. CADN, ARCFPS, C334, Port Said, 30 July 1885, Governor General to Cipollaro, Inspector of Police; and ARCFPS, C334, Port Said, 27 July 1885, Fathalla Darviche to French Consul.

122. DWQ, 2001-019841, Port Said, 19 May 1881, H. Meymar, Director of the Sanitary Office, to Dr. Hassan Bey Mahmoud, President of Sanitary Maritime Council; and 2001-011306, Port Said, 20 November 1887, Ibrahim Tewfik, Governor General, to Mustafa Pacha Helmy, Minister of the Interior.

123. DWQ, 2001-011478, Port Said, 6 August 1888, Ibrahim Tewfik, Governor General, to Riaz Pacha, Ministry of the Interior; and 2001-011478, 27 August 1887, Ministry of the Interior, to Governor General.

124. Huber, *Channelling Mobilities*, 83–84.

125. Tignor, *Egypt*, 111, 224–25; Mostafa Minawi, *The Ottoman Scramble for Africa: Empire and Diplomacy in the Sahara and the Hijaz* (Stanford, CA: Stanford University Press, 2016), 106–7.

126. AMMBPA, HC 34 a Port Said, Journal de Terre Sante, no. 172, "Port-Said 1882," Port Said, 20 August 1882; and Luigi Dori, "Esquisse historique de Port Said: Port Said après l'inauguration du Canal (1869–1900)," *Cahiers d'histoire egyptienne* 8, no. 1 (January 1956): 31.

127. CADN, ARCFPS, C21, Alexandria, 25 August 1882, E. de Vorges to Dobignie, French Consul in Port Said; "Al-Suways," *Al-Ahram*, 21 August 1882, 1; and "Hawadith mahalliya," *Al-Ahram*, 23 August 1882, 1.

128. Robert L. Tignor, "The 'Indianization' of the Egyptian Administration under British Rule," *American Historical Review* 68, no. 3 (1963): 636; and Robert L. Tignor, *Modernization and British Colonial Rule in Egypt, 1882–1914* (Princeton, NJ: Princeton University Press, 1966), 21.

129. Marie-Laure Crosnier-Lecomte, Gamal Ghitani, and Naguib Amin, *Port-Saïd: Architectures XIXe–XXe siècles* (Cairo: IFAO, 2006), 329. This building no longer stands; it was destroyed by the British when they left Port Said in 1956.

130. Tignor, *Egypt*, 231. For dissenting assessments of the historical role of 'Urabi and the 1882 upheaval, see Donald M. Reid, "The 'Urabi Revolution and the British Conquest, 1879–1882," in *The Cambridge History of Egypt*, vol. 2, *Modern Egypt, from 1517 to the End of the Twentieth Century*, (Cambridge: Cambridge University Press, 1998), 2:218; Cole, *Colonialism and Revolution in the Middle East*, 4, 212, 19; and Alexander Schölch, *Egypt for the Egyptians! The Socio-Political Crisis in Egypt, 1878–1882* (London: Ithaca Press, 1981), 311, 314. For references to the statue, see Fu'ad Farag, *Mintaqat qanal as-Suways wa-mudun al-qanal Bur Sa'id, as-Suways, al-Isma'iliya wa-siwaha* (Cairo: Matba'at al-Ma'arif wa-Maktabuha, 1950), 234.

131. AMMBPA, HC 34 a Port Said, excerpt of the letters of the Mother Superior in Port Said to the Mother General, Journal de Terre Sainte, no. 170, Port Said, 21 August 1882.

132. Najm, *Bur Sa'id*, 424–25, 386–390.

133. AMMBPA, HC 34 a Port Said, Journal de Terre Sainte, no. 170, Port Said, 5 July 1882, 12 July 1882; and Journal de Terre Sainte, no. 172, Port Said, 20 August 1882 (excerpts from letters of Mother Superior in Port Said to Mother General in Angers). Cole tends to conflate Port Said and Suez when he argues that, in both places, "more aggressive" stances were taken by Egyptians toward Europeans in 1882. Cole, *Colonialism and Revolution in the Middle East*, 203.

134. Charles Royle, *The Egyptian Campaigns, 1882 to 1885: And the Events Which Led to Them* (London: Hurst and Blackett, 1886), 1:241; Dori, "Esquisse historique de Port Said: Port Said après l'inauguration du Canal (1869–1900)," 29–30.

135. DWQ, 2001-009918, Ismailia, 20 October 1883, Lt. Colonel District Inspector, table showing permanent distribution of Constabulary; 2001-009918, Ismailia, 23 March 1883, Gibbons; and 2001-014127, Port Said, 12 February 1883, Colonel Synge, Gendarmerie Inspector, to General Baker Pacha. The original word in Arabic may have been *mustahfaẓ* (pl. *mustahfaẓāt*) meaning "reserve."

136. DWQ, 2001-009918, Ismailia, 23 March 1883, Gibbons; and 2001-014127, Port Said, 18 April 1883, Gibbons to General Baker Pacha.

137. DWQ, 2001-014127, Port Said, 12 February 1883, Colonel Synge, Gendarmerie Inspector, to General Baker Pacha; 2001-014127, Port Said, 18 April 1883, Gibbons to General Baker Pacha; and 2001-014359, Cairo, 6 September 1886, Inspector General of Cairo Division to Inspector General of Egyptian Police in Cairo. On the disruption put forth by drunken police personnel in Istanbul and Calcutta, see Nurçın İleri, "Rule, Misconduct, and Dysfunction: The Police Forces in Theory and Practice in Fin-de-Siècle Istanbul," *Comparative Studies of South Asia, Africa and the Middle East* 34, no. 1 (2014): 147–59; and Harald Fischer-Tiné, "'The Drinking Habits of Our Countrymen': European Alcohol Consumption and Colonial Power in British India," *Journal of Imperial and Commonwealth History* 40, no. 3 (2012): 387–88.

138. DWQ, 2001-014127, Port Said, 18 April 1883, Gibbons to General Baker Pacha; and 2001-014359, Cairo, 6 September 1886, Inspector General of Cairo Division to Inspector General of Egyptian Police in Cairo.

139. ASMAECI, ARDCE, RC, B39, 1891, Port Said, received 12 January 1891, Italian Consul in Port Said to Italian Consul in Cairo.

140. DWQ, 2001-014127, Port Said, 18 April 1883, Gibbons to General Baker Pacha. In 1884 a decree merged the gendarmerie and the Egyptian police into one force simply called "police." Harold Tollefson, *Policing Islam: The British Occupation of Egypt and the Anglo-Egyptian Struggle over Control of the Police, 1882–1914* (Westport, CT: Greenwood Press, 1999), 13, 7.

141. DWQ, 2001-009918, Ismailia, 23 March 1883, Gibbons; 2001-009918, Port Said, 30 March 1883, Gibbons, Gendarmerie Inspector, to Ministry of the Interior; and 2001-012682, Alexandria, 4 October 1884, L. Colonel Coles, order.

142. ASMAECI, ARDCE, RC, B39, 1891, Port Said, received 12 January 1891, Italian Consul in Port Said to Italian Consul in Cairo.

143. "Satiro," *L'Imparziale*, 26 May 1895, 3.

144. CADC, AEC, CCCPS, vol. 1, Port Said, 20 April 1867, Flesch, Vice-Consul in Port Said, to Governor of the Isthmus.

145. Will Hanley, "Papers for Going, Papers for Staying: Identification and Subject Formation in the Eastern Mediterranean," in *A Global Middle East: Mobility, Materiality and Culture in the Modern Age, 1880–1940*, ed. Liat Kozma, Cyrus Schayegh, and Avner Wishnitzer (London: I. B. Tauris, 2015), 183; and Ministero degli Affari Esteri, *Emigrazione e colonie: Raccolta di rapporti dei RR. agenti diplomatici e consolari* (Rome: Tipografia dell'Unione Cooperativa Editrice, 1906), 2:220–21, 230.

146. ANMT, CUCMS, 1995 060 3599, Ismailia, 20 December 1867, Ritt, Representative of the Director General, to Chief of El Guisr Division.

147. Jonathan Hyslop, "The Politics of Disembarkation: Empire, Shipping and Labor in the Port of Durban, 1897–1947," *International Labor and Working-Class History* 93 (2018): 179.

148. ANMT, CUCMS, 1995 060 3601, El Guisr, 18 July 1861, Chief Engineer of Timsa Division, Report.

149. ASMAECI, ARDCE, RC, B46, F33, Port Said, 9 March 1893, Italian Consul to Italian Consul General in Cairo.

150. Timothy Mitchell, *Rule of Experts: Egypt, Techno-Politics, Modernity* (Berkeley: University of California Press, 2012), 111–12.

151. APF, Fondo S.C., Egitto, Copti, vol. 21, Constantinople, 20 January 1877, F. Lodovico Arc. Di Fiunia, Apostolic Vicar of Syria, Cardinal Le Franch, Prefect, 1155r.

152. Eugène de Régny, *Statistique de l'Égypte* (Alexandria: Impr. Française Mourès & Cie, 1873).

153. Rinaldo De Sterlich, *Sugli italiani d'Egitto: Lettera aperta di Fausto* (Cairo: Imprimerie parisienne J. Cèbe, 1888), 8–9; Ministero degli Affari Esteri, *Emigrazione e colonie: Raccolta di rapporti dei RR. agenti diplomatici e consolari* (Rome: Tipografia Nazionale di G. Bertero, 1893), 562, 572; and Ministero degli Affari Esteri, *Emigrazione e colonie* (1906), 2:219–20, 223.

154. APF, Fondo S.C., Egitto, Copti, vol. 22, Cairo, 29 June 1882, Custodian of Terra Santa to Simeoni, 740v.

155. TNA, FO 846-3, Port Said, 5 November 1875, Alfred Wolff v. James O'Connell.

156. Malte Fuhrmann, "'I Would Rather Be in the Orient': European Lower Class Immigrants into the Ottoman Lands," in *The City in the Ottoman Empire Migration and the Making of Urban Modernity*, ed. Malte Fuhrmann et al. (London: Routledge, 2011), 231. For relevant literature, see Will Hanley, *Identifying with Nationality: Europeans, Ottomans, and Egyptians in Alexandria* (New York: Columbia University Press, 2017), 2–3; Alyssa J. Reiman, "Claiming Livorno: Commercial Networks, Foreign Status, and Culture in the Italian Jewish Diaspora, 1815–1914" (PhD diss., University of Michigan, 2017), 189; Jessica M. Marglin, *Across Legal Lines: Jews and Muslims in Modern Morocco* (New Haven, CT: Yale University Press, 2017); Sarah Abrevaya Stein, *Extraterritorial Dreams: European Citizenship, Sephardi Jews, and the Ottoman Twentieth Century* (Chicago: University of Chicago Press, 2016), 128; Ziad Fahmy, "Jurisdictional Borderlands: Extraterritoriality and 'Legal Chameleons' in Precolonial Alexandria, 1840–1870," *Comparative Studies in Society and History* 55, no. 2 (2013): 305–29; Huber, *Channelling Mobilities*, 279; and Julia A. Clancy-Smith, *Mediterraneans: North Africa and Europe in an Age of Migration, c. 1800–1900* (Berkeley: University of California Press, 2011), 201–2.

157. APF, Fondo S. C., Egitto, Copti, vol. 22, Alexandria, 10 December 1883, Chicaro to Simeoni, 1077r; and vol. 23, Lyon, 2 June 1886, planque, 219r.

158. CADN, ARI, C9, Chalouf, 6 October 1868, Merlé, Campsite Agent, Inventory.

159. ASMAECI, ARDCE, CC, B12, Suez, 8 October 1869, Italian Vice-Consul to Italian Consul in Cairo.

160. CADC, AEC, CCCPS, vol. 1, Port Said, 23 April 1874, Saint-Chaffray, French Consul in Port Said, to French Ministry of Foreign Affairs, 321v–322r; and vol. 1, Port Said, 25 May 1874, Managing Consul in Port Said, to French Ministry of Foreign Affairs, 330r–330v.

161. CADN, ARCFPS, C42, Ismail Pacha Hamdy, Governor General of the Canal and Port Said, to Saint-Chaffray, French Consul in Port Said; ARCFPS, C49, Port Said, 6 November 1878, Massimilano Sever to French Consul; and ARCFPS, C46, Port Said, 19 December 1879, Saint-Chaffray, French Consul in Port Said, to Governor General.

162. DWQ, 2001-011213, Kantara, 6 August 1887, Albino Paoletti to Governor General of the Canal and Port Said.

163. Archivio Centrale dello Stato (hereafter ACS), Casellario Politico Centrale (hereafter CPC), Malatesta Aniello, Cenno biografico, Port Said, 14 September 1899, Italian Consul in Port Said; ACS, CPC, Rome, Ministry of the Interior to Prefecture of Treviso, 9 July 1903; and ACS, CPC, Naples, 19 December 1896, Prefecture of Naples to Ministry of the Interior, Rome.

164. CADN, ARCFPS, C46, Port Said, 6 November 1876, French Consul to Governor General; and ARCFPS, C334, Port Said, 25 February 1884, Fathalla Darviche/Darouiche to French Consul in Port Said.

165. CADN, ARCFPS, C21, Alexandria, 14 April 1864, French Consul to Laroche, French Vice-Consul in Port Said; and ANMT, CUCMS, 1995 060 3601, Ismailia, 10 June 1863, Voisin, Director General, order.

166. ASMAECI, ARDCE, CC, AV, B12, Suez, 10 December 1869, Italian Vice-Consul to Italian Consul in Cairo; and CC, AV, B12, Suez, 15 December 1869, Italian Vice-Consul to Italian Consul in Cairo.

167. CADN, ARCFPS, C21, Alexandria, 14 April 1864, French Consul to Laroche, French Vice-Consul in Port Said; and CADN, ARCFPS, C149, no place, n.d. [September–October 1866], no author to Consul General.

168. Paulo G. Brenna, *L'emigrazione italiana nel periodo antebellico* (Firenze: R. Bemporad & Figlio, 1918), 204.

169. ASMAECI, ARDCE, CC, AV, B12, Suez, 11 Sha'ban 1286 (16 November 1869), Dragoman Maurino, Translation of Memorandum by Governor of Suez.

170. ASMAECI, ARDCE, CC, AV, B12, Suez, 10 December 1869, Italian Vice-Consul to Italian Consul in Cairo.

171. For the ineffectiveness of expulsions in the precolonial Tunisian context, see Clancy-Smith, *Mediterraneans*, 223–25.

172. TNA, FO 846–3, Port Said, 2 January 1875, Royle Margossian v. James Testaferrata; CADN, ARCFPS, C334, Port Said, 24 June 1884, Governor Ibrahim Tewfik to French Consul; and no place, 25 June 1884, no author to Governor General.

173. ASMAECI, ARDCE, RC, B64, Port Said, 25 April 1900, Italian Consul to Italian Consul in Cairo; and RC, B64, Port Said, 8 August 1900, Italian Consul to Italian Consul in Cairo.

174. DWQ, 2001-012682, Alexandria, 8 October 1884, Vice Inspector Galli to Inspector Principal of Police.

175. ASMAECI, ARDCE, RC, B64, Port Said, 25 April 1900, Italian Consul to Italian Consul in Cairo; RC, B64, Cairo, 28 June 1900, Italian Consul to Consul in Port Said; and RC, B64, Port Said, 5 July 1900, Italian Consul to Italian Consul in Cairo.

176. ASMAECI, ARDCE, CC, B12, Suez, 8 October 1869, Italian Vice-Consul to Italian Consul in Cairo; and DWQ, 2001-019915, 12 June 1887, Cole, Director Inspector General of the Cairo Division, Report on the barracks and prison in Port Said.

177. "Espulsioni," *L'Imparziale*, 26 May 1895, 3; and ASMAECI, ARDCE, RC, B64, Port Said, 8 August 1900, Italian Consul to Italian Consul, Cairo.

178. Cole, *Colonialism and Revolution in the Middle East*, 202; and Mario M. Ruiz, "Intimate Disputes, Illicit Violence: Gender, Law, and the State in Colonial Egypt, 1849–1923" (PhD diss., University of Michigan, 2004), 131.

179. ASMAECI, ARDCE, RC, B64, Cairo, 28 June 1900, Consul General to Consul.

180. Tollefson, *Policing Islam*, 100, 115–16; and Francis Galton, "Identification Offices in India and Egypt," *Nineteenth Century: A Monthly Review* 48 (1900): 124. See Carlo Ginzburg, *Clues, Myths, and the Historical Method* (Baltimore, MD: Johns Hopkins University Press, 1989), 120–21.

181. Omnia S. El Shakry, *The Arabic Freud: Psychoanalysis and Islam in Modern Egypt* (Princeton, NJ: Princeton University Press, 2017), 88.

182. Torpey, *Invention of the Passport*, 131.

183. ASMAECI, ARDCE, RC, B64, Port Said, 25 April 1900, Governor General to Italian Consul. A photographer called Massoud had issued a receipt for 12 francs for the photographs of an Italian individual who had been repeatedly removed from Port Said.

184. ASMAECI, ARDCE, RC, B64, Port Said, 25 April 1900, Italian Consul to Italian Consul in Cairo; RC, B64, Cairo, 28 June 1900, Consul General to Consul; and RC, B64, Port Said, 5 July 1900, Italian Consul to Italian Consul in Cairo.

185. DWQ, 2001-012682, Alexandria, 20 October 1884, Vice-Inspector Galli to Inspector Principal.

186. ASMAECI, ARDCE, RC, B64, Port Said, 5 July 1900, Italian Consul to Italian Consul in Cairo.

187. Huber, *Channelling Mobilities*, 285.

188. Lucia Carminati, "Alexandria, 1898: Nodes, Networks, and Scales in Late Nineteenth-Century Egypt," *Comparative Studies in Society and History* 59, no. 1 (2017): 136, 140, 151.

189. Ginzburg, *Clues, Myths, and the Historical Method*, 120; Richard B. Jensen, "The International Anti-Anarchist Conference of 1898 and the Origins of Interpol," *Journal of Contemporary History* 16, no. 2 (1981): 332–33; and Anne McClintock, *Imperial Leather: Race, Gender, and Sexuality in the Colonial Contest* (New York: Routledge, 1995), 123n142. For the Ottoman context, see İlkay Yılmaz, "The Ottoman Empire, Police Photographs and Anthropometry," *European Society for the History of Photography*, no. 31 (2019): 90.

190. Alexander Kazamias, "Cromer's Assault on 'Internationalism': British Colonialism and the Greeks of Egypt, 1882–1907," in *The Long 1890s in Egypt*, ed. Marilyn Booth and Anthony Gorman (Edinburgh: Edinburgh University Press, 2014), 153.

191. "*L'Elettrico* di Port Said," *L'Imparziale*, 14 May 1895, 2.

192. Ginzburg, *Clues, Myths, and the Historical Method*, 120.

193. CADN, ARCFPS, C49, Port Said, 14 December 1876, Berthelot to French Consul; ARCFPS, C49, Ismailia, 10 January 1877, Medamer to French Consul; and ARCFPS, C49, Port Said, 15 July 1878, Jourdan Marcelin to French Consul.

194. DWQ, 5013-003230, Alexandria, 7 December 1878, Gava to Khedive; 5013-003230, Alexandria, 24 September 1878, Luigi Isella to Khedive; 5013-003230, Helwan, 4 March 1891, Ghedalia Rosenthal, to Khedive; and 'UM'A, 5013-005402, Cairo, 24 April 1892, Sylva Amira to Khedive.

195. APF, Fondo S. C., Egitto, Copti, vol. 22, Alexandria, 28 January 1884, Chicaro to Simeoni, 098r; and vol. 22, Alexandria, 10 December 1883, Chicaro to Simeoni, 1077r.

196. Caillard, *Lifetime in Egypt*, 16.

197. Konrad Dryden, *Leoncavallo: Life and Works* (Lanham, MD: Scarecrow Press, 2007), 15. His uncle was active in printing and journalism within Egypt. Luigi Dori, "Italiani in Africa: Tipografi e giornalisti italiani in Egitto," *Africa: Rivista trimestrale di studi e documentazione dell'Istituto italiano per l'Africa e l'Oriente* 14, no. 3 (1959): 147.

198. Panaït Istrati, *Vie d'Adrien Zograffi* (Paris: Gallimard, 1969), 404; and "Ba-Ta-Clan," *L'Imparziale*, 21–22 July 1895, 3.

199. Agostino Bistarelli, *Gli esuli del Risorgimento* (Bologna: Il Mulino, 2011), 245–46. In 1854 he moved with his family to Massawa, where the Piedmontese government appointed him to negotiate with the Ethiopian chiefs.

200. CADN, ARCFPS, C49, Port Said, 2 May 1879, Jules Deslandes to Consul. Deslandes's testimony is incredibly detailed. Born in Paris in 1844, Deslandes had been a soldier of the class of 1864 in the regiment of the "Génie," of engineering. Captured in Metz on October 29, 1870, he went back to Paris, where he took part in the Commune on March 18, 1871. He was arrested on July 11, 1871. In March 1872 the Council of War condemned him to deportation; he was sent to New Caledonia on 6 December 1872. Exactly one year later, on 6 December 1873, he evaded capture and arrived in Port Said on April 11, 1874. No details are provided about his spectacular escape toward Egypt. Having committed no crimes of "common law," he asked to be included in the partial amnesty. He declared that he regretted his past and promised not to commit the same "mistakes" again.

201. "Bullettino dell'operaio," *Il Risveglio*, Siena, 5, no. 8, 25 February 1877, quoted in Leonardo Bettini, *Bibliografia dell'anarchismo: Periodici e numeri unici anarchici in lingua italiana pubblicati all'estero (1872–1891)* (Firenze: CP, 1976), 2:282.

202. Ministero degli Affari Esteri, *Emigrazione e colonie* (1893), 561–63, 573; and APF, Fondo S.C., Egitto, Copti, vol. 23, Lyon, 2 June 1886, planque, folio 219r.

203. Balboni, *Gl'Italiani nella civiltà egiziana del secolo XIX*, 1:424; and Vittorio Briani, *Italiani in Egitto* (Roma: Istituto poligrafico e Zecca dello Stato, 1982), 118.

204. Jens Hanssen, *Fin-de-Siècle Beirut: The Making of an Ottoman Provincial Capital* (Oxford: Clarendon Press; New York: Oxford University Press, 2005), 193; Sibel Zandi-Sayek, *Ottoman Izmir: The Rise of a Cosmopolitan Port, 1840–1880* (Minneapolis: University of Minnesota Press, 2012), 87; and İleri, "Rule, Misconduct, and Dysfunction," 150. For the production and circulation of sensationalistic crime narratives in nineteenth-century Istanbul, see Nurçin İleri, "Between the Real and the Imaginary: Late Ottoman Istanbul as a Crime Scene," *Journal of the Ottoman and Turkish Studies Association* 4, no. 1 (2017): 95.

205. "La sicurezza pubblica," *L'Imparziale*, 22 July 1892, 1.

206. Cole, *Colonialism and Revolution in the Middle East*, 217; Khaled Fahmy, "The Police and the People in Nineteenth-Century Egypt," *Die Welt Des Islams* 39, no. 3 (1999): 375; Kazamias, "Cromer's Assault on 'Internationalism': British

Colonialism and the Greeks of Egypt, 1882–1907," 142, 150; and Khaled Fahmy, *In Quest of Justice: Islamic Law and Forensic Medicine in Modern Egypt* (Oakland: University of California Press, 2018).

207. Noémi Levy-Aksu, "A Capital Challenge: Managing Violence and Disorders in Late Ottoman Istanbul," in *Urban Violence in the Middle East: Changing Cityscapes in the Transformation from Empire to Nation State*, ed. Ulrike Freitag et al. (New York: Berghahn Books, 2015), 57; and El Shakry, *The Arabic Freud*, 88–89.

208. Edwin Arnold, *India Revisited* (London: Trübner, 1886), 27. A similar claim is found in Modelski, with the additional racially informed and ungrounded assumption that "after Britain took charge of law and order in Port Said, the overcolorful elements that had earned a certain reputation for the city in its early years were relegated to the western edge of town," where the so-called Arab village was located. Modelski, *Port Said Revisited*, 131.

209. D. A. Farnie, *East and West of Suez: The Suez Canal in History, 1854–1956* (Oxford: Clarendon Press, 1969), 404.

210. "La verità di Port Said," *L'Imparziale*, 9 May 1895, 3; "Bur Sa'id," *Al-Ahram*, 18 March 1892, 2.

211. ASMAECI, ARDCE, RC, B39, 1891, Port Said, received 12 January 1891, Italian Consul in Port Said to Italian Consul in Cairo. Complaints about nighttime thefts in Port Said appear in "La Verità di Port Said," *L'Imparziale*, 29 June 1895, 2; and "La polizia in bicicletta," *L'Imparziale*, 8 March 1895, 3.

212. Michel Foucault, Paul Rabinow, and Nikolas Rose, "Lives of Infamous Men," in *The Essential Foucault: Selections from Essential Works of Foucault, 1954–1984* (New York: New Press, 2003), 286–87; and Michel Foucault, *Discipline and Punish: The Birth of the Prison* (New York: Pantheon Books, 1977), 213–14.

213. ANMT, CUCMS, 1995 060 3601, Ismailia, 20 August 1866, Voisin, Director, to Governor General.

214. On Barak, "Outsourcing: Energy and Empire in the Age of Coal, 1820–1911," *International Journal of Middle East Studies* 47, no. 3 (2015): 432, 434.

CHAPTER 4. ENTERTAINMENT IN PORT SAID,
A SINK OF IMMORAL FILTH

1. Jabez Burns and Thomas Cook (Firm), *Help-Book for Travellers to the East Including Egypt, Palestine, Turkey, Greece and Italy* (London: Cook's Tourist Office, 1872), 6–7.

2. D. A. Farnie, *East and West of Suez: The Suez Canal in History, 1854–1956* (Oxford: Clarendon Press, 1969), 392; and Mary Brodrick, *A Handbook for Travellers in Lower and Upper Egypt* (London: J. Murray, 1896), 279, 286, 245.

3. Archivio della Sacra Congregazione per l'Evangelizzazione dei Popoli o de "Propaganda Fide," Rome (hereafter APF), Scritture riferite nei Congressi (hereafter Fondo S.C.), Egitto, Copti, vol. 23, Suez, 1 October 1885, Cardinal Moran, Archbishop of Sydney, to Cardinal Prefect, 185v; and Richard Meinertzhagen, *Army*

Diary, 1899–1926 (Edinburgh: Oliver and Boyd, 1960), 12, entry dated 4 April 1899. See Mahmud Salih Ramadan, *Al-Haya al-ijtimaʿiya fi Misr fi ʿasr Ismaʿil* (Alexandria: Munshaʾat al-Maʿrifa al-Haditha, 1977), 307.

4. T. N. Mukharji, *A Visit to Europe* (Calcutta: W. Newman, 1889), 16; Nunda Lall Doss, *Reminiscences, English and Australasian: Being an Account of a Visit to England, Australia, New-Zealand, Tasmania, Ceylon Etc.* (Calcutta: M. C. Bhowmick, 1893), 24–25; and Meinertzhagen, *Army Diary, 1899–1926*, 12–13, entry dated 4 April 1899. See also The British Library, London (hereafter TBL), Oriental and India Office Collections (hereafter OIOC), Mss/Eur/B235/1 Colonel Harry Ross, October–November 1898.

5. Sydney Montagu Samuel, *Jewish Life in the East* (London: C. Kegan Paul, 1881), 13.

6. Avner Wishnitzer, "Yawn: Boredom and Powerlessness in the Late Ottoman Empire," *Journal of Social History* 55, no. 2 (2021): 400–425.

7. "Notizie del giorno," *Il Telegrafo*, 18 December 1893, 1; and "Musée," *La Vérité*, 22 August 1896, 1.

8. TBL, OIOC, Mss/Eur/B235/1 Colonel Harry Ross, October–November 1898.

9. Mountstuart Grant Duff, *Notes from a Diary, Kept Chiefly in Southern India, 1881–1886* (London: J. Murray, 1899), 5; and Ronald Storrs, *The Memoirs of Sir Ronald Storrs* (New York: Putnam, 1937), 19, entry dated 5 October 1904.

10. André Chevrillon, *Dans l'Inde* (Paris: Hachette, 1891), 328, entry dated 7 January 1889.

11. Rudyard Kipling, *The Light That Failed* (London: Macmillan, 1891), 36–37; and Richard Tangye, *Reminiscences of Travel in Australia, America, and Egypt* (London: Sampson Low, Marston, Searle & Rivington, 1884), 228.

12. Grant Duff, *Notes from a Diary, . . . 1881–1886*, 5; Mukharji, *Visit to Europe*, 16; and Meinertzhagen, *Army Diary, 1899–1926*, 12, entry dated 4 April 1899.

13. Doss, *Reminiscences*, 25; and Chevrillon, *Dans l'Inde*, 328, entry dated 7 January 1889.

14. Kipling, *Light That Failed*, 38.

15. Harry Alis, *Promenade en Égypte* (Paris: Librairie Hachette, 1895), 8–9; and Felice Santini, *Intorno al mondo a bordo della regia corvetta "Garibaldi" (anni 1879-80-81-82): Memorie di viaggio* (Rome: Cotta, 1875), 10–11.

16. "F. Fioravanti," *Le Phare de Port-Saïd et du Canal de Suez*, 8 August 1893, 2; Archives Nationales du Monde du Travail, Roubaix (hereafter ANMT), Compagnie universelle du canal maritime de Suez (hereafter CUCMS), 2000 059 005, 1854–1901, Arnoux, Photograph *Commerce Street in Port Saïd*. On the Simon Arzt store, see Marie-Laure Crosnier-Lecomte, Gamal Ghitani, and Naguib Amin, *Port-Saïd: Architectures XIXe–XXe Siècles* (Cairo: IFAO, 2006), 178–79.

17. Ayşe Saraçgil, "L'introduction du café à Istanbul," in *Cafés d'Orient revisités*, ed. Hélène Desmet-Grégoire and François Georgeon (Paris: CNRS éd., 1997), 38. Sariyannis notes that leisure should not necessarily be measured on the basis of,

or in contrast with, work. Marinos Sariyannis, "Time, Work and Pleasure: A Preliminary Approach to Leisure in Ottoman Mentality," in *New Trends in Ottoman Studies: Papers Presented at the 20th CIÉPO Symposium*, ed. Marinos Sariyannis (Rethymno: Crete University Press, 2012), 810. Hanssen highlights the connection between the development of public places for leisure and rising levels of material consumption. Jens Hanssen, *Fin de Siècle Beirut: The Making of an Ottoman Provincial Capital* (Oxford: Oxford University Press, 2005), 198. See also Mark J. Bouman, "Luxury and Control: The Urbanity of Street Lighting in Nineteenth-Century Cities," *Journal of Urban History* 14, no. 1 (1987).

18. Società geografica italiana, *Memorie della società geografica italiana. Indagini sulla emigrazione italiana all'estero fatte per cura della società (1888–1889)* (Rome: Presso la Società geografica italiana, 1890), 4:214.

19. Hanssen, *Fin de Siècle Beirut*, 193; and Malte Fuhrmann, "Beer, the Drink of a Changing World: Beer Consumption and Production on the Shores of the Aegean in the 19th Century," *Turcica* 45 (2014): 89.

20. Sariyannis, "Time, Work and Pleasure," 809; and Edward William Lane, *The Manners & Customs of the Modern Egyptians* (J. M. Dent, 1908), 340, 481.

21. Ilham Khuri-Makdisi, *The Eastern Mediterranean and the Making of Global Radicalism, 1860–1914* (Berkeley: University of California Press, 2010), 91.

22. Émile Erckmann and Alexandre Chatrian, *Souvenirs d'un ancien chef de chantier à l'isthme de Suez* (Paris: Hetzel, 1877), 12; and François Georgeon, "Ottomans and Drinkers: The Consumption of Alcohol in Istanbul in the Nineteenth Century," in *Outside In: On the Margins of the Modern Middle East*, ed. Eugene L. Rogan (London: I. B. Tauris, 2002), 9.

23. Mary Lether Wingerd, "Rethinking Paternalism: Power and Parochialism in a Southern Mill Village," *Journal of American History* 83, no. 3 (1996): 886.

24. Crandall A. Shifflett, *Coal Towns: Life, Work, and Culture in Company Towns of Southern Appalachia, 1880–1960* (Knoxville: University of Tennessee Press, 1991), 52.

25. "Hadatha fi layl al-ams," *Al-Ahram*, 11 April 1895, 3.

26. TBL, OIOC, Mss/Eur/C739 John Wilberforce Cassels, November or December 1879; and Alis, *Promenade en Égypte*, 12.

27. Doctor Hérouard, "Note sur l'état medical de Port-Saïd," *L'Isthme de Suez*, 15–18 July 1865, 220–221.

28. "*La Verità* di Port Said," *L'Imparziale*, 19 July 1895, 2; and "Avis," *Le Phare de Port-Said et du Canal de Suez*, 8 August 1893, 2.

29. Mohamed Anouar Moghira, *L'Isthme de Suez: Passage millénaire, 640–2000* (Paris: L'Harmattan, 2002), 202.

30. On Barak, *Powering Empire: How Coal Made the Middle East and Sparked Global Carbonization* (Oakland: University of California Press, 2020), 66.

31. ANMT, CUCMS, 1995 060 3601, Ismailia, 29 October 1869, Voisin, Director of the Works, to Company Agent Superior in Alexandria; and "Aggressione," *L'Imparziale*, 4 December 1897, 3.

32. Moghira, *L'Isthme de Suez*, 203.

33. François Levernay, *Guide général d'Égypte* (Alexandria: Imprimerie nouvelle, 1868), 287, 271, 255; Barbara De Poli, "Il mito dell'Oriente e la espansione massonica italiana nel Levante," in *Storia d'Italia: La Massoneria*, ed. Gian Mario Cazzaniga (Torino: Einaudi, 2006), 639; Marta Petricioli, *Oltre il mito: L'Egitto degli italiani (1917–1947)* (Milan: Mondadori, 2007), 98; and Aldo Alessandro Mola, "Le logge italiane in Egitto dall'Unità al fascismo," in *Italia e l'Egitto dalla rivolta di Arabi Pascià all'avvento del fascismo (1882–1922)*, ed. Romain H. Rainero and Luigi Serra (Settimo Milanese: Marzorati, 1991), 193. In Suez, the Klysma, with Italian orientation, was established in 1905 and functioned until 1925. Another lodge with Italian orientation in Port Said, the Triangolo, may have only existed in 1918; see Mola, "Le logge italiane in Egitto dall'Unità al fascismo," 204; and Petricioli, *Oltre il mito*, 99.

34. APF, Suez, 13 July 1861, F. Alfonso da Cava to Propaganda Fide, 1142r.

35. Ferdinand de Lesseps, *Lettres, journal et documents pour servir à l'histoire du canal de Suez IV* (Paris: Didier, 1875), 4:24, entry dated 4 March 1861.

36. Archives de la Maison-Mère du Bon Pasteur, Angers (hereafter AMMBPA), HC 43 Port Said, De notre Monastère de Port-Saïd, 1870, 10; and The Franciscan Centre of Christian Oriental Studies, "Cronaca di Santa Caterina: Appendice Seconda," *Studia Orientalia Christiana, Collectanea*, nos. 26–27 (1996): 479.

37. Archives du ministère de l'Europe et des Affaires étrangères, Centre des Archives Diplomatiques de Nantes (hereafter CADN), Archives Rapatriés de l'Agence Consulaire de Ismailia (hereafter ARI), C26, Suez, 9 January 1869, Italian Vice-Consul, decree; ARI, C26, "Le soussigné a l'honneur," *Il Telegrafo*, 18 December 1893, 1; and ARI, C26, "Avis," *Le Phare de Port-Said et du Canal de Suez*, 8 August 1893, 2.

38. Grove Music Online, s.v. "Simsimiyya [Semsemiyya, Sumsumiyya]," by Christian Poché, 2019, www.oxfordmusiconline.com/grovemusic/display/10.1093/gmo/9781561592630.001.0001/omo-9781561592630-e-0000052321;jsessionid=CD88255218F5865482E89CEF0183649A; 'Isam Shattati, *Al-Simsimiyah bayna al-waqi' wa-al-usturah* (Cairo: Al-Hay'ah al-'Ammah li-Qusur al-Thaqafah, 2003), 14; Martin Stokes, "Review of La Simsimiyya de Port-Said," *Music and Anthropology* 7 (2002), www2.umbc.edu/MA/index/number7/stokes/simsi.htm; and Mostafa Mohie, *Biographies of Port Said*, Cairo Papers in Social Science (Cairo: American University in Cairo Press, 2021), 36:1:68–69. On the contemporary history of the *simsimiyya* tradition in Port Said, see Alia Mossallam, "Beyond the Din of the Battle: Stories from the Struggle for Port Said," *Mada Masr* (blog), November 7, 2016, www.madamasr.com/en/2016/11/07/opinion/u/beyond-the-din-of-the-battle-stories-from-the-struggle-for-port-said/; and Mohie, *Biographies of Port Said*, 36:1:65–80.

39. "Ismailia: Correspondance," *Le Bosphore égyptien*, 15 March 1885.

40. L. A. Balboni, *Gl'Italiani nella civiltà egiziana del secolo XIX: Storia-biografie-monografie* (Alexandria: Tipo-litografico V. Penasson, 1906), 1:424; and Ministero degli Affari Esteri, *Emigrazione e colonie: Raccolta di rapporti dei RR. agenti diplomatici e consolari* (Rome: Tipografia dell'Unione Cooperativa Editrice, 1906), 2:277–78.

41. Census Department, *The Census of Egypt Taken in 1907* (Cairo: National Printing Department, 1909), 302–3, 7; and Giuseppe Scalise, *L'Emigrazione dalla Calabria, saggio di economia sociale* (Naples: L. Pierro, 1905).

42. Moghira, *L'Isthme de Suez*, 203. Evidence that it did function is lacking. It may have undergone the same fate as the library that the Dante Alighieri committee in Cairo aimed at establishing in 1898. It is unclear whether the latter project went anywhere, as many of those morally supporting it "forgot" to pay attendant fees. "La Dante Alighieri," *L'Imparziale*, 12 May 1898, 2–3.

43. CADN, ARI, C4, Kantara, 25 January 1866, Chef des bureaux détachés (hereafter CdBD), Inventory. See William Henry Hudson, *A Short History of French Literature* (London: G. Bell, 1919), 40.

44. CADN, Archives Rapatriés du Consulat de France à Port-Saïd (hereafter ARCFPS), C396, Port Said, 8 July 1882, G. Colomb, Inventory.

45. "Libreairie et papeteri," *Il Telegrafo*, 18 December 1893, 1; and Brodrick, *Handbook for Travellers in Lower and Upper Egypt*, 1000.

46. Società Dante Alighieri Historical Archives, Rome (SDA), Port Said, 1924–1926, *Regolamenti del comitato di Porto Said* (Port Said: A. de Giorgio, 1931). See Alyssa J. Reiman, "Claiming Livorno: Commercial Networks, Foreign Status, and Culture in the Italian Jewish Diaspora, 1815–1914" (PhD diss., University of Michigan, 2017), 234–35; and G. B. Danovaro, *L'Égypte à l'aurore du XXème siècle* (Alexandria: J.C. Lagoudakis, 1901), 75. Reiman writes that the Dante Alighieri Society had established committees in Alexandria and Cairo in 1895 and 1898, respectively.

47. Luigi Dori, "Italiani in Africa: Tipografi e giornalisti italiani in Egitto," *Africa: Rivista trimestrale di studi e documentazione dell'Istituto italiano per l'Africa e l'Oriente* 14, no. 3 (1959): 146; "S'adresser à son laboratoire," *Il Telegrafo*, 18 December 1893, 1; and Jules Munier, *La presse en Égypte (1799–1900): Notes et souvenirs* (Cairo: Impr. de l'Institut français d'archéologie orientale, 1930), 9.

48. Zayn al-'Abidin Shams al-Din Najm, *Bur Sa'id: Tarikhuha wa-tatawwuruha, mundhu nash'atiha 1859 hatta 'am 1882* (Cairo: Al-Hay'ah al-Misriyah al-'Ammah lil-Kitab, 1987), 366.

49. Moghira, *L'Isthme de Suez*, 437; and Umberto Rizzitano, "Un Secolo Di Giornalismo Italiano in Egitto (1845–1945)," *Cahiers d'histoire Égyptienne* 8, nos. 2/3 (Avril 1956): 146–47. Specific dates are in Ibrahim 'Abduh, *Tatawwur al-sihafah al-misriyah, 1798–1981* (Cairo: Mu'assasat Sijill al-'Arab, 1982), 356. See Najm, *Bur Sa'id*, 368–75.

50. Munier, *La presse en Égypte*, 9, 45; Najm, *Bur Sa'id*, 376–77; and Dori, "Italiani in Africa," 147. See Frontispiece, *L'Écho de Port Said, journal politique, commercial et littéraire*, 9 August 1893.

51. CADN, ARCFPS, C378, Port Said, 21 February 1894, Miaoulis, Androuliakos, Vindiadis to French Consul in Port Said.

52. Munier, *La presse en Égypte*, 7; and Najm, *Bur Sa'id*, 383. See 'Abduh, *Tatawwur al-sihafah al-misriyah*, 336.

53. Munier, *La presse en Égypte*, 39; and Najm, *Bur Sa'id*, 379. By 1897 it had already been transferred to the capital; see "Al lavoro," *L'Imparziale*, 23 June 1897, 3.

54. "*L'Elettrico* di Port Said," *L'Imparziale*, 6–7 January 1895, 2; Archivio Storico del Ministero degli Affari Esteri e della Cooperazione Internazionale, Rome (hereafter ASMAECI), Archivi Rappresentanze Diplomatiche e Consolari all'Estero (hereafter ARCDE), Rappresentanza Cairo (hereafter RC), Busta (hereafter B) 50, folder 8, Port Said, 8 January 1894, Italian Consul to Italian Consul in Cairo.

55. ASMAECI, ARCDE, RC, B50, folder 8, Cairo, 5 February 1893, Italian Consul to Italian Consul in Port Said. Founded by the abovementioned Cumbo, it would later be directed by Malatesta and Santorelli. Alessandra Marchi, "La presse d'expression italienne en Égypte, de 1845 à 1950," *Rivista dell'Istituto di Storia dell'Europa Mediterranea*, no. 5 (2010): 104. Moghira opts for its French title, *Télégraphe*. Moghira, *L'Isthme de Suez*, 437.

56. "Il nostro fraterno saluto," *L'Imparziale*, 30 April 1895, 2; Rizzitano, "Un Secolo Di Giornalismo Italiano in Egitto (1845–1945)," 142; and Munier, *La presse en Égypte*, 45.

57. "Giornali non pervenutici," *L'Imparziale*, 28 June 1895, 2. It became daily in September 1895. "*Lo Spartaco*," *L'Imparziale*, 13 September 1895, 3. See Balboni, *Gl'Italiani nella civiltà egiziana del secolo XIX*, 1:435; and Anna Baldinetti, *Orientalismo e colonialismo: La ricerca di consenso in Egitto per l'impresa di Libia* (Rome: Istituto per l'Oriente C. A. Nallino, 1997), 154.

58. Dori, "Italiani in Africa," 147; and Rizzitano, "Un Secolo Di Giornalismo Italiano in Egitto (1845–1945)," 14–15, 147.

59. Rinaldo de Sterlich, *Sugli italiani d'Egitto: Lettera aperta di Fausto* (Cairo: Imprimerie parisienne J. Cèbe, 1888), 50. From the holdings of Egypt's national library, it appears that *Al Mu'addab* was founded in Port Said in 1921 and that *Al Thaghr al Sharqi* was founded in Suez in 1923. I was not able to verify whether *Al Suwaysiyya* (1904) and *Timsah* (1907) were published in Suez and Ismailia respectively; they are mentioned in 'Abduh, *Tatawwur al-sihafah al-misriyah*, 338.

60. "*L'Elettrico* di Port Said," *L'Imparziale*, 3 August 1895, 2; and "*L'Elettrico* di Port-Said," *L'Imparziale*, 14 August 1895, 3. This applied more generally within Egypt, where some, for example, saw the *Egyptian Gazette* as an English instrument, while *Le Bosphore egyptien* apparently voiced the opposition to the Anglo-Egyptian government. Alexandria's *Messaggero egiziano*, even though run by a Maltese, was said to represent "Italian interests." De Sterlich, *Sugli italiani d'Egitto*, 48–50.

61. Najm, *Bur Sa'id*, 376.

62. ASMAECI, ARCDE, RC, B50, folder 8, Cairo, 5 February 1894, Ministry of Foreign Affairs to Macciò, Italian Consul General in Cairo; Port Said, 8 January 1894, Italian Consul to Macciò, Italian Consul in Cairo; and Cairo, 5 February 1893, Italian Consul to Italian Consul in Port Said.

63. Najm, *Bur Sa'id*, 71–72, 381–83.

64. Sémaphore, "Anniversaire de la Fondation de Port-Said," *L'Isthme de Suez*, 1 June 1865, 157; and "Bur Sa'id," *Al-Ahram*, 19 April 1892,1.

65. "Lo Cham-el-Nessim," *L'Imparziale*, 17–18–19 April 1898, 3; and. Sylvia Modelski, *Port Said Revisited* (Washington, DC: FAROS, 2000), 143.

66. "L'anniversario del sultano a Port Said," *L'Imparziale*, 3 September 1897, 3; "Bur Sa'id," *Al-Ahram*, 18 March 1892, 2; "I fuochi artificiali pel sultano," *L'Imparziale*, 24 August 1897, 3; and "Anniversario khediviale," *L'Imparziale*, 28 October 1897, 3.

67. "Il 14 marzo a Port Said," *L'Imparziale*, 16 March 1898, 3; "*La Verité* di Port Said scriveva," *L'Imparziale*, 8 August 1900, 3; "Il Giubileo della Regina Vittoria a Suez e a Port Said," *L'Imparziale*, 19 June 1897, 2; "Il genetliaco dell'Imperatore Francesco Giuseppe a Port Said," 20 August 1897, *L'Imparziale*, 3; "XIV Luglio a Port-Said," *L'Imparziale*, 13 July 1895, 3; and "Festa maltese," *L'Imparziale*, 4 September 1895, 3.

68. "XX Settembre a Port-Said," *L'Imparziale*, 7 August 1895, 3; "La festa dello Statuto," *L'Imparziale*, 6–7 June 1897, 3; and Denis Mack Smith, *The Making of Italy, 1796–1870* (London: Macmillan, 1988), 195. Choate writes that only in theory were "the festival of the Constitution [Statuto] and the liberation of Rome [20 September 1870], and the birthdays of the King and of the Queen" to be celebrated by the entire Italian "colonies together [. . .] without distinctions of class or wealth." Mark I. Choate, "Sending States' Transnational Interventions in Politics, Culture, and Economics: The Historical Example of Italy," *International Migration Review* 41, no. 3 (2007): 736.

69. "Bur Sa'id," *Al-Ahram*, 17 March 1892, 2; "In occasione delle feste pasquali," *L'Imparziale*, 16 April 1897, 3; "Le agenzie e i consolati," *L'Imparziale*, 17 April 1897, 3; "Il nuovo anno ebraico," *L'Imparziale*, 28 September 1897, 3; "Il gran digiuno israelitico," *L'Imparziale*, 6 October 1897, 3; and Modelski, *Port Said Revisited*, 142, 146.

70. *L'Imparziale*, 21–22 July 1895, "I Maltesi di Port-Said," 3.

71. "Pensiamo al Cairo!," *L'Imparziale*, 13 December 1892, 1; and "XX September in Port Said," *L'Imparziale*, 24 September 1895, 3.

72. Modelski, *Port Said Revisited*, 146; D. A. Farnie, *East and West of Suez: The Suez Canal in History, 1854–1956* (Oxford: Clarendon Press, 1969), 402; "Da Port Said," *L'Imparziale*, 18 March 1892, 2; and "Bur Sa'id," *Al-Ahram*, 18 March 1892,2. See Murad Faraj, *Diwan Murad* (Cairo, 1912), 1:169–71, cited in Sasson Somekh, "Participation of Egyptian Jews in Modern Arabic Culture, and the Case of Murad Faraj," in *The Jews of Egypt:. A Mediterranean Society in Modern Times*, ed. Shimon Shamir (Boulder, CO: Westview Press, 1987), 137. See also Jacob M. Landau, *Jews in Nineteenth-Century Egypt* (New York: New York University Press, 1969), 35.

73. De Sterlich, *Sugli italiani d'Egitto*, 12–14.

74. Kathleen Neils Conzen et al., "The Invention of Ethnicity: A Perspective from the U.S.A.," *Journal of American Ethnic History* 12, no. 1 (1992): 5, 13; and Vito Teti, "Emigrazione e religiosità popolare," in *Storia dell'emigrazione italiana: Arrivi*, ed. Piero Bevilacqua, Andreina De Clementi, and Emilio Franzina (Rome: Donzelli, 2009), 687–708.

75. Ehud R. Toledano, *State and Society in Mid-Nineteenth-Century Egypt* (Cambridge: Cambridge University Press, 1990), 248. For Cairo's coffeehouses,

see Alon Tam, "Cairo's Coffeehouses in the Late Nineteenth and Early Twentieth Centuries: An Urban and Socio-Political History" (PhD diss., University of Pennsylvania, 2018).

76. "Ce soir," *Le Bosphore égyptien*, 4 December 1884, 3; "Un nostro abbuonato," *L'Imparziale*, 17 September 1892, 3; "Ascensione aerostatica," *L'Imparziale*, 14 February 1895, 3; and "Una gravissima rissa," *L'Imparziale*, 28–29 October 1900, 3. The trend continued after the turn of the century. In Fayyum, south of Cairo, for example, a theater group including a Greek female singer performed around 1919. Her unshaved armpits reportedly scandalized attendees. Anna Cachia and Pierre Cachia, *Landlocked Islands: Two Alien Lives in Egypt* (Cairo: American University in Cairo Press, 1999), 95.

77. Thomas Wentworth Russell, *Egyptian Service: 1902–1946* (London: Murray, 1949), 60; and "Da Mansura," *L'Imparziale*, 29 March 1892, 2.

78. ANMT, CUCMS, 1995 060 3601, Cairo, 22 October 1869, Cherif Pacha, Minister of Foreign Affairs, to French Consul; The National Archives, London (hereafter TNA), Foreign Office (hereafter FO) 423-1, Alexandria, 26 April 1860, Robert G. Colquhoun to Lord J. Russell, Report by J. Coulthard; and CADN, ARI, C1, 1861–1870, El Guisr, 19 May 1862, Viller, Chief Engineer of the Timsah Division, to Voisin, Director General.

79. Amy Fullerton, *A Lady's Ride through Palestine and Syria: With Notices of Egypt and the Canal of Suez* (London, 1872), 11; L. A. Balboni, *Gl'Italiani nella civiltà egiziana del secolo XIX: storia-biografie-monografie* (Alexandria: Tipo-litografico v. Penasson, 1906), 2:144. Primo Levi, in a biographical work, recalls a gold panner from the Doria Riparia valley once declaring, "No other people is as free as we are." Primo Levi, *Il Sistema Periodico* (Turin: G. Einaudi, 2014), 128.

80. Rodolfo De Angelis, *Storia del café chantant* (Milan: Il Balcone, 1946), 13; and ANMT, CUCMS, 1995 060 3601, Alexandria, 30 April 1868, French Consul to Director General.

81. Raoul Lacour, *L'Égypte d'Alexandrie à la seconde cataracte* (Paris: Hachette, 1871), 458; and APF, Fondo S.C., Egitto, Copti, vol. 23, Suez, 1 October 1885, Cardinal Moran, Archbishop of Sydney, to Cardinal Prefect, 186r.

82. L. Lavialle de Lameillère, *Voyage en Égypte et à l'isthme de Suez: Récits d'un père à ses enfants* (Limoges: E. Ardant, 1883), 177.

83. Khuri-Makdisi, *Eastern Mediterranean*, 73.

84. Ministero degli Affari Esteri, *Emigrazione e colonie* (1906), 2:230.

85. "Prendiamo dall'*Elettrico* di Port Said," *L'Imparziale*, 20 July 1892, 3; "Au Théâtre de Port Saïd-les-Bains," *L'Écho de Port Said*, 9 August 1893, 2; CADN, ARCFPS, C328, Port Said, 3 November 1897, French Consul to Governor General; and Luigi Dori, "Esquisse Historique de Port Said: Port Said après l'inauguration du Canal (1869–1900)," *Cahiers d'histoire égyptienne* 8, no. 1 (January 1956): 11, 14.

86. Doss, *Reminiscences*, 24.

87. Ute Sonnleitner, "Moving German-Speaking Theatre: Artists and Movement 1850–1950," *Journal of Migration History* 2, no. 1 (2016): 96.

88. ASMAECI, ARCDE, RC, B98, Suez, 30 July 1902, Deperais to Italian Consul in Cairo, Telegram; RC, B50, Cesare Catastini, "Come l'Italia è rispettata in Egitto," *La Riforma: Giornale dell'antica sinistra parlamentare*, 30 October 1893, 1; and RC, B77, folder 4, Cairo, 23 February 1901, Consul Tugini, received from Italian Consul in Port Said.

89. Tangye, *Reminiscences of Travel in Australia, America, and Egypt*, 228.

90. Santini, *Intorno al mondo a bordo della regia corvetta "Garibaldi,"* 10–11; Karl Baedeker, *Egypt, Handbook for Travellers: Part First, Lower Egypt* (Leipzig: Karl Baedeker, 1878), 424; and APF, Fondo S.C., Egitto, Copti, Vol. 23, Suez, 1 October 1885, Cardinal Moran, Archbishop of Sydney, to Cardinal Prefect, 185r.

91. The cited examples are mentioned in Elizabeth Anne McCauley, *A.A.E. Disdéri and the Carte de Visite Portrait Photograph* (New Haven, CT: Yale University Press, 1985), 106.

92. CADN, ARCFPS, C396, Port Said, 8 July 1882, Madame Veuve Valin, Inventory; and ARCFPS, C396, Port Said, 30 June 1882, Monsieur Isnard, Inventory.

93. TNA, FO 846-4, 1 July 1878, Police v. Battista Bespina; Levernay, *Guide général d'Égypte*, 285; CADN, ARCFPS, C334, Port Said, 22 July 1884, Ludovico Euclitano, "Mecanico [*sic*]," Declaration; and Dar al-Watha'iq al-Qawmiyya, National Archives, Cairo (hereafter DWQ), 2001-025132, Port Said, 24 October 1883, Governor General (attached to Port Said, 30 October 1883, Ibrahim Tewfik, Governor General, Memorandum to Consuls).

94. Toledano, *State and Society in Mid-Nineteenth-Century Egypt*, 242; APF, Fondo S.C., Egitto, Copti, vol. 23, Suez, 1 October 1885, Cardinal Moran, Archbishop of Sydney, to Cardinal Prefect, 185v; and Rudyard Kipling, *Letters of Travel: 1892–1913* (London: Macmillan, 1920), 221, entry dated 1913.

95. TNA, FO 846-3, 24 February 1874, Jacques Ruffi, French Brickmaker v Alex Mitrovich; and CADN, ARCFPS, C46, Port Said, 1 March 1874, Saint-Chaffray, French Consul in Port Said, to Governor of Port Said.

96. CADN, ARCFPS, C334, Port Said, 15 July 1884, Consular Chancelor, Report; 1 June 1865, Sémaphore, "Anniversaire de la fondation de Port-Said," *L'Isthme de Suez*, 158; Henri P. C. Baillière, *En Égypte: Alexandrie, Port-Said, Suez, le Caire; Journal d'un touriste* (Paris: J.-B. Baillière, 1867), 114; and Baedeker, *Egypt, Handbook for Travellers: Part First, Lower Egypt*, 424.

97. Ministero degli Affari Esteri, *Emigrazione e colonie* (1906), 2:280.

98. Karl Baedeker, *Egypt: Part First, Lower Egypt and the Peninsula of Sinai* (Leipzig: Karl Baedeker, 1895), 213; Karl Baedeker, *Egypt: Handbook for Travellers* (Leipzig: Karl Baedeker, 1898), 170; Brodrick, *Handbook for Travellers in Lower and Upper Egypt*, 1000; and Simonini's Hotels, *Port Said: Past and Present* (Cairo: Lencioni, 1937). In 1896 Brodrick-Murray called the latter "Royal Exchange." In the 1895 edition, Baedeker does not mention it at all but does so in the 1898 edition. Damaged during the 1956 war, it would be thereafter torn down. Mabel Loomis Todd, "The Amherst Eclipse Expedition," *Nation* 72, no. 1874 (May 30, 1901): 432; and Crosnier-Lecomte, Ghitani, and Amin, *Port-Saïd: Architectures*, 136.

99. CADN, ARCFPS, C328, Port Said, Mahdar 1378 mukhallafat al'ifranjia sanat 1903, 27 Jumada Awwal 1321 (21 August 1903); and Port Said, Na'ib ma'mur qism al ifranj, Mahdar, 1 September 1903.

100. TBL, OIOC, Mss/Eur/B235/1, Colonel Harry Ross, October–November 1898; Gaston Maspero and Louise Maspero, *Lettres d'Egypte: Correspondance avec Louise Maspero, 1883–1914*, ed. Élisabeth David (Paris: Seuil, 2003), 603; and Henry de Monfreid, *Mer rouge* (Paris: Grasset, 2002), 874.

101. Moustapha Fehmy, Ministry of the Interior, "Police Regulation for Public Establishments," *Journal Officiel du Gouvernement Egyptien*, no. 68 (13 June 1891): 859. Contemporary Western sources make a distinction between "Arabian" cafés, consisting of a single booth with a few cane-bottomed seats where customers could sip coffee, and cafés in the European style, where patrons could drink beer at higher prices. Baedeker, *Egypt, Handbook for Travellers: Part First, Lower Egypt*, 229. For a typology of drinking establishments in contemporary Alexandria, see Will Hanley, "Foreignness and Localness in Alexandria, 1880–1914" (PhD diss., Princeton University, 2007), 45–46. An overview of Ottoman coffeehouses is in François Georgeon, "Les cafés à Istanbul à la fin de l'empire ottoman," in *Cafés d'Orient revisités*, ed. Hélène Desmet-Grégoire and François Georgeon (Paris: CNRS éd., 1997), 45–50. Fuhrmann confirms that before the turn of the nineteenth century a process of hybridization had set in. Fuhrmann, "Beer, the Drink of a Changing World," 107.

102. Cachia and Cachia, *Landlocked Islands*, 45.

103. André Raymond, *Artisans et commerçants au Caire au XVIIIe siècle* (Damas: Institut français de Damas, 1973), 2:387; and Toledano, *State and Society in Mid-Nineteenth-Century Egypt*, 242. See Lane, *The Manners & Customs of the Modern Egyptians*, 340.

104. *Majallah al-qada' wa al-bulis al misri* 1, no. 1, January 1908, 3. See Philippe Gélat, *Répertoire général annoté de la législation et de l'administration egyptiennes: Part I, 1840–1908* (Alexandria: Impr. J.C. Lagoudakis, 1906), 2;813–14.

105. Ziad Fahmy, *Ordinary Egyptians: Creating the Modern Nation through Popular Culture* (Stanford, CA: Stanford University Press, 2011), 33–34, 132, 145; and Beth Baron, *The Women's Awakening in Egypt: Culture, Society, and the Press* (New Haven, CT: Yale University Press, 1994), 90–93.

106. Doss, *Reminiscences*, 25.

107. CADN, ARCFPS, C46, Port Said, 1 March 1874, Saint-Chaffray, French Consul in Port Said, to Governor of Port Said.

108. E. A. Wallis Budge, *By Nile and Tigris, a Narrative of Journeys in Egypt and Mesopotamia* (London: J. Murray, 1920), 77; and Farnie, *East and West of Suez*, 1969, 402.

109. CADN, Cairo, C428, Cairo, 5 February 1890, Zoulfikar, Ministry of Foreign Affairs, to French Consul General.

110. Fehmy, "Police Regulation for Public Establishments," 859.

111. "Da Port Said," *L'Imparziale*, 10 March 1892, 1; "*L'Elettrico* di Port Said," *L'Imparziale*, 16 January 1895, 2; and "Nous croyons devoir appeler," *Le Bosphore égyptien*, 28 November 1884, 3.

112. Fuhrmann, "Beer, the Drink of a Changing World," 100; and Georgeon, "Ottomans and Drinkers," 10.

113. Doctor Aubert-Roche, "Rapport," *L'Isthme de Suez*, 15–19 June 1867, 225–26; Archives du ministère de l'Europe et des Affaires étrangères, Centre des Archives Diplomatiques á la Courneuve, Paris (hereafter CADC), Affaires économiques et commerciales (hereafter AEC), Correspondance Consulaire et Commerciale Port-Saïd (hereafter CCCPS), vol. 1, Port Said, Flesch, French Vice-Consul, Notes on 1867, 1–4.

114. CADN, ARCFPS, C396, Maalaka-Zahle (Syria), 4 July 1882, F. Chevalier, Merchant, to French Consul in Beirut.

115. Georgeon, "Ottomans and Drinkers," 9.

116. Balboni, *Gl'Italiani nella civiltà egiziana del secolo XIX*, 2:145.

117. CADN, ARI, C4, El Guisr, 10 September 1865, Delegate for the succession Bialkowsky, List of missing objects; ARI, C4, [1866], Geyler, CdBD, Report about the month of March; and ARI, C4, Sérapeum, 7 May 1866, List of what Vidal, Telegraph agent, owed to Bazin.

118. Hanan Hammad, *Industrial Sexuality: Gender, Urbanization, and Social Transformation in Egypt* (Austin: Texas University Press, 2016), 75; Omar D. Foda, "The Pyramid and the Crown: The Egyptian Beer Industry from 1897 to 1963," *International Journal of Middle East Studies* 46, no. 1 (2014): 140; and Ralph S. Hattox, *Coffee and Coffeehouses: The Origins of a Social Beverage in the Medieval Near East* (Seattle: University of Washington Press, 1988), 123. Hattox explains that *buza* contains some, though probably very little, alcohol. See also Fuhrmann, "Beer, the Drink of a Changing World," 83.

119. CADN, ARI, C1, Ismailia, 27 July 1863, H/Naker to Voisin, Director General.

120. Società geografica italiana, *Indagini sulla emigrazione italiana*, 4:198; and "Bière française," *L'Écho de Port Said*, 9 August 1893, 4.

121. Fuhrmann, "Beer, the Drink of a Changing World," 86, 83, 89.

122. "Bière alsacienne," *Le Phare de Port-Said et du Canal de Suez*, 8 August 1893, 2.

123. CADC, AEC, CCCPS, vol. 1, Port Said, 20 July 1867, Flesch, French Vice-Consul, to French Ministry of Foreign Affairs. Foda argues that in Egypt in general, even though Greeks and Italians seemed to prevail among the drinking establishments' managers, there was a significant Egyptian contingent among owners. Foda, "Pyramid and the Crown," 144. On the role of Greek tavernkeepers, especially in the Egyptian countryside, see Sayyid 'Ashmawi, "Perceptions of the Greek Money-Lender in Egyptian Collective Memory at the Turn of the Twentieth," in *Money, Land and Trade an Economic History of the Muslim Mediterranean*, ed. Nelly Hanna (London: I. B. Tauris in association with the European Science Foundation, Strasbourg, France, 2002), 256. In nineteenth-century Istanbul, it was Greeks, Armenians, and Jews who reportedly monopolized the retail trade in alcohol. Georgeon, "Ottomans and Drinkers," 8–9, 12.

124. Georgeon, "Les cafés à Istanbul à la fin de l'empire ottoman," 50.

125. Levernay, *Guide général d'Égypte*, 285, 272.

126. Georgeon, "Les cafés à Istanbul à la fin de l'empire ottoman," 50.

127. De Sterlich, *Sugli italiani d'Egitto*, 11–12.

128. CADC, AEC, CCCPS, vol. 1, Port Said, 28 February 1875, Consul in Port Said to French Ministry of Foreign Affairs, 395v; and Port Said, 2 March 1875, "Opposition" to French Minister of Foreign Affairs, 401v–402r.

129. "Per Carnot a Port Said," *L'Imparziale*, 27 June 1895, 3.

130. Panaït Istrati, *Vie d'Adrien Zograffi* (Paris: Gallimard, 1969), 409; ASMAECI, ARCDE, RC, B39, folder 8, Port Said, received on 12 January 1891, Italian Consul in Port Said to Italian Consul in Cairo. It was unclear whether the Italian Consul, who bore witness of the mess, was a customer or simply a passerby.

131. Loomis Todd, "Amherst Eclipse Expedition," 432.

132. Chevrillon, *Dans l'Inde*, 329, entry dated 7 January 1889.

133. CADN, ARCFPS, C149, Port Said, 16 September 1866, Regnier, Campsite Agent, to French Vice-Consul in Port Said; ANMT, CUCMS, 1995 060 3601, Ismailia, 3 December 1867, Ismail Bey, Governor, to Director General; and ASMAECI, ARCDE, CC, AV, B12, Suez, 30 December 1869, Italian Vice-Consul to Italian Consul in Cairo.

134. TNA, FO 846-4, 1 July 1878, Police v. Battista Bespina; and FO 846-4, 26 July 1879, Vignocchi, Sandri, Fergali Mohammed & Cordina.

135. CADN, ARCFPS, C42, Port Said, 7 October 1879, Police Bureau to Ibrahim Rouchdy Pacha, Governor; and DWQ, 2001-013273, Cairo Division, Cairo, 1 March 1885. Sources do not specify Cucco's nationality. In contrast, they do specify that his interlocutor was a "native." Hence, it seems safe to assume that Cucco was not.

136. Alan Mikhail, "The Heart's Desire: Gender, Urban Space and the Ottoman Coffee House," in *Ottoman Tulips, Ottoman Coffee: Leisure and Lifestyle in the Eighteenth Century*, ed. Dana Sajdi (London: Tauris Academic Studies, 2007), 152.

137. Doss, *Reminiscences*, 25; and APF, Fondo S.C., Egitto, Copti, vol. 23, Port Said, 6 December 1889, Fr. Agostino da Tarnac, Superior of the Mm. Oo. Community of Port Said, 654.

138. CADC, AEC, CCCPS, vol. 1, Port Said, 12 September 1867, Forest, Pagnon, Flesch, Minutes, 39r, 41r; vol. 1, Port Said, received on 9 October 1867, Flesch to Ministry of Foreign Affairs, 23r–23v; and vol. 1, Port Said, 13 September 1867, Vice-Consul of France and Managing Consular Agent of Austria in Port Said, Deposition of Michele Arghiro, 27r.

139. Athanasios (Sakis) Gekas, "Class and Cosmopolitanism: The Historiographical Fortunes of Merchants in Eastern Mediterranean Ports," *Mediterranean Historical Review* 24, no. 2 (2009): 103.

140. Similarly, Georgeon claims that Istanbul's coffeehouses were no hotbeds for nationalist agitation. Georgeon, "Les cafés à Istanbul à la fin de l'empire ottoman," 72–73.

141. TNA, FO 846-4, 6 December 1877, First Mate & Chief of Service v. Master Samuel Shean.

142. CADN, ARI, C4, Ouady District, 19 December 1865, Ibrahim Naguib, District Doctor, Certificate; ARI, C4, Tell el Kebir, 31 March 1866, Ibrahim Naguib, to Geyler, CdBD, Attached list; and ARCFPS, C42, Port Said, 4 January 1880, Vignocchi, Police Bureau, to Ibrahim Rouchdy Pacha, Governor General.

143. George W. Balfour, "Clinical Lecture on the Treatment of Delirium Tremens," *Lancet* 113, no. 2892 (1879): 146.

144. DWQ, 2001-025100, Port Said, 13 February 188[?], Ibrahim Tewfik, Governor General of the Canal and Port Said, to Sabet Pacha, Ministry of the Interior, Cairo; CADN, ARCFPS, C334, Port Said, 24 March 1885, Doctor Covidou, Certificate; and TNA, FO 846-3, 14 April 1875, Matilda Carlesso v. Vincenzo Savone.

145. Najm, *Bur Sa'id*, 191. Albeit implicitly, Ramadan also hints at a connection between Greeks, drunkenness, and crime in Port Said; Ramadan, *Al-Haya al-ijtima'iya fi Misr fi 'asr Isma'il*, 307.

146. DWQ, 2001-010948, 19 April 1886, Port Said, Ibrahim Tewfiq, Governor General, to Norrish, British Managing Vice-Consul.

147. A. B. Clot-Bey, *Aperçu général sur l'Égypte: Orné d'un portrait (de Méhémet Ali) et de plusieurs cartes et plans color* (Brussels: Société de Librairie, 1840), 2:267; Harald Fischer-Tiné, "Liquid Boundaries: Race, Class, and Alcohol in Colonial India," in *A History of Alcohol and Drugs in Modern South Asia: Intoxicating Affairs*, ed. Harald Fischer-Tiné and Jana Tschurenev (London: Routledge, 2014), 91–92.

148. "Igienisti all'erta!," *L'Imparziale*, 25 August 1897, 3.

149. Valeska Huber, *Channelling Mobilities: Migration and Globalisation in the Suez Canal Region and Beyond, 1869–1914* (Cambridge: Cambridge University Press, 2013), 72.

150. "Bur Sa'id," *Al-Ahram*, 20 April 1892, 2; and "Bur Sa'id," *Al-Ahram*, 13 April 1895, 1.

151. Philippa Levine, *The British Empire: Sunrise to Sunset* (Harlow, UK: Pearson Longman, 2007), 46–47; and Jan Gothard, "Wives or Workers? Single British Female Migration to Colonial Australia," in *Women, Gender, and Labour Migration: Historical and Global Perspectives*, ed. Pamela Sharpe (London: Routledge, 2011), 146.

152. DWQ, 2001-014127, Port Said, 18 April 1883, Gibbons to General Baker Pacha; 2001-014359, Cairo, 6 September 1886, Cairo Division, Director Inspector General, to Inspector General of the Egyptian Police in Cairo; and 2001-019915, 8 June 1887, Crookshank, Director General of Prisons, to Inspector General of the Police.

153. Richard Tangye, *Reminiscences of Travel in Australia, America, and Egypt* (London: Sampson Low, Marston, Searle, & Rivington, 1883), 230. On Australian soldiers' penchant for drinking and brawling, see Mario M. Ruiz, "Manly Spectacles and Imperial Soldiers in Wartime Egypt, 1914–19," *Middle Eastern Studies* 45, no. 3 (2009): 360.

154. Cesare Catastini, "Come l'Italia è rispettata in Egitto," *La Riforma: Giornale dell'antica sinistra parlamentare*, 30 October 1893, 1; and DWQ, 2001-010948, Port Said, 13 April 1886, Norrish to Governor.

155. "Guerra russo-giapponese a Port-Said," *L'Imparziale*, 24 July 1905, 3; and "Baruffe su baruffe," *L'Imparziale*, 7 August 1905, 3.

156. ANMT, CUCMS, 1995 060 3601, El Guisr, 18 July 1861, Chief Engineer of the Timsah Division, Report.

157. ASMAECI, ARCDE, CC, AV, B12, Suez, 24 October 1869, British Consul, French Consul, Austrian Vice-Consul, Italian Vice-Consul, to Governor.

158. Doss, *Reminiscences*, 25.

159. On the connection between seamen and brawls in coeval Alexandria, see 'Imad Ahmad Hilal, *Al-Baghaya fi Misr: Dirasah tarikhiyah ijtima'iyah, 1834–1949* (Cairo: 'Arabi lil-Nashr wa-al-Tawzi', 2001), 89–90; and Will Hanley, *Identifying with Nationality: Europeans, Ottomans, and Egyptians in Alexandria* (New York: Columbia University Press, 2017), 27–28.

160. DWQ, 2001-025132, Port Said, 24 October 1883, Governor General (attached to Port Said), 30 October 1883, Ibrahim Tewfik, Governor General, Memorandum to Consuls.

161. CADN, ARCFPS, C334, Port Said, 15 June 1885, Secretary of German Consulate, Translation; and "Rissa," *L'Imparziale*, 16 January 1895, 3.

162. Hanssen, *Fin-de-Siècle Beirut*, 203; and Thomas Philipp, *Ǧurǧi Zaidān, His Life and Thought* (Beirut: Orient-Institut der Dt. Morgenländ. Ges., 1979), 147.

163. TNA, FO 846-3, 11 January 1875, Farrugia v. Mifsud.

164. Mario M. Ruiz, "Intimate Disputes, Illicit Violence: Gender, Law, and the State in Colonial Egypt, 1849–1923" (PhD diss., University of Michigan, 2004), 141.

165. DWQ, 2001-010948, 19 April 1886, Port Said, Ibrahim Tewfiq, Governor General, to Norrish, British Managing Vice-Consul.

166. DWQ, 2001-011478, Port Said, 6 August 1888, Ibrahim Tewfik, Governor General, to Riaz Pacha, Ministry of the Interior; and "Filippo Avenatti," *L'Écho de Port Said*, 9 August 1893, 4.

167. TNA, FO 846-4, 8 December 1879, Police through Mr. Vignocchi v. Carmelo Zammit.

168. CADN, ARCFPS, C42, Port Said, 12 September 1885, Ibrahim Tewfik, Governor, to Giullois, French Managing Consul; and DWQ, 2001-013273, Cairo Division, Cairo, 1 March 1885.

169. TNA, FO 846-4, 29 September 1879, Costantino Nicola Vaticiotis v. Spiro Brihazian.

170. Panaït Istrati, *Vie d'Adrien Zograffi* (Paris: Gallimard, 1969), 373. The occurrence reported by Istrati may have been a lie, but it provides what must have been a credible background.

171. Ramadan, *Al-Haya al-ijtima'iya fi Misr fi 'asr Isma'il*, 307–8; and W. N. Willis, *Anti-Christ in Egypt* (London: Anglo-Eastern Publishing, 1915), 150.

172. CADC, AEC, CCCPS, vol. 1, Port Said, 20 July 1867, Flesch, French Vice-Consul, to French Ministry of Foreign Affairs; APF, Fondo S.C., Egitto, Copti, vol. 23, Porto-Said, 6 December 1889, Fr. Agostino da Tarnac, Superior of the Mm. Oo. Community of Port Said, 654.

173. Fehmy, "Police Regulation for Public Establishments," 858.

174. ANMT, CUCMS, 1995 060 3599, Ismailia, 24 May 1867, Director General to Gioia, Chief Engineer of El Guisr Division.

175. CADN, ARCFPS, C42, Port Said, 11 December 1879, Aly Sabit, Sub-Governor General, to Saint-Chaffray, French Consul.

176. DWQ, 2001–025132, Port Said, 24 October 1883, Governor General (attached to Port Said), 30 October 1883, Ibrahim Tewfik, Governor General, Memorandum to Consuls; 2001-025132, Cairo, 4 September 1884, Ministry of the Interior to Governor General; and 2001-025132, Port Said, 10 September 1884, Governor General of the Isthmus, Decree.

177. DWQ, 2001-025132, Port Said, 10 September 1884, Ibrahim Tewfik, Governor General, to Abdel Kader Pacha Hilmy, Minister of the Interior; and 2001-010755, Port Said, 18 September 1884, Tewfik to Hilmy.

178. Fehmy, "Police Regulation for Public Establishments," 859.

179. DWQ, 2001-010948, Port Said, 13 April 1886, Norrish, British Managing Vice-Consul, to Ibrahim Tewfiq, Governor General; and 2001-010948, Port Said, 17 April 1886, Norrish to Tewfiq. A similar claim had already been presented to the governor two years prior, in 1884, by British Consul Barrell; 2001–010948, Port Said, 19 April 1886, Tewfiq to Norrish. Apparently the Ministry of Foreign Affairs yielded to the British consul's request or at least tried to prevent other similar protestations; 2001-010948, 22 April 1886, Ministry of Foreign Affairs to Ministry of the Interior.

180. Ruiz, "Intimate Disputes, Illicit Violence," 141.

181. Hattox, *Coffee and Coffeehouses*, 108; Georgeon, "Les cafés à Istanbul à la fin de l'empire ottoman," 56; and Cemal Kafadar, "How Dark Is the History of the Night, How Black the Story of Coffee, How Bitter the Tale of Love: The Changing Measure of Leisure and Pleasure in Early Modern Istanbul," in *Medieval and Early Modern Performance in the Eastern Mediterranean*, ed. Arzu Öztürkmen and Evelyn Birge Vitz (Turnhout: Brepols, 2014), 261.

182. Ruiz, "Intimate Disputes, Illicit Violence," 139; and Toledano, *State and Society in Mid-Nineteenth-Century Egypt*, 242–43.

183. Hattox, *Coffee and Coffeehouses*, 90; Georgeon, "Les cafés à Istanbul à la fin de l'empire ottoman," 44; and Cengiz Kırlı, "Coffeehouses: Public Opinion in the Nineteenth-Century Ottoman Empire," in *Public Islam and the Common Good*, ed. Dale F. Eickelman and Armando Salvatore (Leiden: Brill, 2005), 76; and Mikhail, "Heart's Desire," 146.

184. CADC, AEC, CCCPS, vol. 1, Port Said, 11 September 1867, Alacacci/Ulacacci/Ullacacci to French Vice-Consul in Port Said.

185. CADN, ARCFPS, C334, Port Said, 12 August 1886, Georges Sucur, Drogman at French Consulate.

186. CADN, ARCFPS, C334, Port Said, 12 August 1886, Hadj Hamza Ben Saleh to French Consul.

187. Tangye, *Reminiscences of Travel in Australia, America, and Egypt*, 228; and Kafadar, "How Dark Is the History of the Night," 261.

188. Foda, "Pyramid and the Crown," 146.

189. Duff, *Notes from a Diary*, 5. Tangye, *Reminiscences of Travel in Australia, America, and Egypt*, 228; Paul Bourde, *De Paris au Tonkin* (Paris: Calmann Lévy, 1888), 14, entry dated 2 January 1884; and Chevrillon, *Dans l'Inde*, 329–30, entry dated 7 January 1889. On female Bohemian orchestras in the late Ottoman Empire, see Malte Fuhrmann, *Port Cities of the Eastern Mediterranean: Urban Culture in the Late Ottoman Empire* (Cambridge: Cambridge University Press, 2020), 315–20.

190. Kipling, *Light That Failed*, 38.

191. Erckmann and Chatrian, *Souvenirs d'un ancien chef de chantier*, 6, 10, 15; and Willis, *Anti-Christ in Egypt*, 141–42, 147. The historical record reveals that there was indeed, at Serapeum in 1866, a male canteen-keeper called Aubry, like the older woman known as "Madame Aubry" presiding over the ramshackle canteen depicted by Erckmann and Chatrian. One can only wonder if the authors and Aubry ever met in person. See CADN, ARI, C4, 8 June 1866, Serapeum, Aubry, Innkeeper, to Geyler.

192. Bourde, *De Paris au Tonkin*, 14.

193. Ruiz, "Intimate Disputes, Illicit Violence," 134.

194. CADN, ARCFPS, C334, Port Said, 11 February 1885, Excerpt of prison register; C334, Port Said, 15 July 1884, Governor General to French Consul; C334, Port Said, 13 July 1882, Frankel Nicolas, Sub Inspector First Class, to Police Inspector; and C334, 29 July 1882, Consular Tribunal to Governor of Port Said.

195. CADN, ARCFPS, C334, Port Said, n.d., Cipollaro, Police Inspector of Port Said and Canal, to Ibrahim Pacha Tewfik, Governor General, Translation.

196. Erckmann and Chatrian, *Souvenirs d'un ancien chef de chantier à l'isthme de Suez*, 37.

197. ASMAECI, ARCDE, RC, B72, folder 1, Suez, 24 October 1897, Maria Ursini to Italian Consul in Cairo, Telegram; and RC, B72, folder 1, Suez, 20 June 1898, Vice-Consul to Consul General in Cairo.

198. DWQ, 2001-015916, no place, 31 August 1880, Neroutsos Bey, Health Service President, Documents sent to the Ministry of the Interior (e.g., Alexandria, 30 August 1880, Dr. Londinsky, to Neroutsos Bey, "Devoirs des inspecteurs de police delegués au service des moeurs" and "Règlement maîtresses des maisons"); and ASMAECI, ARCDE, CC, B30, Port Said, 2 September 1878, Vice-Consul Zerbani to Malmusi, Italian Consul in Cairo.

199. DWQ, 2001-018598, Cairo, 8 January 1891, Riaz, Ministry of the Interior, to Director of Health Service; and *Al-Mu'ayyad*, 27–28 December 1890. Such interpretation is supported by Hilal, who writes that "dance was the apparent profession of most of secret prostitutes as well as of some of the official ones"; Hilal, *Al-Baghaya fi Misr*, 233.

200. "La tratta delle bianche," *L'Imparziale*, 14 June 1904, 3. This point is also made by Bruce W. Dunne, "Sexuality and the 'Civilizing Process' in Modern Egypt" (PhD diss., Georgetown University, 1996), 94; and Toledano, *State and Society in Mid-Nineteenth-Century Egypt*, 237.

201. CADN, ARCFPS, C149, no place, n.d. [after 1883], Project of police regulation in Port Said and Ismailia.

202. Henry de Monfreid, *Mer rouge* (Paris: Grasset, 2002), 868; and Willis, *Anti-Christ in Egypt*, 141–42, 144.

203. Dunne, "Sexuality and the 'Civilizing Process' in Modern Egypt," 145; and Khaled Fahmy, "Prostitution in Egypt in the Nineteenth Century," in *Outside In: On the Margins of the Modern Middle East*, ed. Eugene L. Rogan (London: I. B. Tauris, 2002), 45. Yet some historians claim that "the diffusion of prostitutes in Port Said brought to the diffusion of some venereal diseases such as the syphilis that struck a number of *qawas*." Najm, *Bur Saʿid*, 115. A similar point is made by Serge Jagailloux, *La médicalisation de l'Égypte au xixe siècle: 1798–1918* (Paris: Éd. Recherche sur les civilisations, 1986), 249.

204. DWQ, 4003-037240, Port Said, 8 August 1883, Doctor Robertson to Hospital Director in Port Said & Ismail Pacha Yousry, Health Service Director in Cairo.

205. DWQ, 4003-037240, Health Office in Port Said and Ismailia, 1883–1885, Weekly reports about the inspection of birth and death; and 2001-016382, Cairo, 1 July 1885, Abdel Kader, Decree (based on the suggestion of the Director of the Service of Health and Public Hygiene and approved by the Council of Ministers).

206. ASMAECI, ARCDE, RC, B72, folder 1, Cairo, 5 September 1897, Italian Consul to Italian Minister of Foreign Affairs; Samuel, *Jewish Life in the East*, 16; and Najm, *Bur Saʿid*, 114–15. Lapidus writes that, in Ottoman parlance, "Rum" (literally Roman) designated Greek Orthodox Christians. Ira M Lapidus, *Islamic Societies to the Nineteenth Century a Global History* (New York: Cambridge University Press, 2012), 456.

207. DWQ, 2001-025132, Port Said, 24 October 1883, Governor General (attached to Port Said), 30 October 1883, Ibrahim Tewfik, Governor General, Memorandum to Consuls.

208. DWQ, 2001-010266, Port Said, 24 June 1884, Cadey, to Ibrahim Tewfik, Governor General; and 2001-019970, no place, 2 May 1891, no author to President of Legal Bureau of the State.

209. ASMAECI, ARCDE, RC, B39, Port Said, 16 March 1891, Minutes of meeting among Governor General of Port Said and the Canal, De la Corte (Spain), Leoni (Italy), Rouyer (Denmark), Broadbent (USA), G. Laporte, C. de Toracuchi, Van der Duyn, Hollebecke, W. Macdonald, Ch. Smyrniadis; and RC, B39, Port Said, 4 April 1891, Italian Consul in Port Said, to Italian Consul in Cairo.

210. "*La Verità* rileva lo sconcio," *L'Imparziale*, 4 May 1895, 2; "*L'Elettrico* di Port-Saïd," *L'Imparziale*, 19 July 1895, 2; and Jocelyne Dakhlia, *Lingua franca* (Arles: Actes Sud, 2008), 476. Compare with Alexandria's *Rue des Soeurs* or *Al-Sabʿ Banat*. Hanley, *Identifying with Nationality*, 27–28, 34.

211. Willis, *Anti-Christ in Egypt*, 149.

212. Bruna Bianchi, "Lavoro ed emigrazione femminile (1880–1915)," in *Storia dell'emigrazione italiana: Partenze*, ed. Piero Bevilacqua, Andreina De Clementi, and Emilio Franzina (Rome: Donzelli, 2009), 262. Here Bianchi is mainly referring to women leaving the Italian peninsula and especially its southern regions. See Anthony Santilli, "Penser et analyser le cosmopolitisme: Le cas des Italiens

d'Alexandrie au XIXe siècle," *Mélanges de l'École française de Rome—Italie et Méditerranée modernes et contemporaines*, no. 125-2 (January 15, 2013).

213. Fuhrmann, *Port Cities of the Eastern Mediterranean*, 374.

214. Museo Centrale del Risorgimento di Roma (hereafter MCRR), B1177/2 (2), 30 June 1878, Ezio Galli, Report.

215. The Women's Library, London (hereafter TWL), International Bureau for Suppression of Traffic in Persons (hereafter IBS), Country Files, 4IBS/6, Egypt, FL 113, Cicely McCall, Organizing Secretary of the Central Committee of Egypt, *The International Bureau for the Suppression of Traffic in Women and Children: Its Work in Egypt* (Cairo: Nile Mission Press, February 1930), 4; and Derek Hopwood, *Tales of Empire: The British in the Middle East, 1880–1952* (London: I. B. Tauris, 1989), 66. Excerpt written by Thomas Rapp, British Vice-Consul in Port Said in 1920; managing consul in Cairo in 1922 and in Suez in 1923, in Hopwood, *Tales of Empire*, 84, 86, 87.

216. Harald Fischer-Tiné, "'White Women Degrading Themselves to the Lowest Depths': European Networks of Prostitution and Colonial Anxieties in British India and Ceylon ca. 1880–1914," *Indian Economic and Social History Review* 40, no. 2 (2003): 172–73.

217. Liat Kozma, *Global Women, Colonial Ports: Prostitution in the Interwar Middle East* (Albany: State University of New York Press, 2017), 89–90; and Mary Anne Poutanen, *Beyond Brutal Passions: Prostitution in Early Nineteenth-Century Montreal* (Montreal: McGill-Queen's University Press, 2015), 29.

218. Gur Alroey, "Journey to Early-Twentieth-Century Palestine as a Jewish Immigrant Experience," *Jewish Social Studies* 9, no. 2 (2003): 57.

219. AMMBPA, Annale 9: 1882–1885, Port Said, 11 April 1882, Mother Superior to the General Superior (Mère Générale), 19. On the overrepresentation of Jews, see Kozma, *Global Women, Colonial Ports*, 35.

220. TWL, 4IBS/6, Egypt, FL 113, 30 December 1905, Major Hopkinson to Lord Cromer, Report of Executive and General Committee of the General Meeting.

221. MCRR, B1177/2 (2), 30 June 1878, Ezio Galli, Report; and Malte Fuhrmann, "Down and out on the Quays of İzmir: 'European' Musicians, Innkeepers, and Prostitutes in the Ottoman Port-Cities," *Mediterranean Historical Review* 24, no. 2 (2009): 177–78.

222. Gustave Nicole, *La Prostitution en Égypte, par M. G. Nicole* (Paris: J.-B. Baillière, 1879), 207–8; and TWL, 4IBS/6, Egypt, FL 113, 30 December 1905, Major Hopkinson to Lord Cromer, Report of Executive and General Committee of the General Meeting. See Francesca Biancani, *Sex Work in Colonial Egypt: Women, Modernity and the Global Economy* (London: I. B. Tauris, 2018), 31–33.

223. Willis, *Anti-Christ in Egypt*, 149.

224. ASMAECI, ARCDE, CC, B30, Suez, 9 September 1875, Italian Vice-Consul to Malmusi, Italian Consul in Cairo; and CC, B30, Port Said, 2 September 1878, Italian Vice-Consul to Italian Consul in Cairo.

225. Biancani, *Sex Work in Colonial Egypt*, 33; and Laura Schettini, *Turpi traffici: Prostituzione e migrazioni globali 1890–1940* (Rome: Biblink, 2019).

226. TWL, 4IBS/6, Egypt, FL 113, 30 December 1905, Major Hopkinson to Lord Cromer, Report of Executive and General Committee of the General Meeting.

227. For an analysis of the former, see Lucia Carminati, "'She Will Eat Your Shirt': Foreign Migrant Women as Brothel Keepers in Port Said and Along the Suez Canal: Prostitution as Business and Survival, 1880–1914," *Journal of the History of Sexuality* 30, no. 2 (2021): 161–94.

228. TWL, 4IBS/6, Egypt, FL 113, 30 December 1905, Major Hopkinson to Lord Cromer, Report of Executive and General Committee of the General Meeting.

229. CADN, ARCFPS, C328, Cairo, 18 February 1900, Marie Goujat, to French Consul.

230. Società geografica italiana, *Indagini sulla emigrazione italiana*, 4:227, 62.

231. "Al lavoro," *L'Imparziale*, 23 June 1897, 3. The locution habitually found is "*ufficio di collocamento*." Pea also calls such establishments "*ufficio di sensale*" and "*cancello della servitù*," the latter of which is also used by consular authorities. Enrico Pea, *Vita in Egitto* (Milan: Mondadori, 1949), 60; and Società geografica italiana, *Indagini sulla emigrazione italiana*, 4:62.

232. "Una giovane sposa," *L'Imparziale*, 20 February 1897, 3.

233. Janet Henshall Momsen, *Gender, Migration, and Domestic Service* (London: Routledge, 1999), 5; and Mirjam Milharčič Hladnik, "Trans-Mediterranean Women Domestic Workers: Historical and Contemporary Perspectives," in *From Slovenia to Egypt: Aleksandrinke's Trans-Mediterranean Domestic Workers' Migration and National Imagination*, ed. Mirjam Milharčič Hladnik (Göttingen: VetR Unipress, 2015), 14.

234. Pea, *Vita in Egitto*, 60–61.

235. Ministero degli Affari Esteri, *Emigrazione e Colonie* (1906), 2:169, 171.

236. Paulo G. Brenna, *L'emigrazione italiana nel periodo antebellico* (Firenze: R. Bemporad & Figlio, 1918), 205, 258–61; and Debra L. DeLaet, "Introduction: The Invisibility of Women in Scholarship on International Migration," in *Gender and Immigration*, ed. Debra L. DeLaet and Gregory A. Kelson (New York: New York University Press, 1999), 13–14.

237. Sonya O. Rose, *Limited Livelihoods: Gender and Class in Nineteenth-Century England* (Berkeley: University of California Press, 1992), 152–53; and Sonya O. Rose, "Gender Antagonism and Class Conflict: Exclusionary Strategies of Male Trade Unionists in Nineteenth-Century Britain," *Social History* 13, no. 2 (1988): 205–6.

238. Liat Kozma, *Policing Egyptian Women: Sex, Law, and Medicine in Khedival Egypt* (Syracuse, NY: Syracuse University Press, 2011), xxi.

239. Poutanen, *Beyond Brutal Passions*, 24.

240. Joan Wallach Scott, *Gender and the Politics of History* (New York: Columbia University Press, 1999), 143.

241. T. Holmes, *Heart and Thought. Memories of Eastern Travel* (Bolton, UK: J. W. Gledsdale, 1887), 110; Mukharji, *Visit to Europe*, 16; and Loomis Todd, "Amherst Eclipse Expedition," 432.

242. Farnie, *East and West of Suez*, 395–96; and George W. Steevens, *Egypt in 1898* (New York: Dodd, Mead, 1898), 26.

243. DWQ, 4003-036553, Cairo, 15 February 1900, Ministry of Finances to Under Secretary of Public Works.

244. TBL, OIOC, Mss/Eur/B235/1, Private Papers, Memoirs of Colonel Harry Ross, October–November 1898; and Loomis Todd, "Amherst Eclipse Expedition," 432.

245. CADN, ARCFPS, C328, Port Said, 3 November 1897, French Consul to Governor General.

246. Bernhard Huldermann, *Albert Ballin* (London: Cassell, 1922), entry dated 16 January 1901; and Hopwood, *Tales of Empire*, 16.

247. Brad Beaven, Karl Bell, and Robert James, introduction to *Port Towns and Urban Cultures: International Histories of the Waterfront, c. 1700–2000* (London: Palgrave Macmillan, 2016), 8.

248. Fuhrmann, "Down and out on the Quays of İzmir"; Nile Green, *Bombay Islam: The Religious Economy of the West Indian Ocean, 1840–1915* (Cambridge: Cambridge University Press, 2011), 83; and Harald Fischer-Tiné, "Flotsam and Jetsam of the Empire? European Seamen and Spaces of Disease and Disorder in Mid-Nineteenth Century Calcutta," in *The Limits of British Colonial Control in South Asia: Spaces of Disorder in the Indian Ocean Region*, ed. Ashwini Tambe and Harald Fischer-Tiné (Abingdon, UK: Routledge, 2008), 121–54.

249. Edwin Arnold, *India Revisited* (London: Trübner, 1886), 27; and Loomis Todd, "Amherst Eclipse Expedition," 432.

250. Kipling, *Light That Failed*, 36; and Steevens, *Egypt in 1898*, 27–28.

251. Willis, *Anti-Christ in Egypt*, 139, 150, 151.

252. Percy F. Martin, *Egypt Old & New, a Popular Account of the Land of the Pharaohs from the Traveller's and Economist's Point of View* (New York: Doran, 1923), 90–91, quoted in Farnie, *East and West of Suez*, 401.

253. "Miseria," *L'Imparziale*, 6–7 March 1892, 1; Ministero degli Affari Esteri, *Emigrazione e colonie: Raccolta di rapporti dei RR. agenti diplomatici e consolari* (Rome: Tipografia Nazionale di G. Bertero, 1893), 570.

254. Ministero degli Affari Esteri, *Emigrazione e colonie* (1906), 2:201, 206 and 561–62, 575.

255. Cachia and Cachia, *Landlocked Islands*, 42, 48, 51; and Società geografica italiana, *Indagini sulla emigrazione italiana*, 4:160.

256. DWQ, 5013-005405, Athens, 2 April 1893, Technical Director, Certificate; 5013-005405, Alexandria, 1 June 1893, Pierre Philippides & Family to Khedive; and 5013-005405, 5 June 1893, to Philippides, Reply.

257. Pea, *Vita in Egitto*, 211; and "Meccanico svizzero," *L'Imparziale*, 6 November 1900; and "Signorina tedesca," *L'Imparziale*, 6 November 1900, 3.

258. Khuri-Makdisi, *Eastern Mediterranean*, 5.

259. Ministero degli Affari Esteri, *Emigrazione e colonie* (1906), 2:273–74, 280.

260. Najm, *Bur Saʿid*, 157, 174, 354, 356, 364; Joel Beinin and Zachary Lockman, *Workers on the Nile: Nationalism, Communism, Islam, and the Egyptian Working Class, 1882–1954* (Princeton, NJ: Princeton University Press, 1987), 27–31; John Chalcraft, "The Coal Heavers of Port Saʿid: State-Making and Worker Protest,

1869–1914," *International Labor and Working-Class History* 60 (October 2001): 110–24; and Khuri-Makdisi, *Eastern Mediterranean*, 142.

261. "I'lanat Bur Sa'id," *Al-Ahram*, 1 July 1900,1. In 1924 Port Said still registered the presence of a "great number" of individuals who had come from different towns of Upper Egypt and worked in coal heaving or goods transportation; see "Audat al-'Asabia," *Al-Mu'addab*, 17 February 1924, 4.

262. Ministero degli Affari Esteri, *Emigrazione e colonie* (1906), 2:273–74. ASMAECI, Polizia Internazionale, B25, Alexandria, 23 June 1899, Italian Consulate, Sentence; Archivio Centrale dello Stato, Rome (hereafter ACS), Casellario Politico Centrale (hereafter CPC), Malatesta Aniello, Cenno biografico, London, 7 October 1907, Virgilio; ACS, CPC, Port Said, 14 September 1899, Italian Consul, Letter's excerpt; and ACS, CPC, Naples, 19 December 1896, Prefecture to Italian Ministry of the Interior. For a discussion of the Italian anarchist community in Alexandria, see Anthony Gorman, "'Diverse in Race, Religion, and Nationality . . . But United in Aspirations of Civil Progress': The Anarchist Movement in Egypt 1860–1940," in *Anarchism and Syndicalism in the Colonial and Post-Colonial World, 1870–1940*, ed. Lucien Van der Walt and Steven Hirsch (Leiden; Boston: Brill, 2010), 3–32; Khuri-Makdisi, *Eastern Mediterranean*; and Lucia Carminati, "Alexandria, 1898: Nodes, Networks, and Scales in Late Nineteenth-Century Egypt," *Comparative Studies in Society and History* 59, no. 1 (2017): 127–53.

263. Richard Allen, *Letters from Egypt, Syria, and Greece* (Dublin: Gunn & Cameron, 1869), 26.

264. Bourde, *De Paris au Tonkin*, 14–15; and Willis, *Anti-Christ in Egypt*, 140–41.

265. Avner Wishnitzer, "Into the Dark: Power, Light, and Nocturnal Life in 18th-Century Istanbul," *International Journal of Middle East Studies* 46, no. 3 (2014): 515–16; Kırlı, "Coffeehouses," 89; Georgeon, "Les cafés à Istanbul à la fin de l'empire ottoman," 41, 71; and Hattox, *Coffee and Coffeehouses*, 91, 101, 103. For critique, see Mikhail, "Heart's Desire," 134–35, 156–59.

266. Nga Li Lam, "Women as Pleasure Seekers: Courtesans, Actresses, and Female Visitors in the Amusement Halls of Early Republican Shanghai," *Journal of Urban History* 45, no. 4 (2019): 385; and Harald Fischer-Tiné, "'The Drinking Habits of Our Countrymen': European Alcohol Consumption and Colonial Power in British India," *Journal of Imperial and Commonwealth History* 40, no. 3 (2012): 385.

267. Kırlı, "Coffeehouses," 76; and Tam, "Cairo's Coffeehouses in the Late Nineteenth and Early Twentieth Centuries."

268. Fuhrmann, *Port Cities of the Eastern Mediterranean*, 343.

269. Kipling, *Letters of Travel*, 221, entry dated 1913.

CONCLUSION

1. "*L'Elettrico* di Port Said," *L'Imparziale*, 21 September 1895, 2.

2. Pierre-Henri Couvidou, *Itinéraire du canal de Suez* (Port Said: A. Mourès, 1875), 44.

3. Médiathèque de la Communauté Urbaine d'Alençon, Fonds Adhémard Leclère, Ms.-697/1/f, 17 February 1907, "Voyages sur la route de l'Inde."

4. Henri P. C. Baillière, *En Égypte: Alexandrie, Port-Said, Suez, le Caire; Journal d'un touriste* (Paris: J.-B. Baillière, 1867), 112.

5. "Bur Sa'id. Madiha haditha wa mustaqbalha," *Al-Mu'addab*, 30 August 1925, 3.

6. Harry Alis, *Promenade en Égypte* (Paris: Librairie Hachette, 1895), 10.

7. Casimir Leconte, *Promenade dans l'isthme de Suez* (Paris: N. Chaix, 1864), 95–96.

8. Doctor Aubert-Roche, "Rapport," *L'Isthme de Suez*, 15-19 June 1867, 226.

9. Archives du ministère de l'Europe et des Affaires étrangères, Centre des Archives Diplomatiques á la Courneuve, Paris (hereafter CADC), Affaires économiques et commerciales (hereafter AEC), Correspondance Consulaire et Commerciale Port-Saïd (hereafter CCCPS), vol. 1, Port Said, 26 June 1871, Pellissier, French Consul in Port Said, to Ministry of Foreign Affairs, 158r.

10. Paul Borde, *L'isthme de Suez* (Paris: E. Lachaud, 1870), 32–34; and Raoul Lacour, *L'Égypte d'Alexandrie à la seconde cataracte* (Paris: Hachette, 1871), 458.

11. Couvidou, *Itinéraire du canal de Suez*, 57.

12. Dar al-Watha'iq al-Qawmiyya, National Archives, Cairo (hereafter DWQ), 2001-009687, Port Said, 14 January 1880, Ramacciotti, Cancellieri, Rossi, Boschi, Citty, Lombard, Serris, German, Duc, Stannich, Cendo, to Ibrahim Rouchdy Pacha, Governor General of the Canal and Port Said, Petition.

13. Muhammad Amin Fikri, *Jughrafiyat Misr* (Cairo: Matba'at Wadi al-Nil, 1879), 289.

14. Archivio della Sacra Congregazione per l'Evangelizzazione dei Popoli o de "Propaganda Fide," Rome (hereafter APF), Scritture riferite nei Congressi (hereafter Fondo S.C.), Egitto, Copti, vol. 23, 19 October 1885, Jacobini M. Domenico, Report on meeting in Naples with Cardinal Moran about the plan to erect a new apostolic vicariate in Egypt, 189.

15. 'Ali Mubarak, *Al-Khitat al-tawfiqiyah al-jadidah li-Misr al-Qahirah wa-muduniha wa-biladiha al-qadimah wa-al-shahirah* (Cairo: Matba'at Dar al-Kutub wa-al-Watha'iq al-Qawmiyah, 2004), 1:57. On 'Ali Mubarak and the concern of his age with the Egyptian urban space, see Timothy Mitchell, *Colonising Egypt* (Berkeley: University of California Press, 1988), 63–68.

16. APF, Fondo S.C., Egitto, Copti, vol. 23, Frascati, 17 September 1885, Massaia Cappuccin to Cardinal Prefect, 187r.

17. The National Archives, London (TNA), Foreign Office (FO) 423-1, Alexandria, 26 April 1860, Robert G. Colquhoun to Lord J. Russell, Report by J. Coulthard; and Public Record Office (PRO) 30/22/93/57, 10 April 1865, Sir H. Bulwer to Lord Russell, 336–38.

18. John Steele, Thomas Gray, and Marshall Simpkin, *The Suez Canal: Its Present and Future; a Round-about Paper* (London: Simpkin, Marshall, 1872), 11; and

Thomas Cook (Firm), *Cook's Tourist Handbook for Egypt, the Nile, and the Desert* (London: Thomas Cook, 1876), 229.

19. The British Library, London (TBL), Oriental and India Office Collections (OIOC), Mss/Eur/C739 John Wilberforce Cassels, November or December 1879.

20. CADC, AEC, CCCPS, vol. 1, Port Said, August 1876 (no day), Saint-Chaffray, French Consul in Port Said, to French Ministry of Foreign Affairs, 445.

21. Karl Baedeker, *Egypt, Handbook for Travellers: Part First, Lower Egypt* (Leipzig: Karl Baedeker, 1878).

22. DWQ, 2001-009687, Port Said, 14 January 1880, Petition to Ibrahim Rouchdy Pacha, Governor General.

23. Archives de la Maison-Mère du Bon Pasteur, Angers (AMMBPA), HC 43 Port Said, De notre Monastère de Port-Saïd, 6 January 1881, 16.

24. Archives du ministère de l'Europe et des Affaires étrangères, Centre des Archives Diplomatiques de Nantes (CADN), Archives Rapatriés du Consulat de France à Port-Saïd (ARCFPS), C149, Port Said, 26 December 1887, Delegates of English, Austro-Hungarian, Italian, Hellenic, French and Spanish colonies, to Khedive, Petition of the Port-Saidian Population.

25. John Ninet and Anwar Luqa, *Lettres d'Égypte: 1879–1882* (Paris: Éditions du Centre national de la recherche scientifique, 1979), 128, entry from Zagazig, 11 April 1881; and E. A. Wallis Budge, *By Nile and Tigris, a Narrative of Journeys in Egypt and Mesopotamia on Behalf of the British Museum between the Years 1886 and 1913* (London: J. Murray, 1920), 77, entry dated 1886–1887.

26. Ministero degli Affari Esteri, *Emigrazione e colonie: Raccolta di rapporti dei RR. agenti diplomatici e consolari* (Rome: Tipografia dell'Unione Cooperativa Editrice, 1906), 2:281.

27. Richard Tangye, *Reminiscences of Travel in Australia, America, and Egypt* (London: Sampson Low, Marston, Searle, & Rivington, 1883), 232.

28. "I'lanat Bur Sa'id," *Al-Ahram*, 1 July 1900, 1.

29. Luigi Dori, "Esquisse historique de Port Said: 1900–1914," *Cahiers d'histoire egyptienne* 8 (July 1956): 327; and Lionel Wiener, *L'Egypte et ses chemins de fer* (Brussels: Weissenbruch, 1932), 426–30.

30. Alis, *Promenade en Égypte*, 12.

31. Muhammad Rashid Rida, "Mashru' sikka al-hadid bayna Bur Sa'id wa al-Basra," *Al-Manar* 1, no. 19 (7 Rabi' al-awwal 1316 H [26 July 1898]): 348.

32. "Linea Cairo-Port Said," *L'Imparziale*, 24 November 1900, 3.

33. Michael J. Reimer, "Urban Government and Administration in Egypt, 1805–1914," *Welt des Islams* 39, no. 3 (1999): 312–13; and Relli Schechter and Haim Yacobi, "Rethinking Cities in the Middle East: Political Economy, Planning, and the Lived Space," *Journal of Architecture* 10, no. 5 (2005): 502. See Roger Owen, *The Middle East in the World Economy, 1800–1914* (London: Methuen, 1981), 246; and Charles Philip Issawi, *An Economic History of the Middle East and North Africa* (New York: Columbia University Press, 1982), 5.

34. DWQ, 2001-020454, Cairo, 1 February 1902, President of the Council of Ministers Moustapha Fehmy and President of the Company Prince Auguste d'Arenberg, *Convention relative au chemin de fer a voie normale d'Ismailia a Port Said et au port de Port-Saïd* (Cairo: Imprimerie Nationale, 1902); Archives Nationales du Monde du Travail, Roubaix (ANMT), Compagnie universelle du canal maritime de Suez (CUCMS), 1995 60 090, 4 November 1902, Commission consultative internationale des travaux, Meeting minutes, "Port de Port-Saïd," 12; and Paul Reymond, *Le port de Port-Saïd* (Cairo: Impr. Scribe égyptien, 1950), 109–11, 118.

35. Ministero degli affari esteri, Commissariato dell'emigrazione, "Emigrazione in Egitto," *Bollettino dell'emigrazione*, no. 6 (1902): 65–66; and Ministero degli affari esteri, Commissariato dell'emigrazione, "Egitto," *Bollettino dell'emigrazione*, no. 12 (1902): 79.

36. Panaït Istrati, *Vie d'Adrien Zograffi* (Paris: Gallimard, 1969), 408; and Olivier Hambursin, ed., *Récits du dernier siècle des voyages: De Victor Segalen à Nicolas Bouvier* (Paris: Presses de l'Université Paris-Sorbonne, 2005), 129. See Dori, "Esquisse historique de Port Said: 1900–1914," 327.

37. G. B. Danovaro, *L'Égypte à l'aurore du XXème siècle* (Alexandria: J.C. Lagoudakis, 1901), 126.

38. Khaled Fahmy, "The Essence of Alexandria, Part I," *Manifesta Journal* 14 (January 2012): 68.

39. Kenneth J. Perkins, *Port Sudan: The Evolution of a Colonial City* (Boulder, CO: Westview Press, 1993), 12. See Sylvia Modelski, *Port Said Revisited* (Washington, DC: FAROS, 2000), 57–58; and Valeska Huber, *Channelling Mobilities: Migration and Globalisation in the Suez Canal Region and Beyond, 1869–1914* (Cambridge: Cambridge University Press, 2013), 83, 138.

40. Lucia Carminati, "Port Said and Ismailia as Desert Marvels: Delusion and Frustration on the Isthmus of Suez, 1859–1869," *Journal of Urban History* 46, no. 3 (2019): 622–47; Hélène Braeuner, "À la frontière de l'Égypte: Les représentations du canal de Suez," *In Situ. Revue des patrimoines*, no. 38 (February 12, 2019): 1; and Arnaud de La Grange, Véronique Durruty, and Thomas Goisque, *Nouvelles d'Afrique: à la rencontre de l'Afrique par ses grands ports* (Paris: Gallimard, 2003).

41. Janet Abu-Lughod, "Urbanization in Egypt: Present State and Future Prospects," *Economic Development and Cultural Change* 13, no. 3 (1965): 330. See also Frédéric Bruyas, "Port Said (Egypt), lieu d'articulation du local au mondial. Zone et ville franche: Question d'échelles," *Annales de Géographie* 2000 (2000): 152–71.

42. A. Solletty, "Port-Saïd," *Annales de Géographie* 43, no. 245 (1934): 510.

43. Fu'ad Faraj, *Mintaqat Qanal al-Suways* (Cairo: Matba'at al-Ma'arif wa Maktabuha, 1950), 219; Al-Sayyid Husayn Jalal, *Qanat al-Suways wa-al-tanafus al-isti'mari al-urubbi, 1883–1904* (Cairo: Al-Hay'ah al-Misriyah al-'Ammah lil-Kitab, 1995), 475; and François Zabbal, "Introduction: Dossier Le Canal de Suez, Une Utopie Moderne," *Qantara: Magazine Trimestriel de l'Institut Du Monde Arabe*, no. 106 (January 2018): 27.

44. Charlene L. Porsild, *Gamblers and Dreamers: Women, Men, and Community in the Klondike* (Vancouver: University of British Columbia Press, 1998), 10.

See Marcelo J. Borges and Susana B. Torres, *Company Towns: Labor, Space, and Power Relations across Time and Continents* (New York: Palgrave Macmillan, 2012), 4. On critiques of the use of "cosmopolitanism" as an analytical category in studies of Middle East and Egyptian history, see mainly Henk Driessen, "Mediterranean Port Cities: Cosmopolitanism Reconsidered," *History and Anthropology* 16, no. 1 (2005): 129–41; Will Hanley, "Grieving Cosmopolitanism in Middle East Studies," *History Compass* 6, no. 5 (2008): 1346–67; Ulrike Freitag, "Cosmopolitanism in the Middle East as Part of Global History," *Programmatic Texts*, no. 4 (2010); and Shana Elizabeth Minkin, *Imperial Bodies: Empire and Death in Alexandria, Egypt* (Stanford, CA: Stanford University Press, 2020), 7–8.

45. "Gli stranieri in Egitto," *L'Imparziale*, 9 June 1897, 3.

46. Ministero degli Affari Esteri, *Emigrazione e colonie* (1906), 2:208; 2:233; 2:273–74; and Archivio Storico del Ministero degli Affari Esteri e della Cooperazione Internazionale, Rome (ASMAECI), Archivi Rappresentanze Diplomatiche e Consolari all'Estero (ARDCE), Consolato Cairo (CC), Busta (hereafter B) 12, Suez, 2 January 1868, Enrico Ivaniski to Italian consul in Suez.

47. Mabel Loomis Todd, "The Amherst Eclipse Expedition," *Nation* 72, no. 1874 (May 30, 1901): 432.

48. Gabriel Baer, *Studies in the Social History of Modern Egypt* (Chicago: University of Chicago Press, 1969), 145. Issawi reports that "European migrants and colonists" constituted 28 percent of Port Said's population in 1907. Issawi, *Economic History of the Middle East and North Africa*, 80.

49. Ghislaine Alleaume and Éric Denis, "L'Égypte à l'aube du XX siècle: Pays, bourgs, cités en des temporalités divergentes," in *Urbanité arabe: Hommage à Bernard Lepetit*, ed. Jocelyne Dakhlia (Arles: Sindbad/Actes Sud, 1998), 226.

50. Dori, "Esquisse historique de Port Said: 1900–1914," 330–31; and Perkins, *Port Sudan*, 15–16. In 1930 the appointment of a Japanese man was considered null, as bylaws only mentioned "Europeans." After the Second World War, European member positions were abolished in favor of an exclusively Egyptian representation. See Hilmi Ahmad Shalabi, *Al-Hukm al-mahalli wa-al-majalis al-baladiyah fi Misr mundhu nash'atiha hatta 'am 1918* (Cairo: 'Alam al-Kutub, 1987), 169–88.

51. Robert L. Tignor, *Modernization and British Colonial Rule in Egypt, 1882–1914* (Princeton, NJ: Princeton University Press, 1966), 354; and Derek Hopwood, *Tales of Empire: The British in the Middle East, 1880–1952* (London: I. B. Tauris, 1989), 16–17.

52. "Mushkilat al-batala," *Al-Mu'addab*, 19 March 1926, 1.

53. Jean-Marie G. Le Clézio, "La Port-Saïdienne," in *Nouvelles d'Afrique: À la rencontre de l'Afrique par ses grands ports*, ed. Arnaud de La Grange, Véronique Durruty, and Thomas Goisque (Paris: Gallimard, 2003), 13. For further discussion of Port Said's urban segregation, see Claudine Piaton, "Port Said: Cosmopolitan Urban Rules and Architecture (1858–1930)," in *Revitalizing City Districts: Transformation Partnership for Urban Design and Architecture in Historic City Districts*, ed. Hebatalla Abouelfadl, Dalila ElKerdany, and Christoph Wessling (Cham: Springer,

2017); and Lucia Carminati, "Dividing and Ruling a Mediterranean Port-City: The Many Boundaries within Late Nineteenth Century Port Said," in *Multi-Ethnic Cities in the Mediterranean World: Controversial Heritage and Divided Memories, from the Nineteenth Through the Twentieth Centuries*, ed. Marco Folin and Heleni Porfyriou (London: Routledge, 2021), 2:30–44.

54. Zayn al-'Abidin Shams al-Din Najm, *Bur Sa'id: Tarikhuha wa-tatawwuruha, mundhu nash'atiha 1859 hatta 'am 1882* (Cairo: Al-Hay'ah al-Misriyah al-'Ammah lil-Kitab, 1987), 421–23.

55. Leslie Page Moch, *Moving Europeans: Migration in Western Europe since 1650* (Bloomington: Indiana University Press, 1992), 12.

56. Sarga Moussa and Kaja Antonowicz, *Le voyage en Égypte: Anthologie de voyageurs européens; de Bonaparte à l'occupation anglaise* (Paris: Robert Laffont, 2004), 225; and Hubert Bonin, *History of the Suez Canal Company, 1858–1960: Between Controversy and Utility* (Geneva: Droz, 2010), 14.

57. Elena Chiti, "Quelles Marges Pour Quels Centres? Perceptions Arabes et Européennes d'Alexandrie Après 1882," in *Étudier en liberté les mondes méditerranéens: Mélanges offerts à Robert Ilbert*, ed. Leyla Dakhli and Vincent Lemire (Paris: Publications de la Sorbonne, 2016), 498.

58. C. A. Bayly and Leila Tarazi Fawaz, "Introduction: The Connected World of Empires," in *Modernity and Culture: From the Mediterranean to the Indian Ocean*, ed. C. A. Bayly and Leila Tarazi Fawaz (New York: Columbia University Press, 2002), 9; and Julia A. Clancy-Smith, "Locating Women as Migrants in Nineteenth-Century Tunis," in *Contesting Archives: Finding Women in the Sources*, ed. Nupur Chaudhuri, Sherry J. Katz, and Mary Elizabeth Perry (Urbana: University of Illinois Press, 2010), 37.

59. Gavin D. Brockett, "Middle East History Is Social History," *International Journal of Middle East Studies* 46, no. 02 (2014): 383. I am also inspired by Alan Barenberg, *Gulag Town, Company Town: Forced Labor and Its Legacy in Vorkuta* (New Haven, CT: Yale University Press, 2014), 5–6.

60. Valeska Huber, "Multiple Mobilities, Multiple Sovereignties, Multiple Speeds: Exploring Maritime Connections in the Age of Empire," *International Journal of Middle East Studies* 48, no. 4 (2016): 764.

61. Nancy Y. Reynolds, "City of the High Dam: Aswan and the Promise of Postcolonialism in Egypt," *City & Society* 29, no. 1 (2017): 216.

62. Brad Beaven, Karl Bell, and Robert James, introduction to *Port Towns and Urban Cultures: International Histories of the Waterfront, c. 1700–2000* (London: Palgrave Macmillan, 2016), 5.

POSTSCRIPT

1. "Speech of President Gamal Abd Al-Nasser in the fourth anniversary of the Revolution from Alexandria," 26 July 1956, Bibalex, http://nasser.bibalex.org/Data/GR09_1/Speeches/1956/560726.htm, accessed 2 May 2022.

2. Yoav Di-Capua, "Revolutionary Decolonization and the Formation of the Sacred: The Case of Egypt," *Past & Present* 256, no. 1 (2021): 239–81.

3. Alain Roussillon, "Republican Egypt Interpreted: Revolution and Beyond," in *The Cambridge History of Egypt*, vol. 2, *Modern Egypt, from 1517 to the End of the Twentieth Century*, ed. M. W. Daly (Cambridge: Cambridge University Press, 1998), 339.

4. Mostafa Mohie, *Biographies of Port Said*, Cairo Papers in Social Science (Cairo: American University in Cairo Press, 2021), 36:1:13.

5. Hubert Bonin, *History of the Suez Canal Company, 1858–1960: Between Controversy and Utility* (Geneva: Droz, 2010), 463.

6. William Roger Louis, "The Economic Consequences of Suez for Egypt," in *Suez 1956: The Crisis and Its Consequences*, ed. Roger Owen (Oxford: Clarendon Press; New York: Oxford University Press, 1989), 363–75.

7. Caroline Piquet, *Histoire du canal de Suez* (Paris: Perrin, 2009), 326; and Alia Mossallam, "Hikāyāt Shaʿb—Stories of Peoplehood: Nasserism, Popular Politics and Songs in Egypt, 1956–1973" (PhD diss., London School of Economics and Political Science, 2012).

8. Subhi Al-Sharuni, *Abdel Hadi Al-Gazzar* (Cairo: Elias Modern Publishing House, 2007); and Jessica Winegar, *Creative Reckonings: The Politics of Art and Culture in Contemporary Egypt* (Cairo: American University in Cairo Press, 2008), 269–70.

9. Emad Abou Ghazi and Xavier Daumalin, "Le Canal vu par les Égyptiens," in *L'épopée du Canal de Suez*, ed. Gilles Gauthier (Paris: Gallimard/Institut du monde arabe/Musée d'Histoire de Marseille, 2018), 46–51; and Salma Mobarak and Walid El Khachab, "Le Cinéma de Suez," in *L'épopée du Canal de Suez*, ed. Gilles Gauthier, 146–51.

10. Mohamed Abdel Shakur, Sohair Mehanna, and Nicholas S. Hopkins, "War and Forced Migration in Egypt: The Experience of Evacuation from the Suez Canal Cities (1967–1976)," *Arab Studies Quarterly* 27, no. 3 (2005): 21–22.

11. Mohie, *Biographies of Port Said*, 36:1:22–25, 32, 43.

12. Mohie, 36:1:2, 6, 55, 60.

13. Salma Mobarak, "L'imaginaire cinématographique de Port-Saïd," *Sociétés & Représentations* 2, no. 48 (2019): 98.

14. Frédérique Bruyas, "Aménagement de la ville de Port-Saïd, le point de vue de l'architecte," *Égypte/Monde arabe*, no. 23 (September 30, 1995): 131–68.

15. Ronnie Close, *Cairo's Ultras: Resistance and Revolution in Egypt's Football Culture* (Cairo: American University in Cairo Press, 2019), 7–8, 25–69; and Mohie, *Biographies of Port Said*, 36:1:81–83.

16. Mobarak, "L'imaginaire cinématographique de Port-Saïd," 107.

17. @PSademo Community Organization, Facebook page.

18. Mohie, *Biographies of Port Said*, 36:1:88–90.

19. Claudine Piaton, "Port Said: Decaying Wooden Verandas Tell the Story of a City," *Rawi Magazine*, no. 3 (2011), https://rawi-publishing.com/articles/portsaidverandas/.

20. Hiba Zayadin, "Migrant Workers and the Qatar World Cup," *Jadaliyya*, August 1, 2021, https://www.jadaliyya.com/Details/43169. See Touraj Atabaki, "Far from Home, But at Home: Indian Migrant Workers in the Iranian Oil Industry," *Studies in History* 31, no. 1 (February 1, 2015): 85–114; Andrea Grace Wright, "Migratory Pipelines: Labor and Oil in the Arabian Sea" (PhD diss., University of Michigan, 2015); Pardis Mahdavi, *Crossing the Gulf: Love and Family in Migrant Lives* (Stanford, CA: Stanford University Press, 2020); and Laleh Khalili, *Sinews of War and Trade* (London: Verso, 2020).

BIBLIOGRAPHY

ARCHIVES

Britain

The British Library, London (TBL)
    Oriental and India Office Collections (OIOC)
        Mss/Eur: Western manuscripts
The National Archives, London (TNA)
    Foreign Office (FO)
        FO 78: Political and Other Departments: General Correspondence before 1906, Ottoman Empire
        FO 423: Suez Canal International Commission Correspondence
        FO 846: Consulate, Port Said, Egypt: Consular Court Records
    Public Record Office (PRO)
        PRO 30/22/93: Embassy in Constantinople, Private correspondence
The Women's Library, London (TWL) (former Fawcett Library)
    International Bureau for Suppression of Traffic in Persons (IBS)
        4IBS/6 International Bureau Country Files

Egypt

Dar al-Watha'iq al-Qawmiyya, National Archives, Cairo (DWQ)
    0069: Watha'iq 'Abdin, Documents of 'Abdin
    2001: Diwan al-Dakhiliyya, Ministry of the Interior
    4003: Diwan al-Ashghal al 'Umumiyya, Ministry of Public Works
    5009: Qanat al-Suways, Suez Canal
    5013: 'Usra Muhammad 'Ali, family of Muhammad 'Ali

France

Archives de la Maison-Mère du Bon Pasteur, Angers (AMMBPA)
    Egypt—Port Said
    HC 34 Port Said
    HC 43 Port Said
Archives du ministère de l'Europe et des Affaires étrangères, Centre des Archives Diplomatiques á la Courneuve, Paris (CADC)
    Affaires économiques et commerciales (AEC)
        Correspondance Consulaire et Commerciale Port-Saïd (CCCPS)
    Affaires Politiques (AP)
        Correspondance Politique et Commerciale dite "Nouvelle série," Egypte (CPC)
    Entrées Exceptionnelles et Collections (EEC)
Archives du ministère de l'Europe et des Affaires étrangères, Centre des Archives Diplomatiques de Nantes (CADN)
    Archives Rapatriés de l'Agence Consulaire de Ismailia (ARI)
    Archives Rapatriés du Consulat de France à Port-Saïd (ARCFPS)
Archives Nationales du Monde du Travail, Roubaix (ANMT)
    1995 060: Compagnie universelle du canal maritime de Suez (CUCMS), puis Compagnie financière de Suez
    2000 059: Documents regarding the CUCMS purchased by the board of French archives for the ANMT
Médiathèque de la Communauté Urbaine d'Alençon
    Fonds Adhémard Leclère
Association du Souvenir de Ferdinand de Lesseps et du Canal de Suez (ASFLCS)

Italy

Alinari Archives, Florence
    Favrod Collection
Archivio Centrale dello Stato, Rome (ACS)
    Ministero dell'Interno
        Casellario Politico Centrale (CPC)
Archivio della Sacra Congregazione per l'Evangelizzazione dei Popoli o de "Propaganda Fide," Rome (APF)
    Scritture riferite nei Congressi (Fondo S.C.)
        Egitto, Copti
            Volumes 20–23: 1862–1892
Archivio Storico del Ministero degli Affari Esteri e della Cooperazione Internazionale, Rome (ASMAECI)
    Archivi Rappresentanze Diplomatiche e Consolari all'Estero (ARDCE)
        Consolato Cairo (CC)
            Antico versamento (AV)
            Rappresentanza Cairo (RC)

Personale
    Serie II: Consolati
    Polizia Internazionale (PI)
Istituto per l'Oriente Carlo Nallino, Rome (IPOCAN)
    Bundle "1956 al 1961" [possibly misnamed]
    Bundle "1866–1868"
Museo Centrale del Risorgimento di Roma, Rome (MCRR)
    Carte Mancini, 1884
Società Dante Alighieri Historical Archives, Rome (SDA)
    Port Said, 1924–1926

United Arab Emirates

Akkasah Photography Archive, al Mawrid, New York University Abu Dhabi (NYUAD).
Hisham Khatib Collection

LIBRARIES AND RESEARCH CENTERS

American University in Cairo
Biblioteca civica di Trieste
Biblioteca della Società Geografica Italiana, Rome
Bibliotheca Alexandrina, Alexandria
Bibliothèque nationale de France, Paris
Centre d'études alexandrines, Alexandria, Centre national de la recherche scientifique (CeAlex/CNRS)
Centre d'études et de documentation économiques, juridiques et sociales, Cairo
Dar al-Kutub al-Qawmiyya, National Library, Cairo
Emeroteca dell'Istituto Archeologico Italiano, Cairo
The Franciscan Centre of Christian Oriental Studies, Cairo
Institut français d'archéologie orientale (IFAO), Cairo

PERIODICALS

Al-Ahram
Al-Manar
Al-Mu'addab
Al-Mu'ayyad
Al-Waqa'i' al-misriyya/Journal officiel du gouvernement egyptien
The Egyptian Gazette
Il Telegrafo
La Riforma: Giornale dell'antica sinistra parlamentare

La Riforma Sociale
Le Phare de Port-Said et du Canal de Suez
Le Bosphore égyptien
Le Boulevard
L'Écho de Port Said: Journal politique, commercial et littéraire
L'Eco d'Italia
L'Imparziale
L'Isthme de Suez: Journal de l'union des deux mers
Majallah al-qada' wa al-bulis al-misri
The Spectator
The Times

SELECTED BIBLIOGRAPHY

Due to space constraints, this bibliography only includes those secondary works that bear a relatively strong connection to this book. References to all other secondary and primary sources can be found in the footnotes.

Abou Ghazi, Emad, and Xavier Daumalin. "Le Canal vu par les Égyptiens." In *L'épopée du Canal de Suez*, edited by Gilles Gauthier, 46–51. Paris: Gallimard/Institut du monde arabe/Musée d'Histoire de Marseille, 2018.
Barak, On. *Powering Empire: How Coal Made the Middle East and Sparked Global Carbonization*. Oakland: University of California Press, 2020.
Baron, Beth. *The Orphan Scandal: Christian Missionaries and the Rise of the Muslim Brotherhood*. Stanford, CA: Stanford University Press, 2014.
Bartolotti, Fabien. "Le chantier du siècle: Les entreprises marseillaises et le creusement du canal de Suez (1859–1869)." *Revue Marseille*, no. 260 (2018): 46–48.
Beaven, Brad, Karl Bell, and Robert James. Introduction to *Port Towns and Urban Cultures: International Histories of the Waterfront, c. 1700–2000*, edited by Brad Beaven, Karl Bell, and Robert James, 1–10. London: Palgrave Macmillan, 2016.
Beinin, Joel, and Zachary Lockman. *Workers on the Nile: Nationalism, Communism, Islam, and the Egyptian Working Class, 1882–1954*. Princeton, NJ: Princeton University Press, 1987.
Biancani, Francesca. "Globalisation, Migration, and Female Labour in Cosmopolitan Egypt." In *From Slovenia to Egypt: Aleksandrinke's Trans-Mediterranean Domestic Workers' Migration and National Imagination*, edited by Mirjam Milharčič Hladnik, 207–28. Göttingen: VetR Unipress, 2015.
Bilici, Faruk. *Le canal de Suez et l'Empire ottoman*. Paris: CNRS Editions, 2019.
Bonin, Hubert. *History of the Suez Canal Company, 1858–1960: Between Controversy and Utility*. Geneva: Droz, 2010.
Braeuner, Hélène. "À la frontière de l'Égypte: Les représentations du canal de Suez." *In Situ: Revue des patrimoines*, no. 38 (February 12, 2019): 1–14.

Brown, Nathan J. "Who Abolished Corvee Labour in Egypt and Why?" *Past & Present*, no. 144 (1994): 116–37.

Carminati, Lucia. "Dividing and Ruling a Mediterranean Port-City: The Many Boundaries within Late Nineteenth Century Port Said." In *Multi-Ethnic Cities in the Mediterranean World: Controversial Heritage and Divided Memories, from the Nineteenth through the Twentieth Centuries*, edited by Marco Folin and Heleni Porfyriou, 30–44. London: Routledge, 2021.

———. "'Improvising and Very Humble': Those 'Italians' throughout Egypt That Statisticians and Historians Have Neglected." In *On the Margins of History: Italian Subalterns between Emigration and Colonialism in the Italian Colony in Egypt (1861–1937)*, edited by Costantino Paonessa, 31–52. Louvain: Université catholique de Louvain presse, 2021.

———. "Of Machines and Men: The Uneasy Synergy of Mechanization and Migrant Labor on the Suez Canal Worksites, 1859–1864." In *Oxford Handbook of Modern Egypt*, edited by Beth Baron and Jeffrey Culang. Oxford: Oxford University Press, forthcoming.

———. "Port Said and Ismailia as Desert Marvels: Delusion and Frustration on the Isthmus of Suez, 1859–1869." *Journal of Urban History* 46, no. 3 (2019): 622–47.

———. "'She Will Eat Your Shirt': Foreign Migrant Women as Brothel Keepers in Port Said and along the Suez Canal, 1880–1914." *Journal of the History of Sexuality* 30, no. 2 (May 2021): 161–94.

———. "Suez: A Hollow Canal in Need of Peopling: Currents and Stoppages in the Historiography, 1859–1956." *History Compass* 19, no. 5 (2021): 1–14.

Carminati, Lucia, and Mohamed Gamal-Eldin. "Decentering Egyptian Historiography: Provincializing Geographies, Methodologies, and Sources." *International Journal of Middle East Studies* 53, no. 1 (February 2021): 107–11.

Certeau, Michel de. *The Practice of Everyday Life*. Berkeley: University of California Press, 1984.

Chalcraft, John T. "The Coal Heavers of Port Sa'id: State-Making and Worker Protest, 1869–1914." *International Labor and Working-Class History* 60 (October 2001): 110–24.

Clancy-Smith, Julia A. "Locating Women as Migrants in Nineteenth-Century Tunis." In *Contesting Archives: Finding Women in the Sources*, edited by Nupur Chaudhuri, Sherry J. Katz, and Mary Elizabeth Perry, 35–55. Urbana: University of Illinois Press, 2010.

———. *Mediterraneans: North Africa and Europe in an Age of Migration, c. 1800–1900*. Berkeley: University of California Press, 2011.

Clay, Christopher. "Labour Migration and Economic Conditions in Nineteenth-Century Anatolia." *Middle Eastern Studies* 34, no. 4 (1998): 1.

Corti, Paola. "L'emigrazione temporanea in Europa, in Africa e nel Levante." In *Storia dell'emigrazione italiana: Partenze*, edited by Piero Bevilacqua, Andreina De Clementi, and Emilio Franzina, 213–36. Rome: Donzelli, 2009.

Cresswell, Tim. *On the Move: Mobility in the Modern Western World*. New York: Routledge, 2006.

———. "Towards a Politics of Mobility." *Environment and Planning* 28, no. 1 (February 1, 2010): 17–31.
Crosnier-Lecomte, Marie-Laure, Gamal Ghitani, and Naguib Amin. *Port-Saïd. Architectures XIXe–XXe siècles*. Cairo: IFAO, 2006.
Curli, Barbara. "Dames Employés at the Suez Canal Company: The 'Egyptianization' of Female Office Workers, 1941–56." *International Journal of Middle East Studies* 46, no. 3 (2014): 553–76.
Dakhlia, Jocelyne. *Lingua franca*. Arles: Actes Sud, 2008.
Dalachanis, Angelos. *The Greek Exodus from Egypt: Diaspora Politics and Emigration, 1937–1962*. New York: Berghahn Books, 2017.
De Vito, Christian G. "History without Scale: The Micro-Spatial Perspective." *Past & Present* 242 (2019): 348–72.
Di-Capua, Yoav. "Revolutionary Decolonization and the Formation of the Sacred: The Case of Egypt." *Past & Present* 256, no. 1 (2021): 239–81.
Dori, Luigi. "Esquisse historique de Port Said: Naissance de la ville et inauguration du Canal (1859–1869)." *Cahiers d'histoire égyptienne* 6, nos. 3/4 (October 1954): 180–219.
———. "Esquisse historique de Port Said: 1900–1914." *Cahiers d'histoire égyptienne* 8 (July 1956): 311–42.
———. "Esquisse historique de Port Said: Port Said après l'inauguration du Canal (1869–1900)." *Cahiers d'histoire égyptienne* 8, no. 1 (January 1956): 1–46.
Dunne, Bruce W. "Sexuality and the 'Civilizing Process' in Modern Egypt." PhD diss., University of Michigan, 1996.
El Shakry, Omnia. *The Great Social Laboratory: Subjects of Knowledge in Colonial and Postcolonial Egypt*. Stanford, CA: Stanford University Press, 2007.
Fahmy, Khaled. "The Police and the People in Nineteenth-Century Egypt." *Die Welt Des Islams* 39, no. 3 (1999): 340–77.
Fahmy, Ziad. "Jurisdictional Borderlands: Extraterritoriality and 'Legal Chameleons' in Precolonial Alexandria, 1840–1870." *Comparative Studies in Society and History* 55, no. 2 (2013): 305–29.
Farag, Fu'ad. *Mintaqat qanal as-Suways wa-mudun al-qanal Bur Sa'id, as-Suways, al-Isma'iliya wa-siwaha*. Cairo: Matba'at al-Ma'arif wa-Maktabuha, 1950.
Farnie, D. A. *East and West of Suez: The Suez Canal in History, 1854–1956*. Oxford: Clarendon Press, 1969.
Foda, Omar D. "The Pyramid and the Crown: The Egyptian Beer Industry from 1897 to 1963." *International Journal of Middle East Studies* 46, no. 1 (2014): 139–58.
Foucault, Michel. *Discipline and Punish: The Birth of the Prison*. New York: Pantheon Books, 1977.
———. "Space, Knowledge, and Power." In *The Foucault Reader*, edited by Paul Rabinow, 239–56. New York: Pantheon Books, 1984.
Freitag, Ulrike, Malte Fuhrmann, Nora Lafi, and Florian Riedler. *The City in the Ottoman Empire: Migration and the Making of Urban Modernity*. London: Routledge, 2011.

Frémaux, Céline. "Town Planning, Architecture, and Migrations in Suez Canal Port Cities." In *Port Cities: Dynamic Landscapes and Global Networks*, edited by Carola Hein, 156–73. Abingdon, Oxon: Routledge, 2011.

Frémaux, Céline, and Mercedes Volait. "Inventing Space in the Age of Empire: Planning Experiments and Achievements along Suez Canal in Egypt (1859–1956)." *Planning Perspectives* 24, no. 2 (2009): 255–62.

Fuhrmann, Malte. *Port Cities of the Eastern Mediterranean: Urban Culture in the Late Ottoman Empire*. Cambridge: Cambridge University Press, 2020.

Fuhrmann, Malte, and Vangelis Kechriotis. "The Late Ottoman Port-Cities and Their Inhabitants: Subjectivity, Urbanity, and Conflicting Orders." *Mediterranean Historical Review* 24, no. 2 (2009): 71–78.

Gabaccia, Donna R., and Dirk Hoerder. *Connecting Seas and Connected Ocean Rims: Indian, Atlantic, and Pacific Oceans and China Seas Migrations from the 1830s to the 1930s*. Leiden: Brill, 2011.

Gallant, Thomas. "Tales from the Dark Side: Transnational Migration, the Underworld and the 'Other' Greeks of the Diaspora." In *Greek Diaspora and Migration since 1700: Society, History, and Politics*, edited by Dimitris Tziovas, 17–30. Farnham, UK: Ashgate, 2009.

Gamal-Eldin, Mohamed. "Cesspools, Mosquitoes and Fever: An Environmental History of Malaria Prevention in Ismailia and Port Said, 1869–1910." In *Seeds of Power: Explorations in Ottoman Environmental History*, edited by Onur İnal and Yavuz Köse, 184–207. Winwick, Cambridgeshire: White Horse Press, 2019.

Gekas, Athanasios. "Class and Cosmopolitanism: The Historiographical Fortunes of Merchants in Eastern Mediterranean Ports." *Mediterranean Historical Review* 24, no. 2 (2009): 95–114.

Ghobrial, John-Paul A. "Introduction: Seeing the World like a Microhistorian." *Past & Present* 242 (2019): 1–22.

Ginzburg, Carlo. *Clues, Myths, and the Historical Method*. Baltimore, MD: Johns Hopkins University Press, 1989.

Glick Schiller, Nina, and Noel B. Salazar. "Regimes of Mobility across the Globe." *Journal of Ethnic and Migration Studies* 39, no. 2 (2013): 183–200.

Gobbi, Olimpia. "Emigrazione femminile: Balie e domestiche marchigiane in Egitto fra Otto e Novecento." *Proposte e ricerche: Economia e società nella storia dell'Italia centrale* 34, no. 66 (2011): 7–24.

Gorman, Anthony. "Foreign Workers in Egypt 1882–1914: Subaltern or Labour Elite?" In *Subalterns and Social Protest: History from below in the Middle East and North Africa*, edited by Stephanie Cronin. London: Routledge, 2008.

Gualtieri, Sarah. "Gendering the Chain Migration Thesis: Women and Syrian Transatlantic Migration, 1878–1924." *Comparative Studies of South Asia, Africa and the Middle East* 24, no. 1 (2004): 67–78.

Gutman, David. "Travel Documents, Mobility Control, and the Ottoman State in an Age of Global Migration, 1880–1915." *Journal of the Ottoman and Turkish Studies Association* 3, no. 2 (2016): 347–68.

Haddad, Emily A. "Digging to India: Modernity, Imperialism, and the Suez Canal." *Victorian Studies* 47, no. 3 (2005): 363–96.

Hahn, Sylvia. "Nowhere at Home? Female Migrants in the Nineteenth-Century Habsburg Empire." In *Women, Gender, and Labour Migration: Historical and Global Perspectives*, edited by Pamela Sharpe, 108–27. London: Routledge, 2011.

Hanley, Will. *Identifying with Nationality: Europeans, Ottomans, and Egyptians in Alexandria*. New York: Columbia University Press, 2017.

Hannam, Kevin, Mimi Sheller, and John Urry. "Editorial: Mobilities, Immobilities and Moorings." *Mobilities* 1, no. 1 (2006): 1–22.

Headrick, Daniel R. *The Tentacles of Progress: Technology Transfer in the Age of Imperialism, 1850–1940*. New York: Oxford University Press, 1988.

Herzog, Cristoph. "Migration and the State: On Ottoman Regulations Concerning Migration since the Age of Mahmud II." In *The City in the Ottoman Empire: Migration and the Making of Urban Modernity*, edited by Ulrike Freitag, Malte Fuhrmann, Nora Lafi, and Florian Riedler, 117–34. London: Routledge, 2011.

Hifnawi, Mustafa. *Qanat al-Suways wa-mushkilatiha al-muʻasirah*. 4 vols. Cairo: Al-Hayʾah al-Misriyah al-ʻAmmah lil-Kitab, 2015. First published 1952.

Hilal, ʻImad Ahmad. *Al-Baghaya fi Misr: Dirasah tarikhiyah ijtimaʻiyah, 1834–1949*. Cairo: Al-ʻArabi lil-Nashr wa-al-Tawziʻ, 2001.

Hourani, Albert. "The Syrians in Egypt in the Eighteenth and Nineteenth Centuries." In *Colloque International Sur l'Histoire Du Caire*, vol. 229. Cairo: Ministry of the Arab Republic of Egypt, 1972.

Huber, Valeska. *Channelling Mobilities: Migration and Globalisation in the Suez Canal Region and Beyond, 1869–1914*. Cambridge: Cambridge University Press, 2013.

———. "Multiple Mobilities, Multiple Sovereignties, Multiple Speeds: Exploring Maritime Connections in the Age of Empire." *International Journal of Middle East Studies* 48, no. 4 (2016): 763–66.

Ilbert, Robert. "Qui est Grec? La nationalité comme enjeu en Égypte (1830–1930)." *Relations internationales*, no. 54 (1988): 139–60.

İleri, Nurçın. "Between the Real and the Imaginary: Late Ottoman Istanbul as a Crime Scene." *Journal of the Ottoman and Turkish Studies Association* 4, no. 1 (2017): 95.

———. "Rule, Misconduct, and Dysfunction: The Police Forces in Theory and Practice in Fin-de-Siècle Istanbul." *Comparative Studies of South Asia, Africa and the Middle East* 34, no. 1 (2014): 147–59.

İnalcık, Halil, and Donald Quataert. *An Economic and Social History of the Ottoman Empire, 1300–1914*. Cambridge: Cambridge University Press, 1996.

Isabella, Maurizio, and Konstantina Zanou. *Mediterranean Diasporas: Politics and Ideas in the Long Nineteenth Century*. London: Bloomsbury, 2016.

Ismaʻil Abu Zayd, Rajiyah. *Tarikh madinat al-Ismaʻiliyah: Min al-nashʾah ila muntasaf al-qarn al-ʻishrin*. Cairo: Maktabat al-Adab, 2012.

Issawi, Charles Philip. *An Economic History of the Middle East and North Africa*. New York: Columbia University Press, 1982.

Jakes, Aaron. *Egypt's Occupation: Colonial Economism and the Crises of Capitalism.* Stanford, CA: Stanford University Press, 2020.

Jalal, al-Sayyid Husayn. *Qanat al-Suways wa-al-tanafus al-istiʿmari al-urubbi, 1883–1904.* Cairo: Al-Hayʾah al-Misriyah al-ʿAmmah lil-Kitab, 1995.

Kale, Başak. "Transforming an Empire: The Ottoman Empire's Immigration and Settlement Policies in the Nineteenth and Early Twentieth Centuries." *Middle Eastern Studies* 50, no. 2 (2014): 252–71.

Karabell, Zachary. *Parting the Desert: The Creation of the Suez Canal.* 1st ed. New York: A. A. Knopf, 2003.

Keyder, Çağlar, Eyüp Özveren, and Donald Quataert. "Port-Cities in the Ottoman Empire: Some Theoretical and Historical Perspectives." *Review (Fernand Braudel Center)* 16, no. 4 (1993): 519–58.

Khalili, Laleh. *Sinews of War and Trade.* London: Verso, 2020.

Khuri-Makdisi, Ilham. *The Eastern Mediterranean and the Making of Global Radicalism, 1860–1914.* Berkeley: University of California Press, 2010.

Kitroeff, Alexander. *The Greeks in Egypt 1919–1937: Ethnicity and Class.* London: Ithaca for the Middle East Centre, St. Antony's College, Oxford, 1989.

Kozma, Liat. *Global Women, Colonial Ports: Prostitution in the Interwar Middle East.* Albany: State University of New York Press, 2017.

———. "Women's Migration for Prostitution in the Interwar Middle East and North Africa." *Journal of Women's History* 28, no. 3 (2016): 93–113.

Landau, Jacob M. *Jews in Nineteenth-Century Egypt.* New York: New York University Press, 1969.

Landes, David S. *Bankers and Pashas. International Finance and Economic Imperialism in Egypt.* London: Heinemann, 1958.

Lefebvre, Henri. *The Production of Space.* Oxford: Blackwell, 1991.

———. *Writings on Cities.* Cambridge, UK: Blackwell Publishers, 1996.

Levi, Giovanni. "Frail Frontiers?" *Past & Present* 242 (2019): 37–49.

Lucassen, Leo, and Lex Heerma van Voss. "Introduction: Flight as Fight." In *A Global History of Runaways Workers, Mobility, and Capitalism, 1600–1850,* edited by Marcus Rediker, Titas Chakraborty, and Matthias Van Rossum. Oakland: University of California Press, 2019.

Massey, Doreen B. "A Global Sense of Place." In *Reading Human Geography: The Poetics and Politics of Inquiry,* edited by Trevor J. Barnes and Derek Gregory, 315–23. London: Arnold; New York: Wiley, 1997.

McKeown, Adam. "Global Migration, 1846–1940." *Journal of World History* 15, no. 2 (2004): 155–89.

Mentzel, Peter. "The 'Ethnic Division of Labor' on Ottoman Railroads." *Turcica* 37 (2005): 221–41.

Michel, Nicolas. "La Compagnie du canal de Suez et l'eau du Nil (1854–1896)." In *L'isthme et l'Egypte au temps de la compagnie universelle du canal maritime de Suez (1858–1956),* edited by Claudine Piaton, 273–301. Paris: IFAO du Caire, 2016.

Mikhail, Alan. "Unleashing the Beast: Animals, Energy, and the Economy of Labor in Ottoman Egypt." *American Historical Review* 118, no. 2 (2013): 317–48.

Milharčič Hladnik, Mirjam. "Trans-Mediterranean Women Domestic Workers: Historical and Contemporary Perspectives." In *From Slovenia to Egypt: Aleksandrinke's Trans-Mediterranean Domestic Workers' Migration and National Imagination*, 11–38. Göttingen: VetR Unipress, 2015.

Miran, Jonathan. *Red Sea Citizens: Cosmopolitan Society and Cultural Change in Massawa*. Bloomington: Indiana University Press, 2009.

Mitchell, Timothy. *Colonising Egypt*. Berkeley: University of California Press, 1988.

Mobarak, Salma. "L'imaginaire cinématographique de Port-Saïd." *Sociétés & Représentations* 2, no. 48 (2019): 95–108.

Mobarak, Salma, and Walid El Khachab. "Le Cinéma de Suez." In *L'épopée Du Canal Du Suez*, edited by Gilles Gauthier, 146–51. Paris: Gallimard/Institut du monde arabe/Musée d'Histoire de Marseille, 2018.

Moch, Leslie P. "Domestic Service, Migration, and Ethnic Stereotyping. Bécassine and the Bretons in Paris." *Journal of Migration History* 1, no. 1 (2015): 32–53.

Modelski, Sylvia. *Port Said Revisited*. Washington, DC: FAROS, 2000.

Moghira, Mohamed Anouar. *L'Isthme de Suez: Passage millénaire, 640–2000*. Paris: L'Harmattan, 2002.

Mohie, Mostafa. *Biographies of Port Said*. Cairo Papers in Social Science, vol. 36, no. 1. Cairo: American University in Cairo Press, 2021.

Montel, Nathalie. *Le chantier du Canal de Suez, 1859–1869: Une histoire des pratiques techniques*. Paris: Éd. In forma/Presses de l'École Nationale des Ponts et Chaussées, 1999.

Mossallam, Alia. "Hikāyāt Shaʿb—Stories of Peoplehood: Nasserism, Popular Politics and Songs in Egypt, 1956–1973." PhD diss., London School of Economics and Political Science, 2012.

Moussa, Sarga. "Un Canal Pour Rire. Le Chantier de Suez vu Par Le Charivari." *Sociétés & Représentations* 48, no. 2 (2019): 51–66.

Moya, Jose. "Domestic Service in a Global Perspective: Gender, Migration, and Ethnic Niches." *Journal of Ethnic and Migration Studies* 33, no. 4 (2007): 559–79.

Najm, Zayn al-ʿAbidin Shams al-Din. *Bur Saʿid: Tarikhuha wa-tatawwuruha, mundhu nashʾatiha 1859 hatta ʿam 1882*. Cairo: Al-Hayʾah al-Misriyah al-ʿAmmah lil-Kitab, 1987.

Namnam, Hilmi, and Muhammad Musa Nifin, *Qanat al-Suways, wathaʾiq al-hulm al-misri*. Cairo: Dar al-Kutub wa-al-Wathaʾiq al-Qawmiyah, 2015.

Owen, Roger. *The Middle East in the World Economy, 1800–1914*. London: Methuen, 1981.

Perkins, Kenneth J. *Port Sudan: The Evolution of a Colonial City*. Boulder, CO: Westview Press, 1993.

Petricioli, Marta. *Oltre il mito: L'Egitto degli italiani (1917–1947)*. Milan: Mondadori, 2007.

Piaton, Claudine. "Port Said: Cosmopolitan Urban Rules and Architecture (1858–1930)." In *Revitalizing City Districts: Transformation Partnership for Urban Design and Architecture in Historic City Districts*, edited by Hebatalla Abouelfadl et al. Cham, Switzerland: Springer, 2017.

Piquet, Caroline. *Le canal de Suez, une voie maritime pour l'Egypte et le monde*. Paris: Erick Bonnier Editions, 2018.

———. "The Suez Company's Concession in Egypt, 1854–1956: Modern Infrastructure and Local Economic Development." *Enterprise and Society* 5, no. 1 (2004): 107–27.

Pudney, John. *Suez. De Lesseps' Canal*. New York: Praeger, 1969.

Ramadan, Mahmud Salih. *Al-Haya al-ijtima'iya fi Misr fi 'asr Isma'il*. Alexandria: Munsha'at al-Ma'rifa al-Haditha, 1977.

Reimer, Michael J. *Colonial Bridgehead: Government and Society in Alexandria, 1807–1882*. Boulder, CO: Westview Press, 1997.

Reymond, Paul. *Le port de Port-Saïd*. Cairo: Impr. Scribe égyptien, 1950.

Reynolds, Nancy Y. "City of the High Dam: Aswan and the Promise of Postcolonialism in Egypt." *City & Society* 29, no. 1 (2017): 213–35.

Rossum, Matthias van, and Özgür Balkılıç. "Desertion." In *Handbook: The Global History of Work*, edited by Karin Hofmeester and Marcel Van der Linden. Berlin: De Gruyter Oldenbourg, 2018.

Ruiz, Mario M. "Intimate Disputes, Illicit Violence: Gender, Law, and the State in Colonial Egypt, 1849–1923." PhD diss., University of Michigan, 2004.

Said, Edward W. *Orientalism*. New York: Pantheon Books, 1978.

Schettini, Laura. *Turpi traffici: Prostituzione e migrazioni globali 1890–1940*. Rome: Biblink, 2019.

Sharpe, Pamela, ed. *Women, Gender, and Labour Migration: Historical and Global Perspectives*. London: Routledge, 2011.

Sheller, Mimi, and John Urry. "The New Mobilities Paradigm." *Environment and Planning* 38, no. 2 (2006): 207–26.

Shinnawi, 'Abd al-'Aziz Muhammad. *Al-Sukhrah fi hafr qanat al-Suways*. Alexandria: Munsha'at al-Ma'rifa al-Haditha, 1958.

———. *Qanat al-Suways wa-al-tayyarat al-siyasiyah allati ahatat bi-insha'iha*. Cairo: Ma'had al-Buhuth wa-al-Dirasat al-'Arabiyah, 1971.

Sonnleitner, Ute. "Moving German-Speaking Theatre: Artists and Movement 1850–1950." *Journal of Migration History* 2, no. 1 (2016): 93–119.

Soresina, Marco. "Italian Emigration Policy during the Great Migration Age, 1888–1919: The Interaction of Emigration and Foreign Policy." *Journal of Modern Italian Studies* 21, no. 5 (2016): 723–46.

Sori, Ercole. *L'emigrazione italiana dall'Unità alla seconda guerra mondiale*. Bologna: Il Mulino, 1979.

Tignor, Robert L. *Modernization and British Colonial Rule in Egypt, 1882–1914*. Princeton, NJ: Princeton University Press, 1966.

Todd, David. *A Velvet Empire French Informal Imperialism in the Nineteenth Century*. Princeton, NJ: Princeton University Press, 2021.

Toledano, Ehud R. *State and Society in Mid-Nineteenth-Century Egypt*. Cambridge: Cambridge University Press, 1990.

Torpey, John. *The Invention of the Passport: Surveillance, Citizenship, and the State*. Cambridge: Cambridge University Press, 2018.

Verginella, Marta. "Le Aleksandrinke Tra Mito e Realtà." In *Le Rotte Di Alexandria: Po Aleksandrijskih Poteh*, edited by Franco Però and Patrizia Vascotto, 163–76. Trieste: EUT, 2011.

Yılmaz, İlkay. "Propaganda by the Deed and Hotel Registration Regulations in the Late Ottoman Empire." *Journal of the Ottoman and Turkish Studies Association* 4, no. 1 (May 2017): 137–56.

INDEX

Abbed, Abdalla El, 86
Abbed, Riz El, 86
Abdel Nour, Selim, 124
Abdülmaçit, 32
Abu-Lughod, Janet, 200
Adami, Raffaello, 88
Adele, widow Dumas, 107
agency, 7, 19, 69, 90, 215n31
Agius, Felix, 176
Ahmet: "Hadju" 51; *cawas*, 106; engineer, 174
Aida, 118, 267n23
Akko, xvi*map*, 121, 125
Albertine Statute, 164
Albertini, Madam, 97
Al Busfur. See *Le Bosphore égyptien*
Alcazar Lyrique, 169
alcohol: diversity of preferences, 169–70; excessive drinking and boisterous customers, 157, 173–77; flows of, 169, 201; health issues from, 41; leisure in Port Said, 168, 193, 201; sales from provision boats, 135; smuggling of, 127
Alcoré, Giuseppe, 83
Alessandrini, Ludovico, 62
Alexandria, xvi*map*, 14, 15; anarchists in, 192; British occupation and, 136, 200; changing migration paths, effects of, 26, 47–48, 146; Customs agents at, 127; entertainment in, 163, 165, 166, 169, 170, 176; family connections to, 63, 64, 68, 107; influx of migrants in and through, 4, 35, 46, 50, 53, 57–58, 57*fig*, 64, 73, 92, 93, 99, 110, 116, 125, 141, 143, 145–49, 188, 191, 192, 201; law enforcement and surveillance in, 127, 132, 147–49; prostitution in, 186, 187; railway connections, 28–29, 120; staples from, 41; steamboat connections, 121; as trading port, 47, 120, 196, 197–98, 209; women in, 110, 187–88, 191
Alexandria-Suez railroad, 38–39
Al-'Arish, xvi*map*, 35, 58, 59, 198
Al Firdan, 26, 27, 27*map*; cafés and taverns at, 105; corvée workers at, 36–37; desertion, strikes, and other rebellions, 86; ethnic and racial stereotypes among workers, 82; forced labor at, 38; Serbian worker at, 89; *shaykh* Sidi Ahmed Maghoule at, 38
Al Fowar, 27, 27*map*
Al-Gazzar, Abdel Hadi, 208, 208*fig*
Algeria, xvi*map*, 68, 82, 96, 253n73; Algerians, 144, 146, 180–81. See also *sabir*
Algiers, 14, 20
Al Guisr, 25, 27, 27*map*, 35, 52; brothels at, 97; drinking water resources, 40; entertainment in, 165; forced labor at, 37; gambling in, 103–4; imam at, 25; traders and shopkeepers at, 74
Ali, Hajj, 180–81
almées, 95–96, 116
Al Mourra, 27, 27*map*
al-Sisi, Abdel Fattah, 210
al-Tahtawi, Rifa'a, 25, 31

Amira, Sylva, 150
Anaissa, Hajj, 180
anarchism (anarchists): in Alexandria, 148, 192; Errico Malatesta, 144; section in Port Said, 150; surveillance of, 266n14. *See also* Malatesta, Aniello
Anglo-American Book Depot, 161
animal labor, 35
anthropometric system of identification, 147–49. *See also* Bertillon
Antonia, Austrian waitress, 173, 180
Antonini, Company employee, 165
Antonio, coach driver, 48
Arabic, 13, 34, 82–83, 132, 136, 139, 140, 161, 162, 168, 178, 191, 202
Arabs: ambiguous and overlapping racial and ethnic categories, 80–81, 143; Arab hospital, 121; Arab market, 74; Arab policemen and soldiers, 151, 152, 180; Arab readers, 77; Arab quarter ("village") in Port Said, 27, 28*fig*, 40–41, 62, 66, 67, 74, 97, 98–99, 103, 111, 121, 124, 137–38, 139, 140, 163, 172, 177, 180, 202, 203*map*, 210, 281n208; Arab quarters in other encampments and towns, 74, 97, 103, 111; Arab women and girls, 98, 105–7, 107*fig*, 138, 180, 181, 186; Arab workers, 32, 33–35, 62, 72, 76, 82–83, 86, 87, 91, 100–101, 111, 119, 133, 154, 167, 180, 192, 200; British occupation and, 137–39; as domestic service workers, 94–95; military exhibition of, 118; music of, 159–60; Port Said entertainment and, 165, 171, 172, 177, 181, 186; *shaykh* of, 129; United Arab Republic, 208; workers, division of labor by race and ethnicity, 76–77, 80–81, 82
archaeological sites in Egypt, 65
Armenians, 43, 52, 76, 79, 103, 117
artistic society, 159
Arzt store, 155, 282n16. *See also* Simon Arzt
Aspert, carpenter, 165–66
Aswan, xvi*map*, 73, 206; divisions and hierarchies in, 111–12; High Dam, 208
Asyut, xvi*map*, 37, 73
Aubert-Roche, Louis-Rémy, 80, 96–97
Aubry, canteen-keeper, 296n191
Audibert, Mr., 25

Austria, xvi*map*; Austrians in Port Said entertainment sector, 171, 181; Austrian workers, division of labor and, 76; Jewish migrants from, 185; trade in Port Said, 121; women from, 98, 108, 110–11, 173, 181, 183, 188; workers and other migrants from, 1, 33, 67, 68, 76, 79, 83, 84, 119, 121, 143, 144
Australia, 200; Australian Bar, 175; migrants on their way to, 175; soldiers, 293n153
Austrian Lloyd, 57, 121, 133, 192
Azbakiyya, 166

Baillière, Henri P. C., 25
Bairam, 164
Balcowich, Rosina, 84
Baldini, Dante, 63
banks (financial), 3, 67–68, 198
Baraga, Maria, 59, 85
Barak, On, xi–xii, 4, 41, 78
Barbarins, 80, 94, 182; *shaykh* of, 129
Bari, xvi*map*, 51
Baring, Evelyn. *See* Lord Cromer
Barnolich, Martin, 143
Baron, Beth, xi, 19
bars, 168–70, 193–94; efforts to regulate hours of operation of public establishments, 177–79; excessive drinking and boisterous customers, 173–77; in Cairo, 181; in Izmir, 190; venues, segregation of communities and, 171–73; women and entertainment, 179–89. *See also* breweries; cafés; canteens; taverns; wine
Barsotti, Omero, 161
Barthélemy, Fanny, 172
Bartholdi, Auguste, 15
Bastille Day, 164
bathing, 158, 164, 203*fig*
bathrooms, 122; bathhouses, 193
Baun, Mr., 124, 128
Bavarian, 103, 130
Bayard, Victorine, 106–7
beaches, 158, 158*fig*. *See also* bathing
Bedos, George, 123–24
beer, 74, 165, 170, 172, 192. *See also* alcohol; breweries
Beinin, Joel, 81

Bella Rosa, 183
Ben Ahmed, Hassan, 143–44
Bertillon, Alphonse, 147–48
Bethlehem, xvi*map*, 34
Bialkowski, Mr., 108
Biancani, Francesca, xiii, 186
Bianchi, pianist, 150
Binat, Madam, 181–82
biological determinism, 77–78
Bir Abu Ballah, 27, 27*map*, 36, 40
birth, 63, 66, 95, 142. *See also* children
Bitter Lakes, 27*map*, 4, 100
Blanchet, female café manager, 98–99
Blondel, Sublieutenant, 140
Boesmi, Marco, 124
Bohemia, xvi*map*, 170, 181, 296n189
Bollettino di Porto Said, 162
bondholders, 67–68. *See also* shares
Bon Pasteur nuns, 18, 19, 64, 122, 138, 159, 197
Bornier, Henri de, 102
Bousquet, Frenchman, 167
Bretons, 51–52, 79, 87. *See also* Brittany
breweries, 25, 95, 104, 167, 169, 178, 183, 186; conflated with brothels, 183, 185. *See also* beer; cafés; canteens; taverns; wine
Brindisi, xvi*map*, 49, 57, 116; clandestine migration from, 125
Britain. *See* Great Britain
Brittany, xvi*map*, 51–52, 60. *See also* Bretons
Brocchini, Eugenia, 63
brothels, 23, 96–97, 102, 193; brothel-keepers, 97, 183, 187, 258n132; conflated with breweries, 183, 185; efforts to regulate, 96, 105, 111, 178, 184; in other encampments and towns, 97, 105; in Suez, 97, 183, 186; in Shanghai, 187; Port Said, entertainment in, 154, 172, 177, 178, 180, 183–85, 193; prostitution, poverty or bad morals, 184–89. *See also* prostitution; Rue Babel
Buhayra, 37
Bulbeis, 73
Buoncore, 108
buza, 170, 171, 291n118

Café de Charenton, 173
Café de la Renaissance, 167
Café d'Orient, 169
cafés, 165, 167, 290n101; efforts to regulate hours of operation of public establishments, 177–79; venues, segregation of communities and, 172–73; women and entertainment, 180–81. *See also* bars; breweries; canteens; taverns; *and specific café names*
Cairo, xvi*map*, 4, 13, 15, 26, 30, 31, 32, 41, 112, 195, 200; British occupation of Egypt, 137; canal-based connections, 30, 40; Egyptian National Archive in, 18; influx of migrants in and through, 4, 35, 46, 93, 110, 116, 146, 149–50, 184, 187, 201; law enforcement in, 145, 147, 148, 152; leisure activities, 165, 166, 169, 176, 181, 184; newspapers in, 161, 162, 184, 191, 198; Opera House in, 118; prostitution in, 96, 110, 186; railway connections, 38, 120, 197, 199; smuggling to, 126; women in, 110, 181, 187, 188; labor recruitment in, 34, 188, 191. *See also* Egyptian government
Cairo (newspaper), 162
Calabria, xvi*map*, 50; Calabrian quarter in Ismailia, 111; Calabrians, stereotypes about, 78, 79, 92; Calabrian wet nurses, 191; labor recruitment in, 50–51, 52, 191; unemployed Calabrians at worksites, 92; workers from, 47, 50, 68, 73, 111–12
Calamita, Rosa, 106; Calamita sisters, 48
Calò, Vincenzo, 125
Campi, Antonia, 98
Candida, Antonio, 53
canteens, 54, 102, 157, 296n191. *See also* breweries; cafés; taverns; wine
capitulatory privileges, 4–5, 19, 104, 130, 162, 177, 201
Carjat, Etienne, 101-2, 102*fig*
Carlod, Joseph, 84
Casanova, Joseph, 50, 51, 52
Castello, Nicola, 85
Catholics, 47, 117, 142, 143, 159, 161, 163–64; celebrations, 117, 163; churches and religious accommodations, 66–67, 121, 245n277; interfaith marriages, 63; opinions of music halls and bars and brothels, 97, 177; schools run by, 64,

Catholics (*continued*)
110, 121; sources, 18. *See also* Bon Pasteur nuns; Éugenie
Cazejus, gang head, 89
chain migration, 54, 60, 98, 239n199
Chalouf, 27, 27*map*, 48, 56, 84, 88, 91, 99, 143, 159; brothels at, 97; cholera in, 89; hospital in, 88; music in, 159; railway travel to, 38; taverns in, 104; worker strike at, 90
Charenton: café de, 173; neighborhood in Port Said, 60
Chenet, 52, 55
Cherso, xvi*map*, 53, 59
children, 41, 55, 58, 59, 60, 62, 67, 75*fig*, 100, 106, 110, 119, 144, 163, 174, 181, 241n223; child labor, 64–66, 65*fig*, 244n270; from the Arab quarter ("village"), 138; courbache (whip), use of, 37, 65*fig*; families in Port Said, 63–66; illegitimate births, 95, 142; infantilized workers, 100, 105, 111; mortality of, 65–66, 78; recruitment of, 34, 65, 85. *See also* birth; orphans; schools
China, 3, 124; labor recruitment efforts in, 48–49, 236n167; Chinese objects, 116, 155
cholera, 88–89
Christians, 138, 164, 169; festivals and rituals, 117, 158, 164; Copts, 34, 67, 172; from Syria, 34–35; Maronites, 67. *See also* Catholics; Greek Orthodox
Choupéro, Florio, 79
Cipollaro, chief of the Civil Police, 139
citizenship, 144
Ciurcia, Luigi, 117
Civitavecchia, xvi*map*, 50, 51, 52
Clancy-Smith, Julia, xi, 5, 14, 20, 56, 79
clandestine migration, 55, 68, 124, 125–26, 204, 269–70n61
Clementi, Captain, 140
Clémentine, R., 163
climate, 50, 170; climatology, 77; racial judgements and, 65, 77–78, 81
coal, 12, 97, 268n36; coal-heaving strikes, 192, 200; coaling stations, 15, 115, 120, 190, 192, 197, 221n72; dust from, 154; "duty-free zone" and, 198–99; small provision boats, coal collection and delivery by, 134, 135; workers in coal-heaving, 45, 76, 81, 192
Cobau, Ana (or Anna), 59
coffeehouses, 155, 156–57, 160, 167, 179–80. *See also* cafés
Cohen, Y., 98
Colomb, G., 162
Colomb, Mr., 123
Colombo, 186, 187
Colón (formerly Aspinwall), 16
Commerce street, in Port Said, 67, 98, 128, 155, 156*fig*, 161
Compagnie Universelle du Canal Maritime de Suez (Universal Company of the Maritime Canal of Suez). *See* Company
Company: agreements on use of Egyptian labor force, 28–33; corporate familism of, 260n160; creation of, 1; customs duties owed to, 197–98; desertion and labor unrest, response to, 85–90; Egyptian concession of land to, 44–45; freshwater canals, creation of, 40–41; infantilized male workers and, 102–5; international and on-site labor recruitment, 48–52; invoices for canal inaugural ceremonies, 119; Isthmus of Suez in second half of 1860s, 71–72, 111–13; law enforcement in Port Said and, 128–31; nationalization of, 206–8; newspaper coverage of, 161, 162; newspaper published by, 20; poaching of workers by, 87; Port Said's urban management and, 15, 198–99; position on unemployed and undocumented migrants, 91–94; prostitution, regulation of, 96–97; railway service managed by, 198–99, 199*fig*; recruitment of wives and families, 105–6; recruitment of labor, 33–36, 45, 48–52, 55, 87; reliance on forced labor, 30–33, 36–39, 42–45, 206, 208*fig*; rules and exceptions in engagement of workers, 74–83; scholarship on, 10; setting aside land for religious buildings, 66–67; Suez canal project, first phase of, 1–3; supply shops, 41–42, 73–74; tonnage fees for navigation, 120; use of railway, 38; work encampments, development of, 27, 27*map*; workers as children of a big family, 100–102, 102*fig*;

work-related injury and death, response to, 88. *See also* Lesseps, Ferdinand de
concession: Suez Canal's concessions, 3, 29–30, 40, 43; concessions of land, 44; railway concession, 197
Conti, Clément, 109
Conti, Ms., 25–26
contractors, 26, 33, 34, 36, 41, 50, 66, 67, 69, 87, 92, 108; subcontractors, 51, 52, 82
conversion, 63, 125
coolies, 48–49, 236n167
Corriere del Canale, 162
Corriere di Porto Said, 162
Cortari, Pierre, 84
corvée, 30–31, 32, 36–39, 43, 82, 206, 208, 208*fig,* 227n28; abolition or end in Suez of, 44, 45, 48–49, 234n138; desertion, strikes, and other rebellions, 85–90; worksites, conditions at, 42; Ottoman government opposition to, 43–45
Costa, N., 34
cotton production, 43, 46, 119, 120, 197
courbache (whip), 37, 65*fig*
Courrier de Port-Saïd, 161
Couvreux, Alphonse, 49–50; Couvreux enterprise, 52
Cresswell, Tim, 6
Crete, xvi*map,* 121, 129, 170, 185, 188
crime: Egyptian government efforts to build a modern state, 151–52; gambling as a gateway to, 103, 147; on Isthmus of Suez, 114; law enforcement in Port Said, 128–31, 138–41; in Port Said, 123–26, 128, 293n145; smuggling of goods, 126–28. *See also* criminal prosecution
Crimean War, 41, 170, 185
criminal prosecution, avoidance of, 56
Croce, Saul, 159
Cromer, Lord, 9, 139
Cumbo, Ferdinando, 161, 286n55
Currency, 89–90, 267n24; falsified currency, 55; moneylending, 79
Customs, 8, 114, 126–27, 152, 203*map*; director of, 202; importation taxes, 122, 197, 198–99; railing along Port Said's quay, 270n64; smuggling in Port Said, 8, 126–28
Cyprus, xvi*map,* 41, 170, 191

Dalmatia, xvi*map,* 35, 228–29n43; Dalmatians, 68, 78, 83, 87, 121, 130
Damietta, xvi*map,* 1, 15, 26, 32, 33, 79, 98, 104, 129; changing migration paths, effects of, 47; divisions and hierarchies in, 111–12; drinking water resources, 40; opening of the canal, effect of, 120; railways to, 120, 122; goods, traders and shops, 49, 73, 121, 126, 127; unemployed and undocumented migrants, entry point for, 92–93; worker population in and from, 36, 111
damma, 160
dance, 48, 106, 118, 154, 160, 172, 181–82, 184; "boy dance," 185
Dante Alighieri Society, 161, 285n42
Daqhaliyya, 32, 37, 111
Darviche, Fathalla, 135–36, 144–45
Darwin, Charles, 76–77, 249n28
Dawson City, 15–16. *See also* Yukon
death, 30, 63, 72, 90, 138, 161; alcohol and, 174; cholera epidemic, 88–89; commemoration of, 163, 171; death-related records, 68, 83–84, 142; from disease, 42, 83–84, 90; widowhood, 107; work-related, 88
De Certeau, Michel, 7
Delbergues, Marius, 99
de Lesseps, Ferdinand. *See* Lesseps, Ferdinand de
de Lesseps square in Port Said, 167, 172, 173
delirium tremens, 174. *See also* alcohol; health issues
delta, xvi*map,* 37, 45–46, 81, 87, 141, 160, 165, 197; music in, 160, 165
demographics: Egypt, foreign residents in, 9; population registers, 18, 141; Port Said population, 9, 60, 64, 91, 103, 121–22, 141, 153, 170, 192, 202
Déré, Mr., 109
desalination of water, 40–41, 66
desert, 4, 16, 29, 31, 42, 44, 48, 84, 91, 101, 102*fig,* 111, 119, 197, 199; climate of, 77, 170; surroundings, facilitating crime and vice in Port Said, 127, 140, 152, 154, 170; Western writers on, 200
desertion, 56, 85, 86, 89, 90, 123, 175
Deslandes, Jules, 150, 280n200

differentiated mobility, 6–7, 201, 204
disease, 30, 53, 79, 81, 83–84, 90, 112, 191; cholera epidemic, 88–89; forced hospitalization, 90; venereal diseases, 97, 104, 112, 183, 184. *See also* cholera; syphilis
Djibouti, 15
domestic labor, 60, 91, 94–95, 98–99, 110, 188; speaking broken French, 138; recruitment of, 188
Dot, Emmanuel, 25
Dracoti, Jean, 80
Dredger, 78, 85, 87, 208, 247n1
Dumas, Alexandre, 107
Duse company, 166
Dussaud Brothers, 82
Dutruy, Marie, 99

Eastern Exchange Hotel, 168, 168*fig*, 190, 289n98
economic downturns, 26; Depression of 1873, 3; financial crisis of 1907, 8–9
Effendia, Ahmad, 109–10
Egypt: British occupation of, 136–41, 137*fig*, 200; Depression of 1873 and, 3; financial crisis of 1907, 8–9; foreign residents, data on, 9; Egypt's president Gamal Abdel Nasser, 206; Israel, Sinai and, 209; Isthmus of Suez as an Egyptian Klondike, 45–47; labor recruitment networks in, 51–52; Port Said's status in relation to rest of Egypt, 199–200, 209–10; semicolonial status of, 3; sources for study, 17–21; sovereignty of, 31, 45, 195, 204, 210; time and space in the modern history of, 12–14; in world history, 14–16
Egyptian government: agreements with Company on use of Egyptian labor force, 28–33; building of a modern state, 31, 151–52; concession of land to Company, 44–45; Customs in Port Said, 126–28; efforts to control gambling, 169; efforts to regulate hours of operation of public establishments, 177–79; Health Bureau, 133; law enforcement in Port Said, Company relations and, 130–32; local governors, appointment and power of, 32, 34, 35, 39, 103, 124, 129, 130–32, 136, 138, 143, 148, 151, 174–76, 178–79, 184, 202, 271n87; local governor, on the unemployed, 122–123, 145–46; monitoring the return of unwanted visitors, 147–49; negotiations with Company 1902, Port Said infrastructure, 198–99; opening of the canal festivities (1869), 117–22; Passports Bureau in Port Said, 132–36, 134*fig*; population of Isthmus of Suez, documentation of, 141–49; Port and Lighthouses, regulation of, 133, 135*fig*; public works projects, forced labor and, 30–31; Service of Docks and Entrepôts in Port Said, 133; Suez Canal as public works project, 29–30; unemployed and undocumented migrants, treatment of, 92–94; unemployed persons, removal of, 122–26. *See also* Cairo
Egyptian State Railway, 38–39
Eldorado: cinema in Port Said, 15; Egypt as, 46; theater and gambling hall, 166, 167
electricity, 168; ship navigation and, 189–90
Elias, Démitri, 255n94
Elias, Giorgi, 87
Elias: "the Greek," 18
Elisa, Miss, 98
El Keir, Oum, 98
Emilia: female worker in Port Said, 97; Emilia region, 240n206
Emilia, Maria Giuseppa, 63
Emma, Ida Gemma, 63–64
entertainment: beaches, 158, 158*fig*; dancing, 118, 154, 160, 172, 181–82, 185; drinking establishments, 166–70; efforts to regulate hours of operation of public establishments, 177–79; excessive drinking and boisterous customers, 173–77; festivals and rituals, 163–65; fishing and hunting, 158–59; gambling, 167, 169, 261n174; leisure cultures, diversity of, 156–57; music, 159–60; music halls, 154, 167, 192–93; nighttime navigation, effects of, 189–92; in Port Said, overview of, 153–57, 158*fig*, 165–66, 192–94; prostitution, 183–89, 104; reading and publishing on the Isthmus, 160–63, 169; regattas, 158; lake and sea

baths, 158, 158*fig*; venues, segregation of communities and, 171–73; women and entertainment, 179–89. *See also* bathing; dance; music
environmental determinism, 77–78. *See also* climate
Esna, xvi*map*, 37
Ethiopia, 136, 184, 280n199; Abyssinia, xvi*map*, 181
ethnicity, 76–77, 204, 248–49n19; festivals and rituals, 163–65; rules and exceptions in engagement of workers, 74–83; segregation, 171, 173; wage inequities and, 81
Eugénie: boulevard, 123, 168*fig*; empress of France, 44, 117–18; Quay, 167; Sainte Eugénie church, 66, 163–64
expulsion, 108, 145–146, 152, 278n171; papers of, 143, 147; of prostitutes, 97; re-expulsion, 148; of unemployed workers, 145–47
Ewich, Andrea, 83

Fahmy, Khaled, 31
Fahmy, Ziad, xi, 14
Faraj, Fu'ad, 200, 214n11
Farge, Arlette, 21
Farrugia, bumboatman, 176
Fayyum, 162, 288n76
Felicite, Jeanne, 108–9
Ferrugio, Achille, 63–64
fertility, 62, 77–78, 95. *See also* marriage; women
festivals and rituals, 163–65
Ficcanaso, 162
Fikri, Muhammad Amin, 64, 196
Filele, Abdalla, 124
fishing, 134–35, 135*fig*, 158–59, 198
Folaitis, Minas, 128
Fontane, Mr., 62
food, corvée workers and, 29, 30; cost of, 50, 73–74, 90, 165–66; in Port Said, 41–42, 164; lack of, 32, 89–90, 149; smuggling of, 126; trade in, 47, 49, 121, 135, 140
forced labor. *See* corvée
Forneroni, Anna, 159
Fraissinet, navigation company, 57, 121, 133
France: anthropometric system in, 147–48; French children, 65–66; French workers, division of labor and, 75–76; labor recruitment efforts in, 33, 50, 51–52, 55, 106; nationalization of Suez Canal Company, response to, 207–8; opening of the canal festivities (1869), 117–18; Paris-centered perspectives, 5, 18, 20, 31, 33, 43–44, 106, 112, 150, 167, 184; prostitution, regulation of, 96; tripartite aggression, role in, 207–8; women's work, debate on, 109. *See also* Brittany
Free Officers' Revolution, 206
Freitag, Ulrike, 6
Frigieri, Luigia, 95

Gabaccia, Donna, 17
gambling, 156, 179; efforts to regulate hours of operation, 177–79; in Port Said, 153, 154, 166, 167, 169, 182, 190, 261n174; as a gateway to crime, 103, 147; at the Grand Casino, 171; infantilized male workers and, 102–105; in Suez, 172; venues, segregation of communities and, 171–73
Garibaldi: Inn, 170; café 171
Gasparini, chief of Suez police, 139
Gava, artist, 149
gender, 11–12, 71–72, 200–201, 204; domestic labor market, 94–95; in entertainment, 157, 180, 193; femininity,189; feminization, 94; gendered metaphors, 101; infantilized male workers, 102–5, 112; masculinity, 23, 101–102, 102*fig*, 179–80, 189; mobility and, 6–7, 189, 201, 216n34; respectability, 109–10, 154, 189; women, scattered livelihoods of, 98–100; women and entertainment, 179–89
geographic determinism, 77–78
Georgette, female servant, 106
Germans: children, 65; language, 82, 140; beaten up in Port Said, 175; job-seekers, 191
Geroli, Madame, 94, 113
Geyler, superintendent, 113
Giammugnai, Giuseppe, 62
Gimigliano, 88
Gioia, Edoardo, 37

Giorgini, Domenico, 108
Girardeau, Emile, 51
Giuffrida, Maddalena, 98
Goëlo, Alphonse, 60
Goldman, Leone, 161
Gorin, contracting company, 119
Gouin worksite, 61*fig*
Goujat, Marie, 187
Gouthman, Madame Chai, 98
Grand Hôtel Continental. *See* Hôtel Continental
Graziano, Saverio. *See* Saverio Graziano
Great Britain: Alexandria-Suez railroad, 38–39; British women as domestic servants in Egypt, 188; citizens of Ionian Islands as British subjects, 68; efforts to recruit English police personnel, 139–40; excessive drinking and boisterous behavior by British, 174–75, 176, 179; financial loans to Egypt, 68; London-centered perspectives, 5, 18, 43–44; nationalization of Suez Canal Company, response to, 207–8; occupation of Egypt, 7, 12, 116, 136–41, 137*fig*, 147, 151, 152, 200; official position on forced labor, 31; opening of the canal (1869), 117; Port Said, law enforcement efforts in, 152; Port Said as a link in British maritime chain, 13; Port Said's Protestant community, 67, 163; purchase of Egyptian canal shares (1875), 3; Scottish workers, 61, 140. *See also* Maltese
Greece: labor recruitment efforts in, 35; Morea, tobacco from, 127; opening of the canal festivities (1869), 117; Peloponnese, xvi*map*, 127, 191; travel restrictions on women, 55. *See also* Greeks
Greeks: boats and ships, 129, 134; consul and clergy, 104, 117, 119; contractors, 34; domestic workers, 60, 188; entertainment venues, 103, 165, 166, 168, 169, 171, 172, 175, 261n174, 291n123; Greek quarter ("village") of Ismailia, 97, 105, 111; Greeks, categorization of, 18, 68, 80, 141, 142*fig*, 143, 245–46n287, 252n56, 297n206; Greeks, drinking, 173, 174, 176; Greeks, involved in target shooting, 131; Greeks, stereotypes about, 79, 87, 126, 128, 174, 293n145; language, 74, 82–83, 143, 162; migrant workers, 33, 45, 54, 60, 67, 82, 89, 92, 153, 186; moneylenders, 79; population of Port Said, 1, 64, 66, 42, 121–22, 124; school in Port Said, 62, 64; singer, 288n76; smugglers in Port Said, 126–28; traders and shopkeepers on Isthmus of Suez, 41, 73, 248n9; women and families in Port Said, 60–62, 63, 64, 65, 184, 186; workers, division of labor, 76, 78; unemployed, 119, 201
Greek Orthodox, 66, 68, 117, 164, 171, 184, 266n17, 297n206; Greek chapel, 121; Greek clergy, 117, 119
Greene, Julie, 81
Grinsley, Moussa, 172
Grouaz, trader, 113
Guastalla, Michele, 151
Guerin, Marie Angelique, 95

Hamit, domestic worker, 94
Hamley, William George, 153
Hamza Ben Saleh, Hajj, 180–81, 189
Hardon: Alphonse, 34; enterprise, 1
Harvey, wanderer, 125
Harvey Pasha, 147
Hashish, 104, 127, 168–69; giddiness and, 172
headmen (*shaykhs*), 6, 35, 37–38, 39, 78, 86, 129, 230n74; neighborhood *shaykh*, 180
Health Bureau, 133, 151
health issues, 202; alcohol-related issues, 173–74; children and, 65; cholera epidemic, 88–89; forced hospitalization, 90; health authorities, 65–66, 92, 132, 133, 202; medical care at worksites, 42; migration and, 58, 59; prostitution, regulation of, 97; racial judgments of, 77–78; sexually transmitted diseases, 97, 183, 184; of workers, 41, 79–80, 83–84; work-related injury and death, 88, 101. *See also* Health Bureau; syphilis
Heidegger, workshop, 87
Hejaz, xvi*map*, 143
Helwan, 150, 162

Hezzi, Al Hadja Zarifa Bent Abdelrahman, 98
Hong Kong, 15, 48, 236n167
Horn, J., bookstore, 161. *See also* Anglo-American Book Depot
hospital, 29, 42, 66, 79, 88, 90, 93; "European" hospital and "Arab" hospitals in Port Said, 121, 203*fig*; hospital visit for prostitutes, 184; French hospital in Suez, 89, 94
Hôtel Continental, 168, 171
Hôtel de France et Café, 167, 171, 178
Hôtel de la Division, 167
Hôtel de Paris, 171
Hôtel des Pays Bas, 167, 135*fig*
Hôtel des Voyageurs, 107
Hôtel d'Europe, 98
Hôtel du Louvre, 167
hotels, 73, 135*fig*, 156, 157, 167–68, 171, 168*fig*; at Al Guisr, 165; clients of, 166, 167; employment at, 82, 107, 135; 154, 191; hotelkeepers, conduct of, 107, 117, 127, 145, 187, 266n14; women as owners of, 98. *See also specific hotel names*
housing, 46, 85, 92, 195, 200, 208; Company control of, 29, 42, 60; costs in Port Said, 122
Huber, Valeska, 10, 20, 200
hunting, 158–59; weapons for, 104
Hyènes, 89–90, 95

Ibrahim, Hasan Efendi, 138
Il Telegrafo, 162
inauguration: of Port Said, 213n1; of the Suez Canal, 12, 114–19, 118*fig*, 206; of the "new" Suez Canal 210, 218n49; of the Suez Canal, changes following, 22, 119–20, 152, 153
injuries, work-related, 88. *See also* health issues
International Commission, 3
Isella, Luigi, 149
Islam: Al-Azhar, 31, 117, 266–67n17; conversion to, 63, 125; festivals and rituals, 164; interfaith marriage, 63; mosques and religious accommodations, 66–67. *See also* Muslims; religion

Isma'il, Hamdi (or Bey), 36, 114, 129, 131, 138
Isma'il, ruler, 43, 44, 68, 117, 119, 129, 151
Ismailia, 27, 27*map*; Calabrian quarter in, 111; connection to Port Said, 197–99; drinking water resources, 41; entertainment in, 165; Greek quarter ("village") of, 97, 105, 111; languages in, 74; marriages in, 112; masonic lodges in, 159; music in, 160; opening of the canal, effect of, 120; rail travel to, 38; Rue François-Joseph in, 74; Rue Negrelli in, 99, traders and shopkeepers at, 74. *See also* Lake Timsah; Timsah
Israel, 207–8, 209
Issawi, Charles, 14
Istanbul, xvi*map*, 121, 143, 144, 151, 163, 171; prostitution and, 185–86
Isthmus of Suez: as an Egyptian Klondike, 45–47; British occupation of Egypt, 136–41, 137*fig*, 200; completion of canal, effect on population, 114–22; encampments on, 26–27, 27*map*; expulsion of unemployed workers, 145–47; leading a nomadic life after 1869, 149–52; population, documentation of, 141–49; Port Said, impact on, 8–9; in second half of 1860s, 71–72, 111–13; traders and shopkeepers on, 73–74, 119–20, 132. *See also* desert; entertainment; inauguration; *names of specific locations*
Istrati, Panaït, 1, 24, 225n123
Italy, 173; goods from, 121; Italian school, 63; Italians, stereotypes about, 79; Italian workers, division of labor and, 76; king of, 163; labor recruitment efforts in, 35, 50–51; masonic lodges, ties to, 159; previously gendarmes in, 139; Rome-centered perspectives, 18, 47, 51, 161, 164; travel documents, need for, 54–55; unification of, 164; workers from, 56, 68, 73, 98, 153, 160, 188; unemployed workers from, 119. *See also* Calabria; Piedmont; Tuscany
Ivankovic, Giovanni, 63
Izmir, xvi*map*, 14, 48, 120, 144; entertainment in, 190; migrants from, 106; labor recruitment in, 52

Jabal Janifa, 27, 27*map*, 33
Ja'far Pasha, 32
Jaffa, xvi*map*, 14, 58, 120–21, 185
Jalal, al-Sayyid Husayn, 200
Japan: girl from, 187; man from, 305n50; objects from, 155; Russo-Japanese War, 175
Jauffret, Marius, 161
Jerusalem, xvi*map*, 125, 135, 149, 267n17
Jewish population: anti-Semitic riots, 164; bars for, 171; festivals and rituals, 164; from New York, 155; in Tripoli, 143; prostitution and, 184, 185; school for, 64; stereotypes about, 79; synagogues and religious accommodations, 64, 67; workers, division of labor, 76
Joice-Bey, Mr., 133
Journal du Canal, 161
Junot, coppersmith, 87
Juretich, Andrea, 143

Kasos, xvi*map*, 60, 162, 242n241
Kariclia, match-maker, 188
Kemsé-Abdel-Kérim, 94
Keyser, E. C., 159
Khalil, domestic worker, 94
Khalil, Ahmet Effendi, 174
Khedivial Company, 121
Kherges, Ibrahim, 125
Kipling, Rudyard, 154, 181–82, 194
Klondike, 15–16; Isthmus of Suez as, 45–47, 67
Kyriacopoulo, Mr., 162

labor: child labor, 29, 34–35, 64–66, 65*fig*; completion of canal, effect on workers, 115–16, 119–22; for cotton production, 43; day laborers, 72–73; deaths, data on, 30, 88; desertion, strikes, and other rebellions, 85–90, 192; division of labor in worksites, 72–74; Egyptian labor force agreements with Company, 28–33; extravagant hope for profit, 83–85; forced labor, 30–31, 36–39, 43–45, 82, 206, 208, 208*fig*; infantilized male workers, 102–5; international and domestic recruiting efforts, 48–52; Isthmus of Suez as an Egyptian Klondike, 45–47; opportunities, in second half of 1860s, 71–72, 111–13; opportunities, in early 1890s, 191–92; personal expenses at worksites, 42–43; port cities in world history, 14–16; recruitment of workers, 33–36; rules and exceptions in engagement of workers, 74–83; wages of, 29–30, 34–35, 42–43, 72–73, 81, 87; workers as children of a big family, 100–102, 102*fig*; work-related injury and death, 88; worksites, conditions at, 39–45. *See also* corvée; women
Lake Timsah, 4, 40, 89–90, 120
language: in sources, 17, 19; drogmans, 82–83, 135, 154, 253n76, 274n119; familiar languages, 59, 143; foul language, 146, 179; in Port Said, 74, 83, 93, 140, 154, 161, 254n77; multiplicity of, 74, 82–83, 112, 143, 162, 193; reading and publishing on the Isthmus, 161–62; School of Languages, 31. *See also* sabir
Lapeyrouse, Claudine, 98, 99
La Phare de Port-Saïd et du Canal de Suez, 161–62, 170
Laroche, Félix, 33, 130, 132
Lasseron, company, 41
L'Avenir Commercial de Port-Saïd, 161
La Verità, 162
law enforcement, 128–31, 138–41; efforts to control gambling, 169, 261n174; efforts to regulate hours of operation of public establishments, 177–79; Egyptian government efforts to build a modern state, 151–52; expulsion of unemployed workers, 145–47; monitoring the return of unwanted visitors, 147–49; prostitution, regulation of, 184
Le Bosphore égyptien, 46, 161, 286n60
Lebret, Pierre, 47
Lebrizzi, Nicola, 88
L'Écho de Port Said: Journal politique, commercial et littéraire, 161–62
Ledilly, Jacques, 83–84
Lefebvre, Henri, 14
Leghorn, xvi*map*, 63, 98, 150, 164
Legroud, Mr., 99
Le Havre, 52, 55
Le Journal de Port-Saïd, 161

L'Elettrico: Giornale politico, letterario e commerciale, 162
Le Moniteur de Port-Saïd, 161
Leoncavallo, pianist, 150
Le Pesquer, Joseph, 52, 55
Le Petit Port-Saidien, 162
Le Phare de Port-Said et du Canal de Suez, 161, 162, 170
Lepin, Manuel, 61
Lepori, Giacomo, 26
Le Progrès, 162
Lesegno, xvi*map,* 94
Lesieur, Paul, 167
Lesseps, Ferdinand de, 1, 16, 20, 21, 28, 29, 102*fig,* 115, 160; on almées, 95–96; on families in Port Said, 61–62, 64; on labor force, foreign and domestic, 29, 32–35, 61–62; on land negotiations with Ottoman government, 44–45; de Lesseps square in Port Said, 167, 172, 173; on railway concession, 197, 204; public opposition to corporal punishment, 37; labor recruitment, 33–36, 61–62; statue of, 206, 207fig, 275n130; workers as children of a big family, 100–102, 102*fig*
L'Harmonie, 160
libraries, 160–61
Lifonti, Gildo, 159
L'isthme de Suez, 20, 159
literacy: among migrants, 20, 74, 95, 98, 160; reading and publishing on the Isthmus, 160–63, 169
Lockman, Zachary, 6, 81
Lorient, xvi*map,* 52. *See also* Brittany
Lower Egypt, cotton production in, 43; entertainment in, 165; European "colonies" in, 141; forced labor from, 37; freshwater sources, 40; labor recruitment in, 34, 45, 141. *See also* delta
Lucia, Gaspare, 108
Luciani, Lucia, 94

Maghoule, Sidi Ahmed, 38
Mahmudiyya Canal, 30, 31, 227n31
Maillard, Madame, 182–83
Maison Bazin, 57
Makdisi, Ussama, 19
Malatesta, Aniello, 144, 192
Malia, Eugenio, 133
Malta, xvi*map,* 121, 164, 185; workers from, 79. *See also* Maltese
Maltese, 142*fig*; alcohol use and boisterous behavior, 172, 173, 176; boats, 134; festivals and celebrations, 163–64; as laborers, 1, 36, 53, 59, 82, 150, 199; lace, 155; migrant workers, 1, 36, 53, 59, 65, 150; newspapers by and for, 161–62; offspring of, 66; population in Port Said, 1, 121; stereotypes about, 65, 79, 251n47; workers, division of labor, 82; unemployed, 119
Mamluck, Soleiman el, 143, 147. *See also* Salemi, Francesco
Manovich, Jean, 83
Mansura, 37, 73, 162, 165
Manzala, Lake, 28, 36, 127, 195, 198; goods through, 121, 122, 126
Marcantonio, Ulisse, 148
Marcelin, Jourdan, 149
Marcovich, Milan, 84
Marendi, Raffaele, 56
Mariutti, Pietro, 88
Marmonti, Antonietta, 63–64
marriage, 56, 60, 62–63, 112, 186, 187, 188; match-making, 188. *See also* fertility
Marseille, xvi*map,* 12, 14, 26; changing migration paths, effects of, 14, 48, 186; as port through which migrants and provisions transited, 49, 55, 57, 62, 69, 70*fig,* 107, 120–21, 124, 159, 170; visitors from, 116
Marseille Company of Steam Navigation, 121
Marthoud, Ethienne, 56, 57*fig*
Martin, Jean, 83
masonic lodges, 159
Massey, Doreen, 6
Matay, 165
Maurice, Mr., 93–94
Maxama, 36
Mayond, Joseph, 48
Mazzanti, Mr., 166
Mazzolini, Mr., 162
McKeown, Adam, 9, 69
Meaouad, Ahmed, 182–83

medical care. *See* health issues
Mediterranean Sea, xvi*map*, 2*map*, 8, 13, 193; British and French maritime dominance, 13–14; plans for Suez Canal and, 1, 3–4, 71; migratory paths, 5, 26, 35, 47–48, 53, 56–57, 58, 59, 69, 80, 124, 144, 151, 189, 201, 216n34; Orientalist platitudes across the, 185; people from, 78, 79, 82; Port Said as new hub of connectivity, 8–9, 13, 14, 210; steamship transportation, 56–58, 189; trade routes, 120–21, 196
Megyessi, Charles, 82
Menard, Jules, 51–52
Messageries, 56–58, 124–25
microhistory, 11–12, 205
Mifsud, Maltese drinker, 176
migrants: clandestine migration networks, 125–26; completion of canal, effect on migrant populations, 115–16, 119–22; departure points for, xvi*map*; diversity of migrants in Port Said, 66–70; by the early 1890s, 191–92; Egyptian government's desire to limit migrant labor, 31–33; expulsion of unemployed workers, 145–47; extravagant hope for profit, 83–85; flight from criminal prosecution, 56; flight from military conscription, 56; on Isthmus of Suez, documentation of, 141–49; Isthmus of Suez as an Egyptian Klondike, 45–47; labor recruitment, 33–36, 45, 48–52, 55, 87; leading a nomadic life after 1869, 149–52; literacy rates, 20, 74, 95, 98, 160; multiple identities of, 143–45; musical diversity of, 159–60; need for nationality documents, 4–5; Passports Bureau in Port Said, 132–36, 134*fig*, 135*fig*; Port Said as choice for the study of, 9–12; Port Said as new hub of connectivity, 8–9; Port Said history, overview of, 200–205, 203*fig*; social mobility strategies, 55; sources for study, 17–21; travel documents, need for, 54–55, 58, 92–94; unemployed and undocumented migrants at worksites, 91–94; unemployed persons, after canal opening, 119, 122–26; use of term, 6; women, scattered livelihoods of, 98–100; women and families, 60–66, 61*fig*, 65*fig*. *See also* migratory paths
migratory paths: chain or network migration, 54, 60, 98; changes in, 47–48; clandestine migration networks, 125–26; diversity of migrants in Port Said, 66–70; expulsion of unemployed workers, 145–47; leading a nomadic life after 1869, 149–52; Passports Bureau in Port Said, 132–36, 134*fig*, 135*fig*; Port Said as vital transportation node, 120–22, 204–5; promotion of opportunities in Suez Canal, 52–54; prostitution and, 185–89; steamship transportation and, 56–58; travel documents, need for, 54–55, 58, 92–94. *See also* chain migration; migrants
Mikaël, Jouanin, 87
military conscription, avoidance of, 56. *See also* desertion
Minerva, sea baths, 164. *See also* bathing
Minervine, Maria, 88
Ministry of the Interior, Egypt: efforts to control gambling, 169; efforts to regulate hours of operation of public establishments, 177–79; efforts to regulate return of unwanted visitors, 147–49, 201–2; effort to control traffic in Port Said, 136; law enforcement in Port Said, 151–52; population data from, 141. *See also* Egyptian government
Minkin, Shana, xi, 68
Minya, 37, 111–12, 165
mobility: differentiated mobility, 6–7, 201, 204; Port Said and the alteration of circuits of mobility, 8–9, 14, 22–23, 26; sources for study, 17–21; use of term, 6. *See also* migratory paths
Moch, Leslie P., 53
Mohammad, Hassan, 94–95
Mohie, Mostafa, 209, 210
Monham, captain, 140
Montalti, workshop, 87
Monufiyya, 37, 39
Morand, Jules, 59
Morocco, 49; Moroccan workers, 49, 87, 180; port of Tetuan, 49

Mubarak, 'Ali, 3, 214n9, 196
Mubarak, Hosni, 209
Muhammad 'Ali, 30–31
Mulhouse, xvi*map*, 52
music, diversity of, 159–60, 165–66; during festivities, 163; musical instruments, 66, 159–60; musical societies, 160; music halls, 154, 167, 192–93; pianos and pianists, 150, 159, 181
Muslims, 34, 49; alcohol consumption and, 169–70, 181; schools for, 64; stereotypes about, 49, 77. *See also* Islam
Mustapha, Saad, 76

Naggiar, aly Mohamad el, 146
Nagib, Ibrahim, 76
Najm, Zayn al-'Abidin Shams al Din, 9–10, 202–3, 214n11, 297n203
Nakle, Mr., 165
Naples, xvi*map*, 50, 51, 150, 185–86, 192
Napoleon III, 44–45
Nassar, Mohamed, 133
Nasser, Gamal Abdel, 206
nationalism, 81, 164, 208. *See also* ethnicity
nationality: documents, need for, 4–5, 54–55, 92–94; Passports Bureau in Port Said, 132–36, 134*fig*, 135*fig*, 140; rules and exceptions in engagement of workers, 74–83
Nava, Clotilde, 144
Nefiche, 40, 80, 126
network migration, 54. *See also* chain migration
newspapers, 169; accounts of crime and safety, 151; promotion of opportunities in Suez Canal, 52–54; reading and publishing on the Isthmus, 161–63
New York, 155. *See also* Simon Arzt
Nicaleau hotel, 171
Nicolas, Mikail, 124
nighttime, 73, 103–104, 132, 138, 164, 189–90
Nile River, xvi*map*; canal with Lake Manzala, 36; drinking water resources, 30, 40, 41, 62, 120; irrigation and cotton cultivation, 46; as transportation network, 33, 38, 40. *See also* delta
Nowowolsky, Simcha, 74

Noyer, Victor, 177
Nubar Pasha, 30, 43, 44
Nubia, xvi*map*, 37, 79, 80, 160

Occupation of Egypt, British, 136–41
O'Connell, James, 142
Odessa, xvi*map*, 121, 186, 187
opening of the canal festivities (1869). *See* inauguration
operas and operettas, 118, 166. *See also* Aida; theater; Verdi
opium, 169, 270n70
Orientalism, 19, 77, 101, 102*fig*; *Orientalism* (Said, 1978), 11. *See also* Said, Edward
orphans, 109, 138, 145; orphanage, 64
Ottoman Empire, 3, 4
Ottoman government: capitulatory privileges, nationality documents and, 4–5; demand for neutrality of the canal (1863), 43–45; desire to limit migrant labor, 31–33; opening of the canal festivities (1869), 117; opposition to Company landholdings, 44–45; opposition to use of forced labor, 44–45; travel documents, need for, 54, 133

Pacho, Enrico, 161
Pagnon, hotel, 167
Palestine, 155. *See also specific city names*
Palidi, Jean, 34
Panama canal, 16, 81. *See also* Colón
Paoletti, Albino, 144
passports, need for, 4–5, 54–55, 92–94; Passports Bureau in Port Said, 132–36, 134*fig*, 135*fig*, 140
Pea, Enrico, 191
peddlers, 154–55, 155*fig*, 160, 181
Pelusium, 3–4, 214n11
Peninsular & Oriental steamers, 125
Perkins, Kenneth, 20
Perrusson, workshop, 87
Petanseau, Mr., 25
Philippides, Pierre, 191
photographs: as sources, 19, 61*fig*, 63; visiting card, 57*fig*, 69, 70*fig*; "obscene" photographs, 136, 154; on sale, 154, 155; for surveillance, 147–49
pieceworkers, 50, 72–73

Piedmont, xvi*map*, 94, 170, 280n199; migrants from, 68, 73, 90, 165, 170, 172; Pinerolo and Doria Riparia valley, 165, 288n79; wine from 170. *See also* Lesegno
Police Bureau of Port Said, 128–31
Ponesso, Macario, 88
port cities, importance in world history, 14–16. *See also specific city names*
Portelli, Joseph, 176
Port Said (Būr Saʿīd), xvi*map*, 5fig, 27*fig*, 28*fig*; British occupation of Egypt, 136–41, 137*fig*, 200; brothels at, 97; checkpoints at, 141–49; as choice for study, 9–12; circuits of mobility, alterations of, 8–9, 47–48; crime in, 123–26; diversity of religions in, 66–67; divisions and hierarchies in, 111–12, 202, 203*fig*; drinking water resources, 40–41; Egypt in world history, 14–16; entertainment in (*See* entertainment); expulsion of unemployed workers, 145–47; failure to prosper, 195–97; food resources, 41–42, 73–74; as free trade zone, 209; gambling in, 103–4, 153, 261n174; groundbreaking of the canal and, 1, 3–6, 5f*ig*; history of, overview, 200–205, 203*fig*; infrastructure, Egyptian government efforts to improve, 198–99; law enforcement in, 128–31, 138–41; leading a nomadic life after 1869, 149–52; medical care at, 42; monitoring the return of unwanted visitors, 147–49; nighttime navigation, effects of, 189–92; opening of the canal (1869), 114–22, 118*fig*, 152; Passports Bureau, 132–36, 134*fig*, 135*fig*; poaching of workers, 87; population and demographics, 121–22, 153, 202; in present day, 209–10; prostitution in, 153, 184–89; railway facilities, lack of, 197–99, 199*fig*, 204; as sanctuary for runaways, 146–47; schools in, 64; sea baths, 158, 158*fig*; shelter for workers, 42, 73; smuggling of goods, 126–28; soccer riot (2012), 209–10; sources for study, 17–21; status in relation to rest of Egypt, 199–200, 209–10; unemployed and undocumented migrants, entry point for, 92–94; unemployed persons, after canal opening, 119, 122–26; as vital transportation node, 120–22; women and families in, 60–66, 61*fig*, 65*fig*; work encampments, 27, 27*map*
Ports and Lighthouses, 133
Port Sudan, 15, 20
postcards, 19, 158*fig*, 199 *fig*
Pratolini, Benedditta, 163
press on the Isthmus, 20, 160–63. *See also* literacy; *specific newspaper titles*
prostitution, 95–97, 104, 153, 154, 258n132; efforts to regulate hours of operation of public establishments, 177–79; poverty or bad morals, 184–89; women and entertainment, 183–84. *See also* brothels; gender; Rue Babel; women
provisions, 41–42, 73–74, 90, 119–20, 154–55, 155*fig*, 156*fig*; daily port operations and, 133–36, 135*fig*, 136*fig*; Port Said as a vital transportation node, 121–22; smuggling of goods in Port Said, 126–28; taxes and fees on, 122
public works projects: Egyptian government provisions for, 29–30, 202; power dynamic, local and foreign, 203–4; use of forced labor on, 4, 30–31, 85. *See also* corvée

Qalyubiyya, 37
Qantara, 27, 27*map*, 35, 36, 97, 138–39, 165
Qassasin, 27*map*, 40
Qena, xvi*map*, 37, 143
quarantine, 50, 136, 152, 203*fig*; office in Suez, 63; worksite known as, 119
Quay François Joseph, 137, 137*fig*
Quiquizot, Mr., 99

race, 71–72; ambiguous and overlapping categories of, 80–81; climatology and racial judgments, 77–78; racial hierarchies, creation of, 76–79, 204; rules and exceptions in engagement of workers, 74–83; wage inequities and, 81
railways, 120; in competition with canal transport, 38–39; on Isthmus of Suez, 8–9; to Port Said, 122, 197–99, 199*fig*, 204; for transport of labor force, 38–39

Ramadan, 156, 164
Ra's al 'Ish (Raz-el-Ech), 27, 27*map*
reading on the Isthmus, 160–63
recreational pursuits. *See* entertainment
recruitment; agencies, 188, 299n231; centers of, 52; 65; of children, 34; of Christian workers, 34–35; Company attempts at, 33–36, 45, 48–52, 55, 87; of English police, 139; networks of, 51; into prostitution, 186–87; of stonecutters, 248n6; of wives and families, 105–6. *See also* Company, prostitution
Red Sea, 1, 3, 160; Port Said as new hub of connectivity, 8–9, 192; as trading hub, 120. *See also* Suez
regattas, 158–59, 163
religion: diversity of religions in Port Said, 66–67; festivals and rituals, 163–65; labor recruitment and, 34–35, 49; masonic lodges and, 159; opening of the canal festivities (1869), 117; in Port Said, 121–22; religious archives and sources, 18–19; schools for children in Port Said, 64. *See also* conversion; Catholicism; Christians; Greek Orthodox; Islam; Jewish population; Muslims
remittances, 59, 85, 115
Remy, Louis Edouard, 84
Remy, Mr., 174
research sources, 17–21
Ribao, Petros, 104
Rida, Muhammad Rashid, 198
Ritelli, Ida, 150
rituals and festivals, 163–65
Rizzo, Antonio, 150
Rogel, Réné, 84
Romania: deserters from, 56; girl migrating from, 186; Jewish migrants from, 185; Romanian Wallachia, xvi*map*, 181, 183
Rosina, 186
Rossi, Mr., 173
Rubena, Jusepha, 106
Rue Babel, 185
Rue de la Division, 159
Ruiz, Mario, xi, 176, 179
Russia, xvi*map*, 120; involvement in prostitution networks, 187; Jewish migrants from, 185; Mediterranean ambitions of, 3; pilgrims from, 125; Russian czar, 116; Russian (Navigation) Company, 57, 121, 127, 133; Russo-Japanese War, 175; sailors from, 176;
Ryzova, Lucie, 20

sabir, 82, 253n73. *See also* Algeria
Sadat, Anwar, 209
Said, Edward, 1, 11, 71
Sa'id, Muhammad, 3, 43
Saint-Nazaire, xvi*map*, 51–52
Saka, Abou, 174
Salemi, physician, 90
Salemi, Francesco, 143, 147. *See also* Mamluck, Soleiman el
Salvagno, Maria (or Marieta), 53–54, 59
Saverio Graziano, hiring hall, 188
Savona, xvi*map*, 62
Schettini, Laura, 186
Schoerr, Emile, 56
schools, 62, 63, 64, 66, 110, 121–22, 161, 163, 244n264
Sciva, Raphael, 106
Scopilitis, Vacilis, 177
sea baths, 158, 158*fig*, 164, 203*map*. *See also* bathing
Serali, Regina, 62
Serapeum, 27, 27*map*, 38, 52, 74, 95, 97, 100, 119
Serre, Madame, 98
Serrière, Jacques, 161
Servat, Marie, 99–100
Service of Docks and Entrepôts in Port Said, 133
Seuil, 34, 36–37, 52, 95, 106
sewer, 111. *See also* bathrooms
Shanghai, xvi*map*, 186, 187
shares, 3, 67–68
Sharqiyya, 37, 39, 40, 111–12
Shaykhs. *See* headmen
Shean, Samuel, 173–74
shelter for workers, 42, 73
shopkeepers, 41–42, 73–74, 98, 119–20, 132, 154–55, 156*fig*, 161
Simon Artz, 155. *See also* Arzt store
Simonini, family, 168
simsimiyya, 160
Sinai, 160, 209

Sirsar, Ahmed, 63
skilled workers, 73; promotion of opportunities in Suez Canal, 52–54; rules and exceptions in engagement of workers, 74–83; wages of, 29–30
smuggling of goods, 126–28, 270n64. *See also* hashish; tobacco
soccer riot (2012), 209–10
social mobility, migrant strategies for, 55
Société des Forges et Chantiers, 87, 119
Soldaini, Ms., 98
Soliman Bey, governor, 132
sources, for study, 17–21
Spartaco, 162
Spinelli, Pasquale, 88
Stanich, Giuseppina, 122
statue: of Ferdinand de Lesseps, 206, 207*fig*; of Queen Victoria, 137, 137*fig*, 275n130; statuettes, 74
steamships: daily port operations and, 133–36, 135*fig*, 136*fig*; electricity and nighttime navigation, 189–90; migratory paths and, 56–58; Port Said as preferred port for, 120–21
Steevens, George W., 195
Steiner, Johann, 51
Stephenson, Robert, 38
stereotypes: about Chinese labor, 48–49; about the unemployed and crime, 128; of various worker groups, 78–83, 128. *See also* climate; environmental determinism; ethnicity
Stinger, Thérèse, 109
stoneworkers, 29, 53, 71, 73, 76, 83, 247–48n6
stowaways, 124–25
strike, 85, 89–90; of coal-heavers, 192, 200. *See also* desertion
Sudan, 9, 143; migrants from, 153, 202. *See also* Port Sudan
Suez, xvi*map*, 4, 27*map*, 120, 166; British occupation of Egypt, 138, 139; cafés in, 172–73; changing migratory paths, 47, 59, 62–63, 67–68, 69; drinking water resources, 40; economy after opening of canal, 120; expulsion of unemployed workers, 145–47; gambling in, 103, 105; land concessions to the Company, 44; masonic lodges in, 159; music, diversity of, 159–60; population after opening of the canal, 119–20, 153; prostitution in, 97; railways to, 38; temperatures at, 39; women and entertainment, 183. *See also* Red Sea
Suez Canal, xvi*map*, 1–3, 2*map*, 27*map*, 208*fig*; British occupation of Egypt, 136–41, 137*fig*, 200; canal cites status in relation to rest of Egypt, 199–200, 209–10; as choice for study, 9–12; as competition for rail transport, 38–39; Egyptian labor stipulations, 28–33; groundbreaking of (1859), 3–6, 5*fig*; nationalization of, 200, 206; nighttime navigation, effects of, 189–90; opening of the canal (1869), 114–22, 118*fig*, 152; operations, revenues from, 67–68; as port where East and West merged, 200; sources for study, 17–21; tonnage fees for navigation, 120. *See also* Company
Sufi, 160
sugar cane cultivation, 45–46
supply shops, 41–42, 73–74, 119–20, 132
syphilis, 42, 104, 183, 184, 297n203. *See also* disease
Syria, xvi*map*, 4, 44, 145; hashish from, 127; labor recruitment efforts in, 34–35, 36, 45; migratory pathways, 33, 58–59, 121, 125, 145, 166; prostitution and, 184, 186; racial and national classifications of workers, 80, 82; trade routes of, 41, 121, 170

Tagliacozzo, Eric, 128
Talamasi, Francesco, 63
Tanta, 162, 183
Tanzimat reforms, 32
taverns, 104–5, 119–20, 153, 154, 156, 157, 165, 166–70, 193–94; conflated with brothels, 183; efforts to regulate hours of operation, 104–105, 177–79; excessive drinking and boisterous customers, 168–69, 173–77; venues, segregation of communities and, 171–73; women as owners of, 98; women and entertainment, 106, 179–89. *See also* breweries; cafés; canteens

Tawfiq, Khedive, 136–37
taxation of goods, 122, 197, 198–99, 202
telegraph, 13, 71, 121, 144, 148, 151
temperatures, challenges of, 39; climatology and racial judgments, 77–78
Testaferrata, James, 115, 146
theater, 122, 154, 156, 161, 165, 166; theater workers, 166, 191, 288n76. *See also* Aida; operas and operettas; Verdi
Thomassin, Marie, 100, 109
Timsah, 27, 27*map*, 36, 37, 38, 40, 128. *See also* Ismailia; Lake Timsah
Tinnis, gulf of, 3–4, 214n12
tobacco, 74, 126–127, 154, 168–69, 172, 182, 193. *See also* smuggling
Toledano, Ehud, 19
Töller, Françoise, 98
tonnage fees for navigation of canal, 120
Tosi, theatrical group, 166
Tossoum, 27, 27*map*
tourism, 19, 196, 198
Toussaint, Madame, 107
trade: alcohol, sources of, 169–70; electricity, effect on trade, 189–90; Ismailia as trading hub, 120; Port Said as free trade zone, 209; Port Said as trading hub, 21, 47, 121–22, 141, 196, 197–98, 200; provisions for workers, 73, 74; Red Sea ports and, 120
transportation costs, 49–52, 56–57
Trieste, xvi*map*, 48, 49, 57, 116, 121, 149, 185
Tripoli (Tripolitania), 143, xvi*map*, 4, 14, 143
Tsing, Anna Lowerhaupt, 17
Tunis, 4, 14; Tunisia, xvi*map*, 96; Tunisian, 98, 180
Turkey, xvi*map*, 121, 142, 143 162; goods from, 155; refugees from, 153, 186; Turkish language, 74, 82–83, 140; Turkish subjects, 131, 139, 184; Turkish tobacco, 127. *See also* Ottoman Empire; Ottoman government
Tuscany, xvi*map*; migrants from, 68, 88, 170, 191. *See also* Gimigliano

Ulacacci, Mr., 173
unemployed persons, 9, 111–12, 202; Company position on, 91–94; Damietta as entry point for, 92–93; Egyptian government, treatment of, 92–94; expulsion of, 145–47; stereotypes about, 128; surveillance of, 128–29; suspicions about, 122–26; unemployed after canal opening, 119, 122–26
Union Jack Bar, 175
unskilled workers: rules and exceptions in engagement of workers, 15, 74–83, 115, 150, 236n167
Upper Egypt, xvi*map*; forced labor from, 37; freshwater sources, 40; labor recruitment from, 4, 37, 45, 86, 192; motivation to settle in canal region, 45–46; music from, 160; stereotypes about workers from, 78–79, 80, 81; trade in goods from, 120; women dancers from, 181
'Urabi, Ahmad, 136–37, 138, 275n130
urbanization, opening of the canal, effect of, 120
Ursini, Maria, 183

vagabonds, 91–94, 105; vagrancy and unemployment, 91–92; arrest for vagrancy, 143, 145, 154
Valentino, Jean, 101
Valiandi, Alessandro, 88
Vaticiotis, painter, 177
Verdi, Giuseppe, 118–19. *See also* Aida
Vial, Clemence, 110
Victoria, Queen of England, 137, 137*fig*, 163
violence: against women and girls, 107, 109, 140; anthropometric means of passport control, 147; bar brawls, 157, 174–76; prosecution of, 131, 147; violent death on the worksites, 88, 91, 146, 166; women blamed for, 157. *See* courbache; corvée
Viscardi, Mr., 166
Vranch, Amelia, 144

wages, 29–30, 34–35, 38, 87, 89, 109, 211; day laborers, 72–73; extravagant hope for profit, 83–85; pieceworkers, 72–73; racial inequities in, 76, 81; wage earning by women, 189
Wallachia, xvi*map*; migrants from, 181, 183. *See also* Romania

Warri, Eugene, 52
water, for drinking, 39–41; for navigation, 4, 12; lack of, 16; laundry work, 48, 63, 71, 91, 94, 95, 98, 113, 187, 188; pipes for, 168–69. *See also* desalination of water
wet nurses, 110, 188; Calabrian wet nurses, 191
wine, 74, 118, 119, 168, 170, 176; trade in, 74, 99, 149, 170; wine shops, 123, 168, 172, 177. *See also* beer; taverns
women, 22, 41, 54, 55, 56, 58, 59, 60–61, 61*fig*, 72, 85, 93–94, 99, 105–11, 128, 134*fig*, 153, 175, 202; Arab women and girls, 98, 105–107, 107*fig*, 138, 180, 181, 186; canal wives, 105–11; chain migration and, 54, 60, 94, 98; domestic labor market, 94–95, 99, 188; entertainment and, 157, 179–89, 193; families on Isthmus of Suez, 60–66, 61*fig*, 65*fig*, 112; Isthmus of Suez in second half of 1860s, 71–72, 111–13; Jewish women, 67, 184, 185; laundrywomen, 48, 63, 71, 94, 98, 113, 187, 188; microhistorical and microspatial perspectives of, 11–12; in migration history, 7, 11, 216n34; mobility of, 110–11; money flows and correspondence of, 59; pregnancy and, 60, 95; prostitution, 95–97; prostitution, poverty or bad morals, 184–89, 239n204; recruitment of, 34, 35, 186–88; scattered livelihoods of, 98–100; sources on and by, 11, 19, 53, 54, 55, 59, 60, 108, 224n115; travel restrictions, 55, 134*fig*, 239n204; widows, 62, 63, 88, 98, 100, 107, 108, 167. *See also* gender; marriage
worksites: canal wives, 105–11; children and, 63–66; cholera epidemic, 88–89; conditions at, 39–45; desertion, strikes, and other rebellions, 85–90, 192; division of labor in, 72–74; domestic labor market, 94–95; drinking water, access to, 39–41; extravagant hope for profit, 83–85; food resources, 41–42, 73–74; infantilized male workers, 102–5; medical care at, 42; prostitution at, 95–97, 104; race and gender issues, 71–72; in second half of 1860s, 71–72, 111–13; shelter at, 42, 73; unemployed and undocumented migrants, 91–94; wages and expenses of workers, 42–43; work-related injury and death, 88, 101. *See also names of specific locations*

Yukon, 15, 47

Zagazig, xvi*map*, 38, 40, 73, 89, 126, 129
Zenzafilé, Mikaël, 87

Founded in 1893,
UNIVERSITY OF CALIFORNIA PRESS
publishes bold, progressive books and journals
on topics in the arts, humanities, social sciences,
and natural sciences—with a focus on social
justice issues—that inspire thought and action
among readers worldwide.

The UC PRESS FOUNDATION
raises funds to uphold the press's vital role
as an independent, nonprofit publisher, and
receives philanthropic support from a wide
range of individuals and institutions—and from
committed readers like you. To learn more, visit
ucpress.edu/supportus.